ESSENTIALS OF OCEANOGRAPHY

Harold V. Thurman
Mt. San Antonio College

ESSENTIALS OF OCEANOGRAPHY

3rd Edition

MERRILL PUBLISHING COMPANY
A Bell & Howell Information Company
Columbus | Toronto | London | Melbourne

Cover Image: A false-color view of Earth's biosphere based on data from two different satellite sensors. The ocean image indicates the concentration of marine phytoplankton, with reds and oranges marking areas of highest productivity, such as nutrient-rich polar seas, temperate and tropical continental shelves, and wind-driven upwellings on the western coasts of the Americas and Africa. Decreasing productivity is shown by color changes from yellows and greens to light blue and purple. Unmonitored areas are black, as are areas of polar sea ice. The land image indicates estimated chlorophyll, from dense vegetation (dark green) to barren deserts, high mountains, and arctic regions (yellow). The ocean image is a composite of 31,000 images acquired by the Coastal Zone Color Scanner (CZCS) Nimbus-7 satellite. Land patterns are derived from three years of daily images from the NOAA-7 satellite. Image courtesy of Jane A. Elrod and Gene Feldman, NASA/Goddard Space Flight Center.

Facing Title Page: Photo by Larry Hamill.

Published by Merrill Publishing Company
A Bell & Howell Information Company
Columbus, Ohio 43216

This book was set in Garamond.

Executive Editor: Steve Helba
Production Editor: Carol Driver
Art Coordinator: Ruth Kimpel
Cover Designers: Cathy Watterson, Brian Deep
Text Designer: Cynthia Brunk
Photo Editor: Terry Tietz

Library of Congress Catalog Card Number: 89–61827
International Standard Book Number: 0–675–21117–4
Printed in the United States of America
1 2 3 4 5 6 7 8 9—94 93 92 91 90

Color Plate Credits: All photos and illustrations copyrighted by individuals or companies listed. Plates 1, 7, 8, 23, 24, 25, 26, 27A, Christopher Newbert; plate 2A, Scripps Institute; plate 2B, Dr. Fred N. Spiess/Scripps Institute; plates 3, 9, 10, 11, 15, 16, 18B, 19B, 20B, 28, Robert W. Tope/Merrill Publishing; plate 4, from C. L. Parkinson, J. C. Comiso, H. J. Zwally, D. J. Cavalieri, P. Gloersen, and W. J. Campbell, 1987, *Arctic Sea Ice, 1973–1976: Satellite Passive-Microwave Observations,* NASA SP-489, National Aeronautics and Space Administration, Washington, D. C., 296 pp.; plate 5, Dr. Charles McLain/Rosensteil School of Marine and Atmospheric Science, University of Miami; plates 6, 21, from H. R. Gordon, "Phytoplankton Pigments from *Nimbus* 7 Coastal Zone Scanner: Comparisons with Surface Measurements," *Science* 210: 63–66, 1980; plate 12, Marie Tharp; plate 13, UNESCO; plate 14, Jet Propulsion Laboratory; plate 17, NASA, Goddard Space Flight Center; plates 18A, 19A, 20A, NOAA/NESDIS; plate 22, Paul C. Fiedler/National Marine Fisheries Service; plate 27B, Gregory Silber/EarthViews; plate 29A, Craig Matkin/EarthViews; plate 29B, Bernie Tershy and Craig Strong/EarthViews.

Figure Credits: Figures 3–9 and 3–10 from "Remote Acoustic Detection of Turbidity Current Surge," Hay, A. E., et al., *Science* Vol. 217, pp. 833–835, 27 August 1982. Copyright 1982. Table 11–1 from Table 3, "Average Beach Face Slopes Compared to Sediment Diameters" (p. 127) from *Submarine Geology,* third edition by Francis P. Shepard. Copyright © 1948, 1963, 1973 by Francis P. Shepard. Reprinted by permission of Harper & Row, Publishers, Inc. Figure 13–8 from Sverdrup/Johnson/Fleming, *The Oceans: Their Physics, Chemistry, and General Biology,* © 1942, renewed 1970, pp. 817, 818. Reprinted by permission of Prentice-Hall, Inc., Englewood Cliffs, NJ. Figure 14–9A from "Nitrogen Fixation by Floating Diatom Mats: A Source of New Nitrogen to Oligotrophic Ocean Waters, " Martine, L., et al., *Science* Vol. 221, pp. 152–154, 8 July 1983. Copyright 1983. Figures 15–4B, C, D, and 16–25 courtesy of James M. King, Graphic Impressions, P. O. Box 21626, Santa Barbara, CA 93121. Appendix V from *Marine Biology,* Thurman, H. V., and Webber, H. H., Merrill Publishing Co., 1984.

PREFACE

Essentials of Oceanography was originally written to provide for the needs of general oceanography courses in quarter-system colleges and community colleges. It contained 18 chapters containing the topics usually covered in such a course.

Because the previous editions were well received, we have made few major organizational changes for the third edition. We have clarified the content wherever possible to help the reader. Another of our primary objectives has been to keep the book as up-to-date as possible. With these objectives in mind, we have also redrawn a number of line drawings. The color inserts have been expanded, with the plates arranged in four central themes: an introductory overview, marine geology, physical oceanography, and marine biology. The images have been carefully integrated into a narrative that explains the significance of each as it relates to the central theme. Both this narrative approach and the placement of the inserts at appropriate locations in the text provide the student with a context for understanding the relevance of each color photograph or line drawing to the specific subject matter.

The Introduction provides a short overview of our reasons for studying the ocean. It includes a discussion of our need to better understand the effects of economic activities such as fishing and petroleum exploitation; the close relationship between physical and biological aspects of the marine environment that exist in marine ecology; the interactions between the atmosphere and ocean as they relate to climate, ocean waves, and currents; and the effects of disposing of our wastes in the ocean.

Chapter 1 provides a background that explains how we got where we are in the study of the oceans, while Chapter 2 describes the place of Earth and its oceans in the cosmos.

Chapter 3, "Marine Provinces," provides a brief discussion of the major bathymetric provinces and their major features as a framework for the discussions of global plate tectonics in chapter 4 and marine sediments in chapter 5.

The study of physical oceanography begins with chapter 6, "The Nature of Water." The important properties of water are summarized in table 6–1 for those who do not wish to consider them in depth, but a comprehensive discussion is also provided for each of the properties. Physical phenomena are discussed in chapters 7 through 10. Chapter 7 provides an overview of the renewable energy resources of the ocean and a more complete look at ocean thermal energy conversion (OTEC). The potential for using wind and currents as sources of energy is discussed in chapter 8; wave potential is covered in chapter 9, and the use of tidal energy is considered in chapter 10.

Those regions of most direct interaction between the continents and the oceans are covered in chapter 11, "The Shore," and chapter 12, "The Coastal Ocean."

Marine biology is first considered in chapter 13, "The Marine Habitat," which provides an ecological overview. Chapter 14, "Biological Productivity—Energy Transfer," introduces marine algae and continues the discussion of broad ecological concepts related to biological energy transfer. An ecological presentation of the variety of marine animals is contained in chapters 15, "Animals of the Pelagic Environment," and 16, "Animals of the Benthic Environment." For those wishing to discuss the taxonomic relationships among phyla, a phylogenetic tree with a brief discussion of relationships is provided in appendix V, and a listing of taxonomic phyla and classes of the five kingdoms identified in chapter 14 is found in appendix VI.

Finally, chapters 17 and 18 cover the primary human exploitation of marine resources, fisheries, and mariculture and the negative effects of human activities on the ocean by means of marine pollution. Chapter 17 provides an up-to-date discussion of fisheries that emphasizes the fundamental concepts of the ecosystem approach to the assessment of fisheries and updates the status of mariculture. Chapter 18 provides an overview of marine pollution that includes the methods of assessing marine pollution and coverage of pollution involving plastics, petroleum, sewage, radioactive waste, halogenated hedrocarbons, and mercury.

Each chapter concludes with a summary and a list of discussion questions and exercises designed to reinforce the student's understanding of important concepts and facts presented in the chapter. To broaden the student's knowledge beyond what can be covered in one chapter of a textbook, related readings from *Sea Frontiers* and *Scientific American* are listed and briefly described.

To assist the student further, there are appendices I "Scientific Notation," II "The Metric System and Conversion Factors," III "Periodic Table of the Elements," IV "The Geologic Time Table," V "The Phylogenetic Tree," and VI "Taxonomic Classification of Common Marine Organisms." An extensive glossary provides quick reference to the definition of many important concepts and terms.

An Instructor's Manual provides not only answers to the end-of-chapter questions but also special-feature articles, covering a wide range of interesting topics, that may be duplicated and distributed to students. These features could be used simply to add breadth and interest to the coverage of the material in a chapter or serve as a basis for student reports.

It is recommended that chapters 3, 4, 6, 7, 8, 9, 10, 11, 12, 13, and 14 be covered completely in most courses—even 10-week-quarter presentations. Teachers of 10-week courses may wish to confine the attention of their students in chapter 6 to table 6–1 and the section "Salinity of Ocean Water." Depending on whether the emphasis in a one-quarter course is on physical or biological oceanography, chapter 7 may be added to increase geological coverage, whereas chapters 15 and 16 would round out biological coverage. In most semester courses, it is likely that the text can be used in its entirety. For the unusual course that extends beyond the semester time frame, the text will serve well as the framework around which a more in-depth coverage using outside references or materials provided by the teacher is achieved.

ACKNOWLEDGMENTS

Many people contributed to the successful completion of this edition, from the reviewers, who provided the focus of the revision, to the excellent production staff at Merrill Publishing. For their reviewing input, which greatly strengthened the text, we wish to thank Charles E. Herdendorf, Huron OH; W. H. Hoyt, University of Northern Colorado; Lyle T. Hubbard, University of Tennessee; Jack H. Hyde, Tacoma Community College; Lawrence McAdam, Seminole Community College; Robin Newbold-Sturm, San Juan Capistrano; Paul Pinot, Colgate University; C. Nicholas Raphael, Eastern Michigan University; Erwin Seibel, San Francisco State University; Yin S. Soong, Millersville University; and Kelly Williams, University of Dayton.

We also express our great appreciation to the individuals, institutions, and government agencies that provided information, figures, and photographs. It is cooperation from such sources that greatly simplifies the production of texts like this.

Major contributors from the Merrill production staff to the development of the book were Carol Driver, production editor; Cynthia Brunk, text designer; Cathy Watterson, cover designer; and Ruth Kimpel, art coordinator. Their efforts are of the most professional quality and greatly appreciated.

CONTENTS

Appendixes

Glossary   373

Index   389

Color Plates

ESSENTIALS OF OCEANOGRAPHY

PLATE 1

We humans are drawn to the ocean by our attraction to our close relatives, whales and other marine mammals, living at its surface. The humpback whale has aroused an especially large amount of interest with its "songs" and spectacular breaching behavior.

Some of the whales, all of which are marvelously streamlined, make remarkably deep dives and flaw-lessly track prey in water that is totally dark.

Sperm whales are known to dive to depths in excess of 2 km (1.2 mi). They are aided in making these long, deep, dives by a much greater volume of blood per unit of body mass than other mammals.

Many whales, including the sperm whale, locate and track their prey in the deep, dark waters by using echolocation—bouncing sound waves off the prey.

PLATE 2

A

B

But we also are driven to learn more of our distant relatives, invertebrates, that inhabit the ocean to its greatest depths. Of particular interest are relatively large bottom-dwelling animals living in association with hot-water springs found along the rift valleys near the axes of submarine mountain ranges. These animals are independent of the microscopic plant life that lives in the sunlit waters and supports the vast majority of marine animals.

These vent communities are supported by bacteria that produce food in total darkness by extracting chemical energy from the hydrogen sulfide gas that is dissolved in the vent water.

The first of these vent communities was discovered in 1977. Off the coast of Ecuador and near the equator, they were recorded by a photographic sled towed through the region of a temperature anomaly at the ocean floor. This initial find was in a rift valley at the axis of a submarine mountain range—the Galapagos Rift.

The more prominent members of this deep assemblage are large tube worms living in tubes over 1 m (3.3 ft) long, as well as clams and mussels up to 25 cm (9 in) long. None of these larger animals possess guts, and they may receive most of their nutrition from a symbiotic relationship with the chemosynthetic bacteria living within their tissue.

Other members of the hydrothermal vent communities may filter the bacteria from the water or graze on bacterial mats growing on the rocky ocean floor.

Subsequent investigations have located vents called black smokers that spew out dark clouds of metallic compounds at temperatures of more than 300°C

(572°F) and biological communities associated with numerous hydrothermal vents in the Pacific and Atlantic oceans.

Cold-water seeps have also been found to support biological communities based on a basic bacterial food supply. The cold-water communities have all been located near the margins of continents. They are known to exist at the base of the Florida Escarpment in the Gulf of Mexico, off the coast of Oregon, and near Japan.

Seeping hydrocarbons support a biological community on the continental slope off the coast of Louisiana.

In addition to the unusual biological phenomena associated with the hydrothermal vents, these emissions precipitate metallic deposits that may be the ultimate source of metal-rich ores mined on the continents.

After the metal deposits form at the axes of the submarine mountain ranges, they are transported away by a slowly moving ocean floor. Eventually, the ocean floor may encounter a continent and descend beneath it. During this process, the metallic compounds are distilled out of the oceanic rocks and emplaced in the overlying rocks of the continent.

PLATE 3

PLATE 4

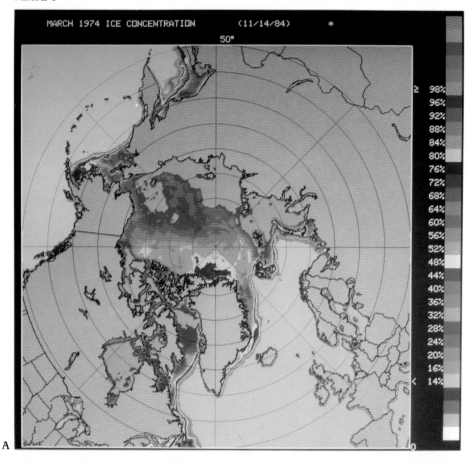

A

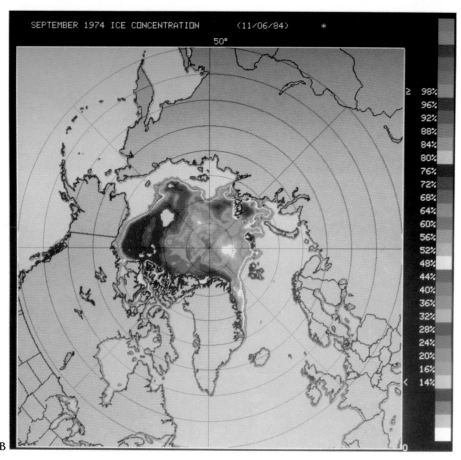

B

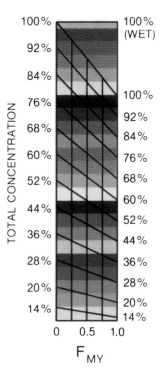

During recent years, much attention has been directed at the polar oceans, where sea ice covers great expanses of ocean surface each winter and melts away in summer months. The seaward edge of this ice is a zone of intense biological activity.

A permanent polar ice accumulation exists in the Arctic Ocean. During the winter, sea ice in the form of pack ice forms on the open sea to extend the polar ice to the south. Fast ice also forms along the shores to the south. Each summer the pack ice and fast ice melt away.

These 1974 images showing percent of sea-ice concentration are accurate to within 15% for first-year sea ice (pack ice and fast ice) and 25% for multiyear ice (polar ice cap). The scales of ice concentration along the right margins of each image are for first-year ice. The scale above can be used for first-year ice (left side) or multiyear ice (right side). The images were developed from the ESMR (Electrically Scanning Microwave Radiometer) aboard the Nimbus 5 satellite.

Sea-surface temperature data gathered by a NOAA satellite was processed to produce this false-color image of the northwest Atlantic Ocean. The warm Gulf Stream waters are shown as orange and red. The colder nearshore waters are blue and purple. Warm water from south of the Gulf Stream is transferred to the north as warm core rings (yellow) surrounded by cooler (blue and green) water. Cold nearshore water spins off to the south of the Gulf Stream as cold core rings (green) surrounded by warmer (yellow and red) water.

The rings form when meanders close to trap warm or cold water within them. The warm rings contain shallow, bowl-shaped masses of warm water about 1 km (0.62 mi) deep with diameters of about 100 km (62 mi). The cold rings have cones of cold water that extend to the ocean floor. They may be more than 500 km (310 mi) across at the surface. The diameter of the cone increases with depth.

PLATE 5

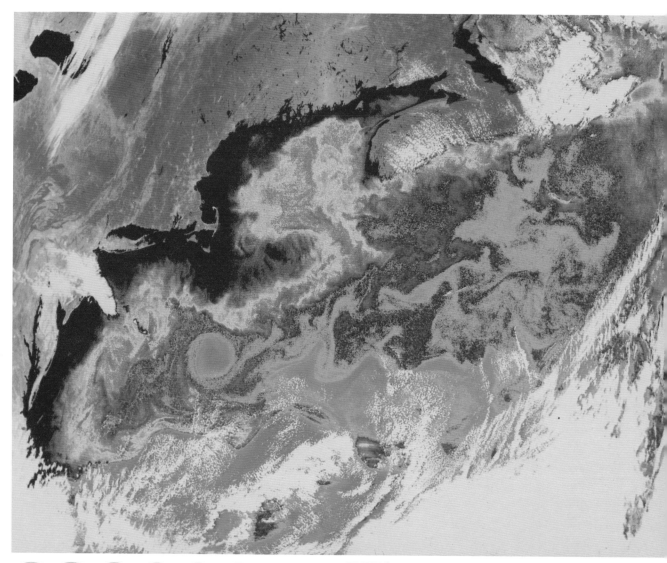

PLATE 6

In shallow coastal waters, where biological activity reaches its greatest development, there is a correlation between decreased temperature of surface water and high levels of microscopic plant production. Remote sensing by satellites may eventually make it possible to develop a global view of this relationship.

This false-color image of chlorophyll pigment concentration in the northwest Atlantic Ocean shows the low level of biological productivity of the Gulf Stream in blue. Increasing levels of productivity are shown as green, yellow, red, and dark red.

A striking feature of this image is the low level of productivity indicated by the blue color of the warm core ring south of Cape Cod. Note that it is surrounded by more productive green nearshore water. The white is cloud cover.

PLATE 7

C oral reefs serve as biological oases in the relatively unproductive
tropical ocean. The coral reef environment includes a great variety of marine life with the most colorful and varied marine communities found in the oceans. Here we see the tentacles of coral polyps
extended to remove microscopic food particles from ocean water.

PLATE 8

Coral reefs are found only in warm water, and reef-building corals have microscopic algae living in their tissues to aid them in secreting the calcium carbonate structure that supports them.

However, more than half of the limestone of coral reefs is produced by calcium carbonate-secreting algae that thrive in the nutrient-rich waters of the reef. Actually, the plant mass of a coral reef is three times as great as the animal mass.

The clown fish, which lives unharmed among the stinging tentacles of the sea anemone, may pay for this protection by bringing food to the anemone. This symbiotic relationship that benefits both participants is called mutualism. Two other types of symbiotic relationships that may be found in coral reefs and the ocean as a whole are commensalism and parasitism. In commensalism, one organism benefits while the other is unaffected. Parasitism involves the benefit to one at the expense of the other.

One of the most striking features of the coral reef is the great variety of brilliantly colored fish. The reason for such an audacious display of color has not been discovered. But for whatever reason it exists, no visual experience on this planet will ever produce the thrill and wonderment of one's first view of this display.

INTRODUCTION

We hope the world will be a better place for our having been here. As good world citizens, we have that responsibility. If you approach the study of the oceans with such a goal in mind, I'm sure the oceans and the world will benefit from your presence.

When I first began writing texts in the early 1970s, I wanted to help students develop an appreciation for the oceans by learning about oceanic processes, both physical and biological—that is, develop an ecological awareness. Some of you may even find a life-long interest in the ocean and continue to study it formally or informally.

I'd like to impart to you what I feel are the basic facts behind oceanic science. First, the physical nature of the ocean and its life forms are dramatically interrelated, a mystery that we are only now beginning to unravel. Second, people are migrating to the coasts, and this trend will have a negative impact on coastal ecology. Third, our technology can be used to increase the threat to our oceans or lessen lethal damage, as we dictate.

INTERRELATIONSHIPS OF THE OCEAN

Since prehistoric times, people have used the ocean as a means of transportation and as a source of food, but the importance of its processes has been studied technically only since 1930. The impetus for this study began then in a search for petroleum, continued with the emphasis on ocean warfare during World War II, and more recently has been expressed in the concern for ocean ecology. Of course, fishermen have always known to go where the physical processes of the ocean offer good fishing. But how life interrelates with ocean chemistry, geology, and physics was more or less a mystery until scientists in these disciplines began to investigate the ocean and use high technology.

For example, a group of geologists investigating global plate tectonics decided in 1977 to research hot water anomalies in the deep eastern Pacific Ocean. (The process of tectonics, confirmed in 1968, involves the creation of new ocean floor along underwater moun-

FIGURE I–1
Tube worm from a Hydrothermal Vent. Dr. Robert D. Ballard of Woods Hole Oceanographic Institution examines one of the large tube worms collected from the Galapagos Rift hydrothermal vents near the equator. The vent biocommunity from which this worm was collected is located about midway between the South American mainland and the Galapagos Islands at a depth greater than 2500 m (8200 ft). (Photo taken by Jack Donnelly and provided courtesy of WHOI.)

POPULATION IMPACTS ON COASTAL ECOLOGY AND THE OPEN OCEAN

Population studies now show that three-fourths of all Americans live within 80 kilometers (50 miles) of the coast or the Great Lakes. This migration to the coasts will further mar a delicate balance between the ocean and the shore, resulting in more harbor and channel dredging; sea disposal of sewage, industrial waste, and dredge spoils; the use of ocean water for cooling of power plants and industries; and the filling in of marshes (figure I–2). But the marshes are vital to the cleansing of runoff waters and the maintenance of coastal fisheries. In the open ocean, deep ocean mining and nuclear waste disposal are planned. How do we deal with the increased demands on the environment; how do we regulate the ocean's use?

Further, the National Academy of Sciences in 1980 predicted a rise in sea level. As carbon dioxide and other gases that increase the atmosphere's ability to hold heat reach higher concentrations, worldwide temperature increases may occur. This "greenhouse effect" could hasten sea-level rise and damage marine ecological systems. Even considering present erosion and deposition, geologists would like to have us move back from the shore as much as possible and leave it to nature. Engineers, on the other hand, say any coastal region can be stabilized for the lifetime of a development; in that case much public money will be spent

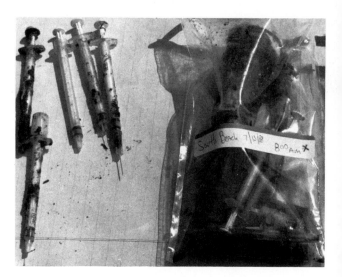

FIGURE I–2
Staten Island Pollution. Shown here is some of the medical waste, including syringes, found on the South Beach of Staten Island, New York, during the summer of 1988. On July 12, all the municipal beaches on Staten Island and in Brooklyn were closed when sewage poured into New York Harbor after a power failure at a sewage treatment plant. (Photo courtesy of UPI/Bettmann Newsphotos.)

tain ranges.) When they searched the hot-water vents along a range, they discovered many large tube worms, clams, mussels, and crabs. Normally the deep ocean floor supports relatively few, small forms of life because there isn't much food. But the vents provided energy for bacteria to support abundant life. This is a striking example of how the physical nature of the ocean determines the distribution of life within it (figure I–1).

Scientists have always been interested in how the atmosphere and the ocean affect each other. Our present technology uses satellites to view ocean circulation and life distribution. Within the next decade we will be gathering data from every ocean, every day. Such simultaneous data will show how the water reacts to wind speed and air temperature and how the air responds to oceanic changes.

FIGURE I–3
Coastal Property Damage. The debris shown is all that Hurricane Camille left of a 32-unit apartment building in Gulfport, Mississippi. (Photo courtesy of NOAA.)

stabilizing the coastal property of a few (figure I–3). Perhaps if the sea level does rise, engineers will be too busy protecting present development and will abandon plans to develop new areas!

RATIONAL USE OF TECHNOLOGY

We look at these issues and see a cause for concern, but certainly not despair. We have a vast resource that hasn't yet been lethally damaged. We have been able to inflict only minor damage here and there along its margin. But as our technology makes us more powerful, the threat of irreversible harm becomes greater—or less, depending on how we use our tools.

Which path will we take? All of you will need to evaluate carefully those you elect to public office, and some of you will have direct responsibility for making the decisions that affect our environment. It is my hope that you, as a new student of the marine environment, will gain enough knowledge from your teacher and this course to help our nation make rational use of the oceans in your lifetime.

1
HISTORY OF OCEANOGRAPHY

The robot *Jason Jr.* explores the prow of the *Titanic.* Photo courtesy of Woods Hole Oceanographic Institution.

With each passing year, human interaction with the ocean increases. The fishing effort is stepped up, and the search for petroleum and minerals intensifies. With this increasing rate of ocean exploitation, we are obliged to increase our knowledge of its effect on the marine environment. The following is a summary of events that have brought us to our present state of knowledge and a discussion of how the pressing need for increased knowledge of the oceans may be met through the application of newly developed technology.

THE CONCEPT OF GEOGRAPHY DEVELOPS

Early History

Humankind probably first viewed the ocean as a source of food. At some later stage in the development of civilization, vessels were built to move upon the ocean's surface, thereby making it a wide avenue over which distant societies could begin to interact.

The first Westerners to develop the art of navigation were the Phoenicians, who as long ago as 2000 B.C. were investigating the Mediterranean Sea, the Red Sea, and the Indian Ocean. The Phoenicians established trade with the Eastern cultures of Southeast Asia and discovered the Canary Islands. The first recorded circumnavigation of Africa in 590 B.C. was made by the Phoenicians, who had also sailed as far north along the European coast as Great Britain.

The world as it was viewed by the Greeks is presented for us on a map constructed in 450 B.C. by the geographer Herodotus in which he depicts the Greek civilization centered on the Mediterranean Sea (figure 1–1). To the north, east, and south lie the three continents—Europe, Asia, and Libya—bordered by three major seas—Mare Atlanticum to the west, Mare Australis to the south, and Mare Erythraeum to the southeast. The northern and northeastern margins of the continents of Europe and Asia are indicated as unknown, but the Greeks believed that the oceans represented a margin of water that surrounded all three continents.

FIGURE 1–1
The World of Herodotus. The world
according to Herodotus, 450 B.C. (From
Challenger report, Great Britain, 1895).

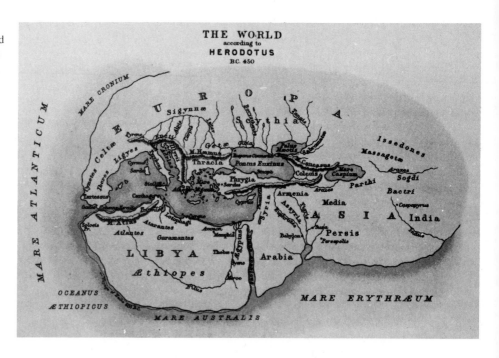

The Greek astronomer-geographer Pytheas sailed north-ward to Iceland in 325 B.C. and worked out a simple method for determining latitude, which involved measuring the angle between the lines of sight from an observer to the North Star and the northern horizon. Using astronomical measurements, he also proposed that the tides were a product of lunar influence.

Eratosthenes (276–192 B.C.), a Greek librarian at Alexandria, was the first known determiner of the world's circumference. He determined that the earth's meridianal circumference was 40,000 km (24,840 mi) which compares very well with the 40,032 km (24,875 mi) determined by the more precise methods in use today.

The Roman, Strabo (63 B.C.-A.D. 24), 30), made a contribution more directly related to the nature of the oceans. He had observed the volcanic activity that is characteristic of the Mediterranean area and concluded that, as a result of this activity, the land periodically sank and rose, causing the sea to invade and then recede from the continents. He further observed that the rivers flowing across the continent erode material from the surface and transport it to the oceans.

In approximately A.D. 150, Ptolemy produced a map of the world that represented the Roman knowledge at that time. He introduced lines of longitude and latitude on his map. Mariners could determine the latitude of any point on the surface of the earth using the method introduced by Pytheas, but it was impossible for them to accurately determine longitude. Ptolemy's map indicated, as did the earlier Greek maps, the continents of Europe, Asia, and Africa and showed the Indian Ocean to be surrounded by a partly unknown landmass. Unlike Herodotus, Ptolemy considered the major oceans to be seas similar to the Mediterranean, having boundaries defined by unknown landmasses.

The Middle Ages

After the fall of the Roman Empire, the Mediterranean area was dominated by Arab influence, and the writings of the Greeks and Romans passed into the hands of the Arabs, to be forgotten by the Christians in Europe. Subsequently, the Western concept of geography degenerated considerably, and one notion envisioned the world as a disc with Jerusalem at the center.

The Arabs, meanwhile, were trading extensively with East Africa, Southeast Asia, and India and had learned the secret of the monsoons. They took advantage of these winds, making their trade voyages easier. During the summer seasons, when the monsoon winds blew from the southwest, ships laden with goods for trade would leave the Arabian ports and sail eastward across the Indian Ocean. The return voyage would be timed to take advantage of the northeasterly trade winds that occurred during the winter season, making the transit relatively simple for their sailing vessels.

In Europe, the nautical inactivity of the southern Europeans was offset by the vigorous exploration of the Vikings of Scandinavia. Late in the 9th century, aided by a period of climatic warming, the Vikings conquered Iceland. In 981 Erik the Red sailed westward from Greenland and discovered Baffin Island. In 995 Erik's son, Leif Eriksson, discovered what was then called Vinland, and he spent the winter in that portion of North America we call Newfoundland (figure 1–2).

Age of Discovery

During the 30-year period from 1492 to 1522, known as the Age of Discovery, the Western world came to a full realization of the vastness of the earth's water-covered surface. The continents of North and South America were discovered. The globe was circumnavigated, and it was learned that human populations existed elsewhere in the world. Human cultures were found throughout the newly discovered continents and islands, although they were vastly different from those with which the voyagers were familiar.

Precipitating these voyages was the capture of Constantinople, the capital of eastern Christendom, in 1453 by Sultan Mohammed II. This event isolated the Mediterranean ports from the riches of the East Indies and caused the Western world to search for a new trading route to this area. As a result of the Arab occupation of Constantinople, the ancient knowledge of the Greeks and Romans was carried out of that city and became available to the powers of southern Europe who were interested in reestablishing trade connections with the East Indies.

Prince Henry the Navigator, of Portugal, had established a marine observatory to improve the Portuguese sailing skills, but its attempts to reestablish trade by ocean routes met with failure for years.

One of the greatest obstacles to an alternate trade route was the need to travel around the tip of Africa. Cape Agul-

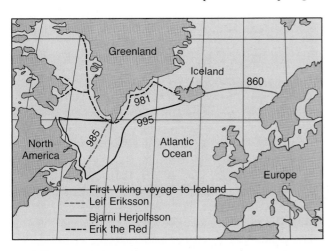

FIGURE 1–2

The Vikings Reach the New World. The expansion of Viking influence spread from Europe and probably reached Iceland by 860. Erik the Red was banished from Iceland and sailed to Greenland in 981. He returned to Iceland in 984 and led the first wave of Viking colonists to Greenland. Bjarni Herjolfsson sailed from Iceland to join the colonists but sailed too far south and is thought to be the first Viking to have viewed what is now called Newfoundland. Bjarni did not land but continued trying to find his way to Greenland. He eventually did land at the settlement Herjolfsness, named for his father, who waited for him there. In 995, Leif Eriksson, son of Erik the Red, bought Bjarni's ship and set out for the land Bjarni had seen to the southwest. Leif set up a camp at Tickle Cove in Trinity Bay and named the land Vinland after the grapes that were found there.

has was finally rounded by Bartholomeu Diaz in 1486. He was followed in 1498 by Vasco da Gama, who continued his trip around the tip of Africa to India.

The idea for a voyage such as the one Columbus undertook was initiated by the Florentine astronomer Toscanelli. He wrote to the king of Portugal, suggesting that a course be charted to the west in an attempt to reach the East Indies by crossing the Atlantic Ocean. Columbus later contacted Toscanelli and was given a copy of the information indicating that India could be reached at a distance that would have carried him just west of the continent of North America.

After his well-known difficulties in originating the voyage, Columbus set sail with 88 men and three ships August 3, 1492, from the Canary Islands. During the morning of October 12, 1492, the first land was sighted, which is generally believed to have been Watling Island. Recent research by the National Geographic Society has convinced them Columbus landed at Samana Cay, 120 km (75 mi) to the southeast. Because Columbus had greatly underestimated the distance to the East Indies via the Atlantic Ocean, he was convinced that he had arrived at these islands. Upon his return to Spain and the announcement of his discovery, additional voyages were planned, and the Spanish and Portuguese continued to explore the coasts of North and South America.

The Atlantic Ocean became familiar to the European explorers, but the Pacific Ocean was not seen by them until 1513 when Vasco Núñez de Balboa attempted a land crossing of the Isthmus of Panama and sighted the Pacific Ocean from atop a mountain.

The culmination of this period of discovery was the circumnavigation of the globe by Ferdinand Magellan (figure 1–3). In September of 1519, Magellan left Sanlucar de Barrameda, Spain, and traveled through a passage to the Pacific at 52°S latitude, now named the Straits of Magellan. After discovering the Philippines on March 15, 1521, Magellan was killed in a fight with the inhabitants of these islands. Sebastian del Caño completed the circumnavigation by taking one of the ships, *Victoria,* across the Indian Ocean and back to Spain in 1522. On this trip Magellan had attempted to measure the depth of the Pacific Ocean by a weighted line, but he was unable to reach bottom.

After these voyages, the Spanish initiated many others for the purpose of removing the gold that had been found in the possession of the Aztec and Inca cultures in Mexico and South America. While the Spanish concentrated on plundering the Aztecs and Incas, the English and Dutch plundered the Spanish. The political dominance of Spain came to an end with the defeat of the Spanish Armada by the British in 1588. With the squelching of an attempted invasion of the British Isles, the English became the dominant maritime power, and they remained so until early in the 20th century.

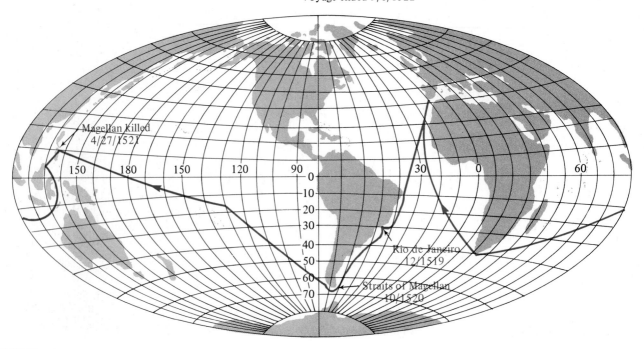

FIGURE 1–3

Voyage of Magellan. Searching for a western route to the East Indies, Magellan culminated the Age of Discovery by beginning a voyage that ended for him with death in the Philippines. San Juan Sebastian del Caño returned to Spain in the *Victoria*, one of five ships that began the voyage, to complete the first circumnavigation of the earth. (Base map courtesy of National Ocean Survey.)

THE SEARCH TO INCREASE SCIENTIFIC KNOWLEDGE OF THE OCEANS

Captain James Cook

The next major focus of interest in the oceans was more scientific, as the English determined that increasing their knowledge of the oceans would help them maintain their maritime superiority. The more successful of the early voyages that were initiated to learn more about the physical nature of the oceans were conducted by the English navigator James Cook (1728–79), son of a farm laborer (figure 1–4).

From 1768 until his death in 1779, Captain Cook undertook three voyages of discovery. He searched for the continent Terra Australis and concluded that if it existed, it lay beneath or beyond the extensive ice fields of the southern oceans. Captain Cook discovered the South Georgia and South Sandwich islands and the Hawaiian Islands, where he was killed after searching for a northwest passage from the Pacific Ocean to the Atlantic Ocean. Cook also developed a diet that prevented his crew from contracting the disease scurvy, which is caused by a vitamin C deficiency.

Cook's expeditions added greatly to the scientific knowledge of the oceans. He determined the outline of the world's largest ocean and was the first person known to cross the Antarctic Circle. Cook also led the way in sampling subsurface temperatures, measuring winds and currents, sounding, and in collecting important data on coral reefs. By proving the value of John Harrison's chronometer as a means of determining longitude, Cook made possible the first accurate maps of the earth's surface.

An American contemporary of Captain Cook was an early contributor to greater understanding of the ocean's surface currents. Benjamin Franklin, deputy postmaster general for the colonies, determined that it took mail ships coming from Europe by a northerly route 2 weeks longer to reach the colonies than it did ships that came by a more southerly route. To find out why this occurred, he asked for information concerning the movement of surface waters from captains who came into port. Franklin inferred that there was a significant current moving in a northerly path along the eastern coast of the United States. The current moved out in a more easterly path across the North Atlantic. He concluded that this east-flowing current, which the ships traveling a northerly route from Europe had to

FIGURE 1–4
Captain James Cook, RN. (From U.S. Navy photograph.)

FIGURE 1–6
Lieutenant Matthew F. Maury—A Photograph of an Engraving by Lemuel S. Punderson after a Daguerrotype, Autographed and Inscribed by Lt. Maury. (Released Naval Historical Center photograph.)

combat, was responsible for increasing the time of their voyage. He subsequently published (in 1777) a map of the Gulf Stream based on these observations (figure 1–5).

Matthew Fontaine Maury

An even greater contribution was made by Matthew Fontaine Maury (1806–73). This career officer in the U.S. Navy was placed in charge of the Depot of Naval Charts and Instruments after suffering an injury early in his career. In the depot were log books containing a large amount of information about currents and weather conditions in various parts of the oceans. Maury's systematic analysis of these logs produced a compilation of wind and current patterns that proved very useful to the navigators of the 19th century. Maury helped organize the first International Meteorological Conference in Brussels in 1853 for the purpose of establishing uniform methods of making nautical and meteorological observations at sea. This standardization greatly increased the dependability of such data, which Maury summarized in *The Physical Geography of the Sea (1855)*. Maury is often referred to as the father of oceanography (figure 1–6).

Charles Darwin

In the early 19th century, an English naturalist named Charles Darwin (1809–82) entered the scientific scene. Darwin's interest, investigating the whole of nature, led him to make one of the most outstanding contributions to the field of biology. Much of the background on which he based his conclusions concerning evolution by natural selection was gained from observations made during his voyage aboard HMS *Beagle*.

The *Beagle* sailed from Devensport on December 27, 1831, under the command of Captain Robert Fitzroy with

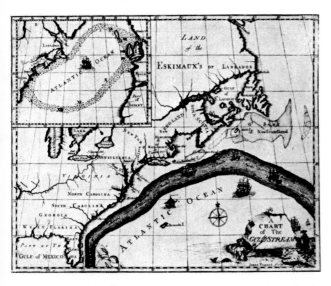

FIGURE 1–5
Chart of the Gulf Stream Compiled by Benjamin Franklin in 1777. (Courtesy of U. S. Navy.)

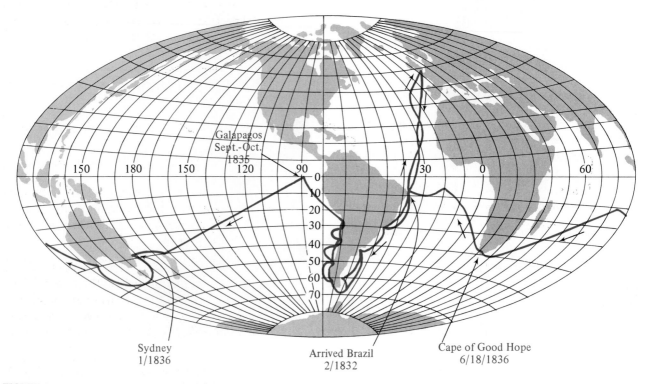

FIGURE 1–7
Voyage of the Beagle. Sailing as naturalist aboard the HMS *Beagle,* Charles Darwin gathered the evidence that allowed him to develop his theory of biological evolution through natural selection. (Base map courtesy of National Ocean Survey.)

the major objective of completing the survey of the coast of Patagonia and Tierra del Fuego.

Darwin spent the next 5 years aboard the *Beagle,* which continued to sail around the world for the purpose of carrying on chronometric measurements (figure 1–7). The voyage allowed the young naturalist the opportunity to study the plants and animals throughout the world. He studied closely the changes that occurred in the animal populations living in different environments and concluded that all animals change slowly over the immense period of time represented by the geologic past.

Darwin's observations led him to conclude that birds and mammals must have evolved from the reptiles and that the similar skeletal framework of the human, the bat, the horse, the giraffe, the elephant, the porpoise, and other vertebrates required that they be grouped together. Darwin felt that the superficial differences that could be observed between populations were the result of adaptation to different environments and modes of existence.

On his return to England, Darwin published his controversial book, *The Origin of Species,* which dealt not so much with the origin of life but with the evolution of living things into their many forms. Although Darwin's ideas were highly controversial, many can be indisputably proven by an impressive array of scientific facts.

The Rosses—Sounders of the Deep

Two of the earliest successful sounders of deep oceans were Englishmen, Sir John Ross and his nephew Sir James Clark Ross. Baffin Bay in Canada was explored by Sir John Ross in 1817 and 1818, and he was able to measure depths. Ross also collected samples of bottom-dwelling organisms and sediments with a device of his own design called a *deep-sea clamm.* He recovered starfish and worms in mud samples from a depth of 1.8 km (1.1 mi).

Sir James Clark Ross extended the soundings to greater depths on voyages to the Antarctic from 1839–43 (figure 1–8). A 7-km (4⅓-mi) sounding line was used on these voyages, and even this great length was at times insufficient to reach the bottom of the ocean. Sir James observed that the animals he recovered from the cold waters of the Antarctic were the same species his uncle had recovered from the Arctic area, and they were found to be very sensitive to temperature increase. This discovery led him to the conclusion that the waters making up the deep ocean must be of uniformly low temperature.

Edward Forbes

Also making an important contribution to the study of marine biology was a British naturalist, Edward Forbes (1815–54). Forbes was interested in determining the vertical distribution of life in the ocean, and after repeated observations, he divided the sea into specific life-depth zones. He came to the conclusion that plant life is limited to the zone near the surface. Forbes also found that the animal concentration was greater near the surface and decreased with increasing depth until only a small trace of life, if any, remained in the deepest waters.

Forbes's followers either misinterpreted his statement of the condition of life's distribution in the ocean or decided to apply their own logic to it, and they concluded that no life existed in the deep ocean, as life would be impossible under the conditions of high pressure and absence of light and air in this environment. Ignoring the findings of the Rosses in the high southern and northern latitudes, they persisted in their belief that no life existed in the abyssal ocean.

The Challenger Expedition

The controversy over the distribution of life in the ocean was one of the factors that helped stimulate interest in the first large-scale voyage with the express purpose of studying the subject.

In 1871, the Royal Society recommended that funds be raised from the British government for an expedition to investigate the following subjects:

1. Physical conditions of the deep sea in the great ocean basins;
2. The chemical composition of seawater at all depths in the ocean;
3. The physical and chemical characteristics of the deposits of the sea floor and the nature of their origin;
4. The distribution of organic life at all depths in the sea, as well as on the sea floor.

A staff of six scientists under the direction of C. Wyville Thompson left England in December, 1872, aboard HMS *Challenger*. A 2306-tn corvette that had been refitted to conduct scientific investigation, the *Challenger* returned in May 1876, after having traversed large portions of the Atlantic and Pacific oceans (figure 1–9). The achievements of the *Challenger* expedition, which covered 127,500 km (79,223

FIGURE 1–8
Ross Expedition. Members of the Sir James Clark Ross expedition to Antarctica after landing on Possession Island. The penguins seem to be at least as curious about the visitors as the members of the expedition are about them. (Reprinted by permission of Bettman Archive.)

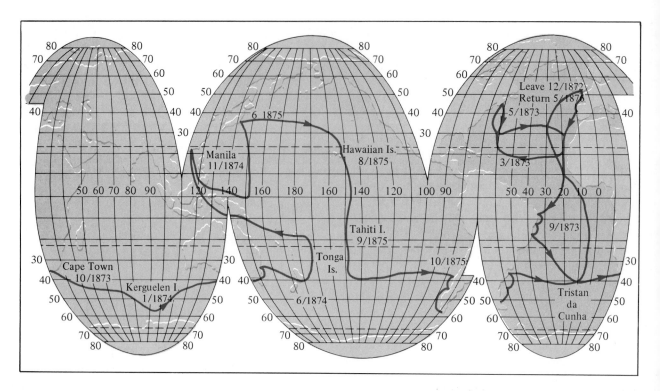

FIGURE 1–9

Voyage of HMS *Challenger*. The route traveled by HMS *Challenger* during the first major oceanographic voyage (December 1872 to May 1876). (Base map courtesy of The National Ocean Survey. Right, HMS *Challenger* from the *Challenger* Report, Great Britain, 1895.)

mi), included 492 deep-sea soundings, 133 bottom dredges, 151 open-water trawls, and 263 serial water temperature observations. The more outstanding aspects of the voyage were the netting and classification of 4717 new species of marine life and the measurement of a water depth of 8185 m (26,850 ft) in the Marianas Trench. Seventy-seven samples of ocean water collected during the Challenger expedition were analyzed in 1884 by C. R. Dittmar. This first refined analysis contributed greatly to our understanding of ocean salinity, which will be discussed in chapter 6.

Fridtjof Nansen

One of the most unusual of the voyages was initiated by a Norwegian, Fridtjof Nansen (1861–1930). Dr. Nansen developed a great interest in exploring the North Atlantic and the Arctic area. To aid him in his planning, the results of earlier attempts to explore the Arctic area during the latter part of the 19th century were compiled. Of particular interest to Nansen was the ill-fated American voyage of George Washington DeLong and his crew on the *Jeanette* in 1879. This expedition had attempted to sail through the Bering Strait to Wrangell Island, from which it planned to go overland to the pole. Scientists believed that Wrangell Island

was possibly the southern tip of a peninsula extending southward from the great, but yet undiscovered, Arctic continent.

The *Jeanette* became stuck in the ice on September 6, 1879, and drifted north of Wrangell Island, proving that the island was not a peninsula of the Arctic continent. After two years of drifting, the *Jeanette* sank off the New Siberian Islands. DeLong and many of his crew perished in the Lena Delta region of eastern Siberia.

In 1884, a number of articles that could be traced to the *Jeanette* were found frozen in the pack ice off the eastern coast of Greenland. On the basis of this evidence, Nansen considered that the articles must have drifted from the wreck of the *Jeanette* to the observed location by following a path that would have taken them over or near the North

Pole. He saw no reason why this ice flow that transported these articles could not be used to transport an expedition across the North Pole as well.

Eventually Nansen was able to raise sufficient funds for building the *Fram,* a ship that would be so designed that the expanding ice would not crush it but rather would force it up to the surface, free of the grip of the growing ice sheets.

Provisions for 13 men for 5 years were stored away on the small ship, and the crew set sail from Oslo on June 24, 1893. On September 21, after fighting their way along the northern coast of Siberia, they had not yet sighted the New Siberian Islands when the ice captured them at a latitude of 78°30′N. They were 1100 km (683 mi) from the North Pole. They had now begun what would be a long and lonely endeavor (figure 1–10).

On November 15, 1895, the *Fram* reached its northernmost point, 394 km (244 mi) from the pole. On August 13, 1896, the little *Fram* broke ice and was again unbound in the open sea. She had drifted a total of 1658 km (1028 mi) during her 3-year entrapment.

The drift of the *Fram* proved that no continent existed in the Arctic Sea and that the ice that covered the polar area throughout the year was not of glacial origin but a freely moving pack ice accumulation that had been formed directly upon the ocean surface. During the voyage, the crew had found that the depth of this ocean exceeded 3000 m (9840 ft) and that there was a rather surprising body of relatively warm water with temperatures as high as 1.5°C (35°F) between the depths of 150 and 900 m (about 500–3000 ft). Nansen correctly described this water as being a mass of Atlantic water that had sunk below the less saline Arctic water.

Nansen's realization of the need for more accurate measurements of salinity and temperature led him to develop the once widely used bottle for sampling the ocean depths, the *Nansen bottle.* His observations of the direction of ice drift relative to the wind direction helped V. Walfred Ekman, a Scandinavian physicist, to develop the mathematical explanation of this phenomenon. Nansen, who was awarded the Nobel Peace Prize in 1922, Ekman, and other Scandinavian scientists led the way in the development of the field of physical oceanography in the early 20th century.

20TH CENTURY OCEANOGRAPHY

Voyage of the *Meteor*

Oceanography entered a new era with the voyage of the *Meteor,* 1925–27. This German expedition was the first to use an echo sounder, making it possible to obtain a continuous depth recording as a vessel proceeded along its course. This represented a great advancement over reliance on scattered soundings and interpolation of the water depth between these points of known depths. The *Meteor*

FIGURE 1–10
Voyage of the Fram. The course followed by the *Fram* after becoming frozen in ice near the New Siberian Islands, September 21, 1893. On August 13, 1896, the little ship was released by the ice off Spitsbergen.

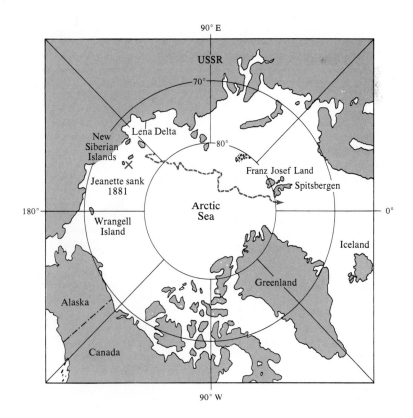

gathered data day and night for 25 months and concentrated on increasing the knowledge of the South Atlantic Ocean. It was this expedition that first revealed the true ruggedness of the ocean floor.

Many contributions from a number of nations have been made to the understanding of the oceans subsequent to the voyage of the *Meteor.* We will not discuss the details of these voyages, but we must emphasize that the knowledge we now possess concerning the world's oceans is still far from complete. Efforts that will be needed to bring it to a more complete state will, of necessity, continue for years into the future.

Oceanography in the United States

Although the first large-scale contribution to oceanography by a citizen of the United States resulted from the efforts of Lieutenant Matthew Maury of the U.S. Navy and culminated in his book *The Physical Geography of the Sea,* it was not until later in the 19th century that the United States became active in organizing voyages for the purpose of increasing our knowledge of the ocean. This phase of involvement actually began in 1877 with the voyages of the *Blake* under the direction of Alexander Agassiz. This young scientist, the son of the Swiss naturalist Louis Agassiz, contributed greatly to the development of oceanographic research in the United States. Following his voyages, many more have been initiated with the prime objectives of increasing the scientific knowledge of the world's ocean. Rather than detail these individual voyages, the following paragraphs will discuss briefly some projects undertaken by the major institu-

FIGURE 1-11

Dr. Maurice Ewing. Dr. Maurice Ewing (1906–74) pioneered seismic techniques for studying sediments on the ocean floor. He was director of Lamont-Doherty Geological Observatory at Columbia University from 1949 to 1972. (Photo courtesy of Columbia University.)

tions in the United States that are concerned with the investigation of the world's ocean.

The U.S. Navy Oceanographic Office has the primary objective of improving the combat readiness of the fleet, and basic research performed toward this objective is a part of its assignment. The results of their research, as well as that

FIGURE 1-12

Floating Instrument Platform (FLIP). FLIP, a 108-m (356-ft) Floating Instrument Platform, designed and developed by the Marine Physical Laboratory at Scripps Institution of Oceanography. In the vertical position (inset) it gives scientists an extremely stable platform from which to carry out underwater acoustic and other types of oceanographic research. FLIP has no motive power and must be towed to a research site in the horizontal position. Once on station, the ballast tanks are flooded and the vessel flips vertically, leaving 17 m (56 ft) of the platform above water. The work completed, water is forced from the tanks and FLIP resumes the horizontal position. (Courtesy of Scripps Institution of Oceanography, University of California, San Diego.)

of the Office of Naval Research and the U.S. Coast Guard, are of value and basic importance to all oceanographers.

On February 10, 1807, the U.S. Coast and Geodetic Survey was created by an act of Congress after President Jefferson requested a survey of the coast of the United States. This organization, now the National Ocean Survey, has the primary responsibility for bathymetric (ocean depth) surveys in the waters adjacent to the United States and its territories. Although the work done by this agency is primarily that of underwater mapping, the vessels that are involved are outfitted to conduct a wider scope of oceanographic research, gathering a wide range of oceanographic data.

The Commerce Department's National Oceanic and Atmospheric Administration (NOAA) works to ensure wise use of ocean resources through its National Ocean Survey, National Marine Fisheries Service, and Office of Sea Grant.

Not directly under the control of the federal government, but very important in the achievement of research of national interest, are three marine laboratories that have pioneered much of the marine research in the United States. The oldest of these institutions is Scripps Institution of Oceanography of the University of California at San Diego. Originated at La Jolla, California, in 1903, Scripps investigates a wide spectrum of marine problems.

Woods Hole Oceanographic Institution originated at Woods Hole, Massachusetts, in 1930 as a private nonprofit organization devoted to the scientific study of the world ocean. The ship *Atlantis II* operated by this institution was one of the first modern ships designed to conduct oceanographic research.

The Lamont-Doherty Geological Observatory of Columbia University was founded at Torrey Cliffs Palisades, New York, in 1949 under the direction of Dr. Maurice Ewing. He was one of the great pioneers of geophysical methods for investigation of the geology of the ocean bottom (figure 1–11).

These institutions have led the way in 20th-century oceanographic exploration. Their development of new devices for the study of the ocean depths has been sparked by the desire to unravel more of the mysteries hidden there. One of these imaginative devices is the Floating Instrument Platform (FLIP) shown in figure 1–12. It is designed for making acoustical and other open-ocean measurements that require a stable platform. Camera systems and magnetometer instrument capsules have been developed for free-fall emplacement on the ocean floor. They can be left to operate for days or months. Automatic timers or acoustical signals release these devices to float to the surface for recovery.

The need for cooperative efforts was recognized in the Deep Sea Drilling Project, which was being carried out under the leadership of the Scripps Institution of Oceanography, and through the cooperation of a number of educational institutions that have developed oceanographic

FIGURE 1–13

Old and New Challengers. Models of HMS *Challenger* and D/V *Glomar Challenger* are shown with the tracks of Phase 1 of the Deep Sea Drilling Project in the background. For propulsion, HMS *Challenger* relied on her sails and auxiliary steam power of 1234 hp. D/V *Glomar Challenger* can produce 8800 continuous or 10,000 intermittent horsepower for propulsion and to operate the drilling equipment. Where HMS *Challenger* shortened sails and used steam power to remain over a site, D/V *Glomar Challenger* has dynamic positioning. The first *Challenger* was 61 m (200 ft) long and displaced 2306 tn. Glomar *Challenger* is 122 m (400 ft) long and displaces 10,500 tn. The scientific impact of both vessels has been revolutionary. While HMS *Challenger* is generally thought of as having begun the science of deep ocean marine geology, the results of D/V *Glomar Challenger's* drilling and coring have truly revolutionized the science. (Courtesy of the Deep Sea Drilling Project, Scripps Institution of Oceanography, University of California, San Diego.)

TABLE 1–1

Some Important Achievements in the History of Human Descent into the Ocean.

360 B.C. Aristotle records the use of air trapped in kettles lowered into the sea by Greek divers in his *Problematum*.

330 B.C. Alexander the Great descends in a glass berrylike bell during the siege of Tyre.

1620 Dutch inventer Cornelius van Drebbel tests the first submarine in the Thames River. Drebbel descended to a depth of 5 m (16.5 ft) with King James I of England on board.

1690 Sir Edmund Halley descends in a lead-weighted diving bell. Air was replenished from barrels lowered into Thames River.

1715 Englishman John Lethbridge tests a leather-covered barrel of air with viewing ports and waterproof armholes to a depth of 18 m (60 ft). This design represented the first armored diving suit.

1776 Colonist David Bushnell's *Turtle* attempts the first military submarine attack on the British ship HMS *Eagle*. The attempt failed, but the British moved the fleet to less accessible waters.

1800 Robert Fulton builds the submarine *Nautilus*. It was the first of several to bear this name and was powered, when submerged, by a hand-driven screw propeller.

1837 Augustus Sieve invents a prototype of a modern helmeted diving suit.

1913 German company Neufeldt and Kuhnke manufactures an armored diving suit with articulating arms and legs.

1934 William Beebe and Otis Barton descend in a bathysphere to a depth of 923 m (3027 ft) off Bermuda.

1943 Jacques-Yves Cousteau and Émile Gagnan invent the fully automatic, compressed-air Aqualung.

1960 Jacques Piccard and Donald Walsh descend to a depth of 10,915 m (35,801 ft) in the bathyscaph *Trieste*.

1962 Hannes Keller and Peter Small make an open-sea dive from a diving bell to a depth of 304m (1000 ft) using a special gas mixture. Small died during decompression.

1962 Albert Falco and Claude Wesly live for 1 week under conditions of saturation in Cousteau's Continental Shelf Station off Marseilles, France.

1964 *Sealab I*—U.S. Navy team led by George F. Bond stays 11 days at 59 m (193 ft) off Bermuda.

1966 Submersibles *Alvin, Aluminaut,* and *Cubmarine* with the aid of unmanned remote controlled vehicle *CURV* locate and recover an atomic bomb from water over 800 m (2624 ft) deep off Palomares, Spain.

1969 Armored diving suit *Jim* is built to operate at depths below 304 m (1,000 ft).

1969–70 *Tektite I* and *II* accommodate 12 scientific teams from 14 to 60 days at a depth of 15 m (50 ft) off the U.S. Virgin Islands.

1976 Jacques Mayol dives to 100 m (328 ft) while holding his breath for 3 minutes and 40 seconds.

1981 Diving scientists at Duke University simulate a dive to 686 m (2250 ft) in a pressure chamber.

---

programs. In November 1963, the director of the National Science Foundation announced the establishment of a national program for sediment coring. Scripps Institution of Oceanography and three other major oceanographic institutions—the Rosenstiel School of Atmospheric and Oceanic Studies at the University of Miami, Florida; Lamont-Doherty Geological Observatory of Columbia University; and Woods Hole Oceanographic Institution in Massachusetts—united to form the Joint Oceanographic Institutions for Deep Earth Sampling (JOIDES). They were later joined by the departments of oceanography of the University of Washington, Texas A&M, the University of Hawaii, Oregon State University, and the University of Rhode Island.

A ship capable of drilling the ocean bottom while floating 6000 m (3.7 mi) above at the ocean surface had to be designed specifically to carry on this program. The *Glomar Challenger* was constructed and launched March 23, 1968 (figure 1–13). The first leg of the Deep Sea Drilling Project commenced on August 11 of that year. The *Glomar Challenger* is 122 m (400 ft) long and 20 m (66 ft) in beam, and its displacement is 10,500 tn loaded. The vessel is a self-sustained unit capable of remaining at sea for 90 days.

Initially the oceanographic research program was financed by the U.S. government, but in 1975 it became international. Financial and scientific support is now received from West Germany, France, Japan, the United Kingdom, and the USSR. In 1983 the Deep Sea Drilling Project became the Ocean Drilling Program (ODP) with the broader objective of also drilling the continental margins. This new phase of operations is supervised by Texas A&M University. Accompanying this change in emphasis was the decommissioning of *Glomar Challenger* after a highly successful 15-year role in which it made possible a major advancement in the field of earth science—confirmation of sea-floor spreading, a process in which new ocean floor is being

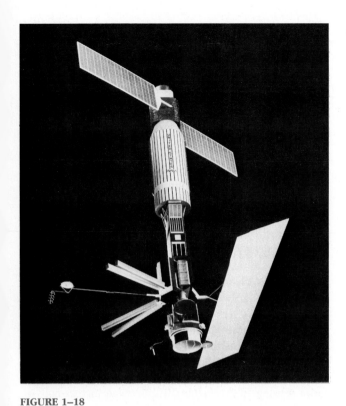

FIGURE 1–18

Seasat A. *Seasat A,* the first oceanographic satellite, was launched July 7, 1978. It shorted out October 10, 1978, but provided sufficient data to prove the value of satellite remote sensing to the study of the oceans. (Credit: Jet Propulsion Laboratory, NASA.)

In 1609, Hugo Grotius, the Dutch jurist and statesman whose writings were important in the formulation of international law, established in *Mare liberum* the doctrine that the ocean is free to all nations. Although controversy continued over whether nations could control portions of the oceans, Cornelius van Bynkershonk produced a satisfactory solution to this problem in *De dominio maris,* published in 1702. It provided for national domain over the sea for the distance that could be protected by cannons on the shore. The first official dimension of this *territorial sea* was probably established in 1672 when the British determined that they would exercise control over a strip of water 1 league (3 nautical miles) wide along the shores of the Commonwealth—the territorial sea.

Because of the rapidly developing technology for drilling beneath the ocean, the first United Nations Conference on the Law of the Sea held in Geneva in 1958 produced a treaty relating to the continental shelf. It stated that prospecting and mining of minerals on the shelf are under the control of the country that owns the nearest land. Unfortunately, the seaward limit of the continental shelf was not well-defined in the treaty.

The second United Nations Conference on the Law of the Sea was held in Geneva in 1960 and achieved little. The third had its first meeting in New York in 1973. Subsequent meetings were held at least once a year.

The third conference came to a disappointing conclusion in April of 1982, when a large majority—130 to 4, with 17 abstentions—voted in favor of a new Law of the Sea treaty. Most of the developing nations who could benefit significantly from the provisions of the treaty voted for its adoption. The opposition was led by the United States, who, along with Turkey, Israel, and Venezuela, voted against it. These countries support private companies planning seabed mining operations and believe certain provisions of the treaty will make it unprofitable. Among the abstainers were the Soviet Union, Great Britain, Belgium, the Netherlands, Italy, and West Germany, all of whom are interested in seabed mining. Because only 30 nations have ratified the treaty, its ability to facilitate the exploitation of seabed resources will be severely hampered.

The primary features of the treaty are as follows:

1. *Coastal Nations Jurisdiction:* A uniform 12-mi territorial sea and a 200-mi Exclusive Economic Zone were established. The coastal nation has jurisdiction over mineral resources, fishing, and pollution within the EEZ and beyond it, up to 350 mi from shore, if the continental shelf (defined geologically) extends beyond the EEZ.

2. *Ship Passage:* The right of free passage on the high seas is maintained; it is also provided for within territorial seas through straits used for international navigation.

3. *Deep-Ocean Mineral Resources:* Private exploitation of seabed resources may proceed under the regulation of the International Seabed Authority (ISA), within which will be a mining company chartered by the United Nations, called Enterprise. Mining sites will be operated in pairs: one chosen by the private mining concern, the other by Enterprise, and both approved by ISA. Revenues from the private concern's chosen site will go to that concern, and those from the other site will be divided among the developing nations.

4. *Arbitration of Disputes:* A Law of the Sea tribunal will perform this function.

SUMMARY

In the Western world, the **Phoenicians** were the first great navigators. Later the Greeks, Romans, and Arabs made significant contributions that lay hidden behind the curtain of the Middle Ages for over 1000 years. The **Age of Discovery** renewed the Western world's interest in exploring the unknown. It began with the voyage of Columbus in 1492 and ended in 1522 with the circumnavigation of the earth by a voyage initiated by **Ferdinand Magellan**.

Captain James Cook, who sailed in the late 18th century, was one of the first navigators to sail the oceans with the primary purpose of learning their natural history. He also made the first accurate maps using the **chronometer** developed by John Harrison. **Matthew Fontaine Maury** was the first American to make a large-scale contribution to the study of the oceans with publication of *The Physical Geography of the Sea* in 1855. By the mid–19th century, the English biologists **Charles Darwin, Sir John Ross, Sir James Ross,** and **Edward Forbes** had made significant contributions to increasing our knowledge of life in the oceans.

The voyage of **HMS *Challenger,*** from December 1872 through May 1876, represented the first major full-scale oceanographic voyage. A feel for the significance of this exploration can be gained by contemplating the fact that 4717 new species of marine life were identified during the voyage. An unusual voyage aboard the *Fram,* which spent 3 years entrapped in Arctic ice was initiated by the Norwegian explorer **Fridtjof Nansen** in 1893. It was with this voyage that the first knowledge of the oceanography of the Arctic Ocean was gained. From 1925 through 1927 during a period of 25 months, the German ship *Meteor* crisscrossed the South Atlantic Ocean, recording the first large-scale continuous depth record with an echo sounder. This voyage first revealed the true ruggedness of the ocean floor.

Recent oceanographic investigations have been numerous. In the United States, the early ones were conducted by government agencies and three pioneering oceanographic institutions—**Scripps Institution of Oceanography, Woods Hole Oceanographic Institution,** and **Lamont-Doherty Geological Observatory**. They have since been joined by many educational institutions. Major projects such as the **Deep Sea Drilling Project** and **Geochemical Ocean Sections** have resulted from this widespread cooperation among institutions and nations.

The use of manned submersibles such as *Alvin* and *Sea Cliff II* has contributed significantly to new discoveries in the deep ocean. Remote unmanned systems such as **RUM III** and the **Seabeam/*Argo-Jason*** systems will be able to replace more expensive submersibles for much deep ocean research. Many new techniques have been developed for studying the ocean. **Ocean Acoustical Tomography** and **remote sensing** from satellites will become increasingly important.

Because of the imminence of extensive exploitation of the ocean, the United Nations has conducted conferences on the **Law of the Sea** to develop international agreement on how the resources of the ocean can be shared and managed. These efforts were certainly set back, and may have failed, in April 1982, when developed and developing nations could not agree. The treaty voted on then may not be ratified by enough developed nations to significantly increase current levels of seabed resource exploitation.

QUESTIONS AND EXERCISES

1. Construct a time line that includes the major events of human history that have resulted in a greater understanding of our planet in general and the oceans in particular.

2. List the Mediterranean cultures that appear on your time line in the probable order of their dominance in the field of marine navigation.

3. Using a diagram, describe the method used by Pytheas to determine latitude in the Northern Hemisphere.

4. How did the Greek concept of geography provided by Herodotus differ from that of the Roman geographer Ptolemy?

5. While the Arabs dominated the Mediterranean region during the Middle Ages, what were the most significant events taking place in northern Europe?

6. List some of the major achievements of Captain James Cook.

7. What was Matthew Fontaine Maury's major contribution to an increased knowledge of the oceans?

8. Define in your own words the process of evolution by natural selection.

9. What was the controversy that developed due to the observations of the Rosses and Edward Forbes?

10. List some major achievements of the voyage of HMS *Challenger.*

11. What new knowledge of the Arctic polar region did the voyage of the *Fram* provide?

12. The voyage of which ship first revealed the true relief of the ocean floor?

List the names and locations of three pioneering oceanographic research institutions.

Describe the functions of the Floating Instrument Platform (FLIP), the Remote Underwater Manipulator (RUM III), and Seabeam/*Argo-Jason*.

Name the major ocean research programs designed to drill into the ocean floor and study ocean circulation.

16. Discuss possible reasons the less developed nations might prefer the concept that the open ocean is the common heritage of all nations, while the more developed nations might prefer the concept that the open ocean's resources belong to those who recover them.

REFERENCES

nendinger, E. 1982. Submersibles:past—present—future. *Oceanus* 25:1, 18–35.

ley, H. S., Jr. 1953. The voyage of the *Challenger. Scientific American* 188:5, 88–94.

rgese, E. M., and Ginsberg, N. 1986. *Ocean Yearbook 6.* Chicago: The University of Chicago Press.

omie, W. J. 1962. *Exploring the secret of the sea.* Englewood Cliffs, N.J.: Prentice Hall.

xbury, A. C. 1971. *The earth and its oceans.* Reading, Mass.: Addison-Wesley.

Knauss, J. A. 1974. Marine science and the 1974 Law of the Sea Conference—science faces a difficult future in changing Law of the Sea. *Science* 184:1335–41.

Mowat, F. 1965. *Westviking.* Boston:Atlantic-Little, Brown.

Ryan, P. R. 1986. The *Titanic* revisited. *Oceanus* 29:3, 2–17.

———. 1985. The *Titanic:* lost and found. 1985. *Oceanus* 28:4, 1–112.

SUGGESTED READING

a Frontiers

xer, S. G. 1981. The continent that wasn't there. 27:2, 108–14. A history of the search for Ptolemy's *Terra Australis Incognita,* the unfound southern continent.

arlier, R. H., and Charlier, P. A. 1970. Matthew Fontaine Maury, Cyrus Field, and the physical geography of the sea. 16:5, 272–81.

biography of Maury with emphasis on his accomplishments as superintendent of the Department of Charts and Instruments. His role in laying the first transatlantic cable and a discussion of his most famous work, *The Physical Geography of the Sea,* are included.

gle, M. 1986. Oceanography's new eye in the sky. 32:1, 37–43.

otographs taken by Paul Scully-Power, a U. S. Navy oceanographer, from the space shuttle add to our knowledge of ocean currents and waves.

effelbein, B. 1983. Law of the Sea treaty: What does it mean to nonsigners (like the United States)? 29:6, 358–66.

overview of the Law of the Sea Convention and a discussion of the implications of the treaty not being signed by all nations.

Clintock, J. 1987. Remote sensing: adding to our knowledge of oceans—and Earth. 33:2, 105–113.

e role of remote sensing, primarily in aiding us in understanding weather and climate, is discussed.

Rice, A. L. 1972. HMS *Challenger*—midwife to oceanography. 18:5, 291–305.

An interesting account of the achievements recorded during the first major oceanographic expedition.

Schuessler, R. 1984. Ferdinand Magellan: The greatest voyager of them all. 30:5, 299–307.

A brief history of the voyage initiated by Magellan to circumnavigate the globe.

Van Dover, C. 1987. *Argo* Rise: outline of an oceanographic expedition. 33:3, 186–194.

A description of the first scientific use of the *Argo* system to survey hydrothermal vent activity on the East Pacific Rise off the coast of Mexico.

Scientific American

Borgese, Elizabeth Mann. 1983. The law of the sea. 248:3, 42–49.

One hundred nineteen nations sign a convention aimed at establishing a new international scheme of regulating the oceans.

Herbert, Sandra. 1986. Darwin as a geologist. 254:5, 116–123.

Before publication of *Origin of species,* Charles Darwin considered himself to be primarily a geologist. This article outlines his contributions in this field with special attention to his theory of coral reef formation.

2

THE ORIGIN OF THE EARTH, ITS OCEANS, AND LIFE IN THE OCEANS

It can be a wonderful and rewarding experience to learn about the world's oceans. The wonder is heightened by adding this new knowledge to a general understanding of the place of the oceans in the cosmos.

A great deal of speculation is involved in these considerations, but science has not made advances by restricting itself to that which is already known. We should, however, begin our discussion with those aspects of the universe which we can readily observe and describe. Then we will discuss some theories of the origin of the universe and our planet.

THE UNIVERSE WE SEE

As astronomers looked into the sky surrounding the planet Earth, they gradually developed an understanding of Earth's position in what is now called the *solar system* (figure 2–1). The solar system is composed of the sun and its nine planets, which occupy a disc-shaped portion of the universe about 13 billion kilometers (8 billion miles) in diameter. Although this seems to be an enormous structure, we find that it represents a very minute portion of the total universe. Our sun is one of perhaps 100 billion billion, or 10^{20}, stars that constitute the known universe. In terms of distances with which we are familiar, the solar system is incomprehensibly isolated. The distance from the sun to the nearest star is 40.7 thousand billion kilometers (25 thousand billion miles).

While the planets revolve around the sun, the entire system is moving at about 280 km/s (174 mi/s) around the center of the Milky Way, the galaxy or the group of stars to which our solar system belongs. Estimates of the dimensions of the Milky Way indicate that it is a disc-shaped accumulation of stars about 100,000 light-years in diameter and 10,000–15,000 light-years thick near the center, thinning gradually toward the edges. A *light-year* is a unit of measure used by astronomers to describe distances within the universe and is equal to the distance traveled by light at a speed of 300,000 km/s (186,000 mi/s) during a period of one year— almost 10 trillion (9.8×10^{12}) km (6.2 trillion miles).

Cape Kiwanda, Oregon. Photo by Ray Atkeson.

FIGURE 2–1

The Solar System. *A*, the orbits of the planets of the solar system are drawn to scale. *B*, relative sizes of the sun and the planets are shown.

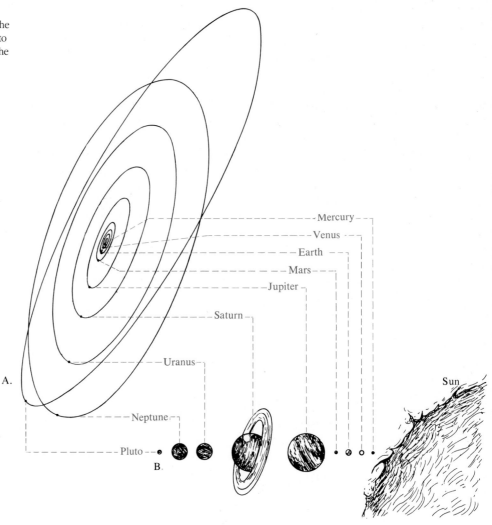

Within the Milky Way are some 100 billion (10^{11}) stars, tens of millions of which very probably possess families of planets, millions of which may be inhabited by intelligent creatures, according to the estimates of some astronomers.

All the stars you see with your naked eye belong to the Milky Way (figure 2–2). If you have exceptional vision and live in the Northern Hemisphere, you may see a hazy patch of light in the constellation Andromeda that represents another galaxy within the universe. This galaxy is some 1.5 million light-years away. It appears that galaxies are exceedingly numerous, and within a distance of a billion light-years from our galaxy, there are at least 100 million galaxies large enough to be observed by our telescopes.

By observing light energy that radiates from these distant galaxies, astronomers have been able to determine that most are moving away from us. The velocity with which the galaxies recede from our system is proportional to their distance from it. Velocities of nearly 240,000 km/s (149,000 mi/s) have been calculated for the most distant galaxies as they recede from the Milky Way. This is in excess of three-fourths of the speed of light!

One may be left with the impression that all the galaxies are being repelled by our solar system, but a more reasonable consideration would be that galaxies are moving away from one another as would fragments that result from an explosion. The universe appears to be rapidly expanding, its component galaxies moving ever farther apart. If all these galaxies are moving away from one another in a manner similar to that of fragments created by explosion, we might ask if they all belonged to one large mass. We, of course, do not know. But if that was the case, the time required for them to have reached their present distribution would have been at least 15 billion years.

ORIGIN OF THE EARTH

As insignificant as our planet is in the overall composition of the cosmos, it is nonetheless of interest to us to learn what we can of its origin. Although no complete understanding of the origin of the earth has yet been developed, it is possible to outline the probable sequence that led to its present form. As galactic matter revolved around its center of rotation, the stars began to form as individual glowing masses due to the concentration of space particles under the force of gravity. These concentrations may have occurred in local eddies and over a vast period of time took their present form. Many of the resulting stars developed assemblages of planets that revolved around them.

Such a general sequence may have occurred with the formation of our sun, and in its early stages it may have had a size at least equal to the diameter of the solar system that we observe today. As this large accumulation of gas and space dust began to contract under the force of gravity, a small percentage of it was left behind in smaller eddies with features similar to the large eddies in which the stars developed in association with the rotation of the galaxies.

As this material was left behind, it flattened itself into a disc that became increasingly more dense. The stage was set for the formation of the suite of planets that revolve around the sun. Because of the increase in density of this disc, it became gravitationally unstable and broke into separate small clouds. These were protoplanets, which later consolidated into the present planets.

Protoearth was a huge mass, perhaps 1000 times greater in diameter than the earth today and 500 times more massive. The heavier constituents of the planet Earth, as with all the other protoplanets, were migrating toward the center to form the heavy core surrounded by lighter materials. Similarly, the satellites we observe around the planets, such as our moon, began to form. Our moon is relatively large compared to the size of the earth. In its early formation it was much closer to the earth than it is today.

During this early formation of the protoplanets and their satellites, the sun eventually began to "shine" as it condensed into such a concentrated mass that forces were set up within its interior, creating energy through a process known as the *carbon cycle*. When temperatures reach tens of millions of degrees, hydrogen atoms, in the presence of carbon, are converted to helium atoms, and large amounts of energy are released. In addition to the light which is so obviously emitted by the sun, there is an intense emission of ionized, or electrically charged, particles. In the early stages of the development of our solar system, this emission of ionized particles served to blow away the nebular gas that remained from the formation of the planets and their satellites.

Meanwhile, the cold protoplanets nearer the sun were being warmed by the solar radiation, and their atmospheres began to boil away. The combination of emission of ionized solar particles and the internal warming of the planets resulted in a drastic decrease in the size of the planets closer to the sun. Since the gas envelope that surrounded the planets was composed mostly of hydrogen and helium (we may consider this to be the composition of the earth's first atmosphere), the energy that was available easily allowed atoms of this small size to escape from the gravitational attraction of the planets. As the protoplanets continued to contract, the heat produced deep within their cores as a product of spontaneous disintegration of atoms (radioactivity) became more intense.

The evidence indicates that the earth ultimately became molten, allowing the heavier components to concentrate in the core. The lighter of the materials segregated according to their densities in concentric spheres around this core. Surrounding the core, which contains abundant iron and nickel, is a zone composed of heavy iron and magnesium silicates—the mantle. The thin crust overlying the mantle ranges in thickness from 4 to 10 km (2.5–6.2 mi) for the basaltic rock underlying the oceans to a 35- to 60-km (22–37 mi) thickness for the lighter granite of the continents.

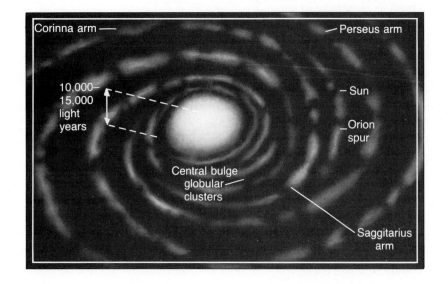

FIGURE 2–2
The Milky Way. The sun and its planets are located about one-half the distance from the center of the Milky Way to the edge. The Milky Way completes a rotation every 200 million years. To maintain its position in the Milky Way, the solar system travels around the center of the galaxy at a velocity of about 280 km/s (149,000 mi/s).

Corinna arm

Perseus arm

10,000–15,000 light years

Sun

Orion spur

Central bulge globular clusters

Saggitarius arm

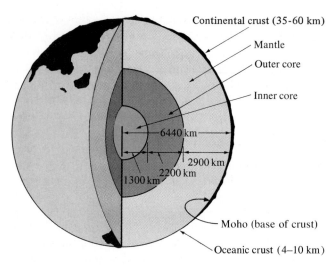

FIGURE 2–3
The Earth's Structure. The three major subdivisions of the earth—core, mantle and crust—are shown in the cross section.

The crust is separated from the mantle by a distinct density change called the *Mohorovičić discontinuity* (Moho), after the Yugoslavian seismologist who discovered its existence (figure 2–3).

ORIGIN OF THE ATMOSPHERE AND OCEANS

The solidification of the earth's crust marks the beginning of geologic history, and evidence indicates that this event occurred about 4.6 billion years ago (figure 2–4). At this time the earth had lost all but a small fraction of its original gas envelope. It would seem improbable that there would have been much of an atmosphere surrounding the earth at this time, if we consider the tremendous amount of energy that was being radiated into space from the earth as it cooled from its molten condition.

All this heat energy was made available to the molecules of the atmosphere, and they undoubtedly would have been able to achieve a velocity well above that required to escape the earth's gravitational field. The lighter the molecule and the higher the energy of that molecule, the greater is the probability that a gas composed of such molecules will escape from the earth's surface. It has been calculated that when the average molecular velocity in an atmosphere exceeds one-fourth of the planet's escape velocity, which for the earth is 11.4 km/s (6.9 mi/s), the atmosphere will be lost into space within a relatively short period of time. At present hydrogen and helium are rapidly lost from the earth's atmosphere, while gases such as oxygen and nitrogen, composed of heavier molecules, are retained.

The Atmosphere Forms

Where did the material to make up the present atmosphere and ocean come from? Although the earth's surface had become relatively cool by this stage in its development, there was still much volcanic activity according to the geologic record. As volcanic gases bubbled up from the earth's upper mantle, they formed a new atmosphere as soon as the earth had cooled enough to retain one. This atmosphere was far different from the one that the earth had lost and from the one it presently has. It contained little free oxygen and was composed of large percentages of carbon dioxide and water vapor. As the earth cooled to the point at which it could retain these gases, a vast envelope of clouds surrounded the planet, causing the earth to reflect about 60 percent of all the sunlight that was directed toward it. Most of the energy it absorbed from the sun remained in the thick envelope of gases and did not reach the surface of the solid earth.

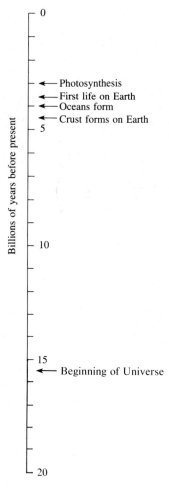

FIGURE 2–4
Major Events in the Evolution of the Earth.

Water vapor released from the surface of the earth was captured in the thick envelope and was forced to remain there as a vapor until sufficient cooling of the earth's crust would allow it to condense and begin to accumulate as a liquid. Eventually, cooling of the crust allowed the water vapor to condense and fall under the force of gravity to the earth's surface. Finally, with further decrease in the surface temperature, permanent accumulation of water began in depressions on the earth's irregular surface (figure 2–5).

The Oceans Evolve

These accumulations were the initial oceans on planet Earth, and it is assumed that this water has been permanently in existence since its formation. The primary crystalline rocks that made up the earth's crust were weathered and eroded. The inorganic compounds of which they were composed were dissolved in the waters that fell relentlessly upon their surfaces and were carried into the newly forming oceans.

The process of cooling the earth's surface, which allowed the water vapor to finally condense and accumulate to form the earth's first oceans, was accelerated by water's property of high heat capacity. The heat energy that prevented the water from accumulating on the earth's surface was being absorbed by that massive cloud of water vapor and carbon dioxide that surrounded the earth's solid surface. As the individual molecules of water absorbed heat at the earth's surface and moved through the cloud to its outer edges where they were frozen into ice crystals, large quantities of heat were released into outer space. We might say that water vapor was working hard to achieve its ultimate goal of establishing for itself a permanent resting place on the earth's surface.

FIGURE 2–5
Formation of the Earth's Oceans.
A, widespread volcanic activity released H_2O and smaller quantities of CO_2, Cl_2, N_2, and H_2, which produced a water vapor atmosphere that also contained carbon dioxide (CO_2), methane (CH_4), and ammonia (NH_3). *B*, as the earth cooled, (1) the water vapor condensed and (2) fell to the earth's surface. There it accumulated to (3) form the oceans.

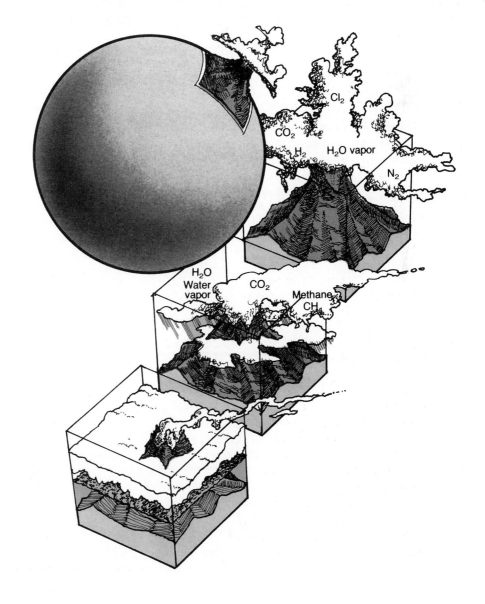

THE CHEMICAL BALANCE SHEET

As chemical weathering of the primary rocks that composed the original earth's crust was made possible by the appearance of water, the process of chemical weathering should be considered. As a result of chemical weathering, the elements that were contained in the primary crystalline rock were freed, dissolved in the water, and carried off to accumulate in the ocean. As the water carried these dissolved elements and compounds into the ocean, it also carried a great load of particulate material that had been weathered and eroded from the crystalline rocks and deposited this material as sediment. The components of the primary crystalline rock that were freed by chemical weathering must be found (1) dissolved in the ocean, (2) as components of the earth's atmosphere, or (3) chemically bound with the sediments that have been carried into the sea by the rivers (see figure 2–6).

The estimates that have been made for these various amounts, of course, will not be exact. The estimate for the ocean is probably one of the most accurate, since the ocean volume is relatively well known and its composition is comparatively uniform because it is well mixed. The estimates are less accurate for crystalline rock, which is quite heterogeneous, and for the sediments, which are of various types. Nevertheless, geochemists have drawn up balance sheets for the various elements, and those calculated by different individuals working on the problem agree fairly well. For those elements that are most common in the crystalline rocks, the geochemical equations balance well within the limits set by the accuracy of the estimates.

Excess Volatiles

There are some components that do not balance in any of the attempts that have been made. These elements and compounds appear to be much more abundant in the at-

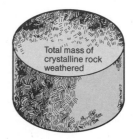
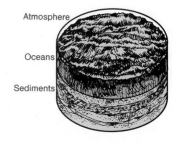

The total mass of crystalline rock components freed by chemical weathering must be found as components of the atmosphere, oceans, or sediments.

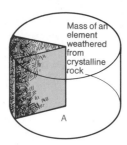
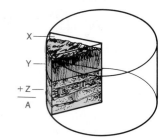

When considering one element represented by the dark wedge in the "pie" to the left, (A), the sum of the masses of that element found in the atmosphere, (x), oceans, (y), and sediments, (z), should equal the mass weathered from the crystalline rock, $A = x + y + z$

FIGURE 2–6
Geochemical Balances.

mosphere, the ocean, and the sediments than can be accounted for by the volume of crystalline rocks that have been estimated to have undergone chemical weathering. All these substances that are found in excessive amounts compared to the products of chemical weathering are volatile (gaseous) at or slightly above the average temperatures found at the earth's surface and are therefore called *excess volatiles*. Most abundant of the excess volatiles are

TABLE 2–1
Excess Volatiles. The volatile elements and compounds that are found in the atmosphere, ocean, and sedimentary rocks in far greater amounts than have been made available by chemical weathering of primary crystalline rocks are called *excess volatiles*. They are believed to have been released from the molten interior of the earth by volcanic activity. Note the absence of free oxygen.

	Mass of Elements or Compounds in 10^{20} g					
	H_2O	C as CO_2	Cl	N	S	H, B, Br, Ar, F, etc.
Found in:						
Atmosphere and ocean	14,600	1.5	276	39	13	1.7
Buried in sedimentary rocks	2,100	920	30	4	15	15
Total	16,700	921.5	306	43	28	16.7
Supplied by:						
Weathering of crystalline rocks	130	11	5	0.6	6	3.5
Excess volatiles (unaccounted for by rock weathering)	16,570	910.5	301	42.4	22	13.2

Courtesy of the Geological Society of America. After Rubey, 1951.

water vapor and carbon dioxide; also important are quantities of chlorine, nitrogen, sulfur, hydrogen, and fluorine (see table 2–1).

Volcanic activity is presently producing these gases and venting them into the earth's atmosphere. A great percentage of the gases that surface as a result of volcanic activity may simply be recycled volatiles that have been dissolved in groundwater and carried down from the earth's surface, but it has been determined that a small percentage of the total volatiles is always composed of new gases that have for the first time been released from the crystalline rock of the earth's crust. When molten magma cools, solid minerals are formed and some of the gaseous constituents of the magma are freed. These gases, freed during the crystallization process, represent the new material coming to the earth's surface from the mantle.

We have geologic evidence in the form of fossilized algae, assumed to have grown in a marine environment, that the oceans have existed on the earth's surface for well over 3 billion years. If we consider the rate of discharge of excess volatiles by hot springs in the United States to be equal to the average rate of production of excess volatiles throughout the last 3 billion years, we find that enough water vapor would have been produced to fill the oceans to more than 100 times their present volume. We must remember that much of this water that comes to the surface is water that is being recycled; it does not represent new water released from the crystallization of magma. Only 1 percent of this water would require such an origin to account for the present volume of the earth's oceans from this source.

DEVELOPMENT OF THE OCEANS AND THEIR BASINS

When the earth began to form its solid crust about 4.6 billion years ago, the mass of the oceans was zero, while at present it is 1.4×10^{24} g. We do not know what the rate of outgassing was throughout this period of 4.6 billion years. However, if we assume it to have been equal to the present rate, we can visualize the formation of the earth's oceans as a gradual process.

Does continual outgassing mean that the oceans are gradually covering more and more of the earth's surface? Not necessarily, since the area that the oceans cover is directly determined by the volume that the basin in which they form is able to accept. The crust, which we now describe as continental, may not have been present during the initial solidification of the earth's crust and may well have formed gradually as the oceans themselves were formed. Since the volume of the ocean basins is directly related to the difference in thickness between the oceanic and the continental crusts, and if we assume that the con-

tinental material has gradually accumulated, we can see that the capacity of the ocean's basins could have gradually increased along with the volume of water that was being produced over this period of time. Thus, it is quite possible that the oceans have had a distribution very similar in total surface area to that of the present oceans for a long period of geologic time and that the major change in the ocean's character has been an increase in depth. Throughout the geologic time during which the oceans and continents were increasing their volumes, the ocean basins were getting deeper to accommodate the increasing volume of water that was being produced in association with the formation of the continental crust. This process will be discussed further in chapter 4.

Considering the history of ocean salinity, we might ask if the oceans have possessed a relatively uniform salinity throughout their history or if they are getting more or less saline. By far the most important component of salinity is the chloride ion, Cl^-, which is produced by the same process that produces the water vapor forming the oceans. So our considerations boil down to one question: has the relative amount of water vapor to chloride ion production remained constant throughout geologic time, or has it varied? We have no indication that there has been any fluctuation in this ratio throughout geologic time. We must then consider, on the basis of the present evidence, that the oceans' salinity has been relatively constant, while their volume and the volume of the crystalline rock that makes up the continental crust has increased throughout the 4.6 billion years since the earth's crust first began to form.

LIFE FORMS IN THE OCEANS

We did not include free oxygen in the excess volatiles that were produced at the earth's surface by volcanic activity. If such a gas were produced, it is certain that it could not remain free in the earth's atmosphere because of the iron found in volcanic rock. Any oxygen that surfaced during volcanic activity would have combined with this iron. As a result of this condition, we may consider that the earth's early atmosphere did not contain this gas that makes up almost 21 percent of our present atmosphere.

We depend on the oxygen in our atmosphere for protection from the ultraviolet radiation that is constantly bombarding our planet. Much of the energy represented by this ultraviolet radiation is exhausted in our upper atmosphere converting the free oxygen (O_2) into ozone (O_3), thus allowing very little of the ultraviolet energy to reach the surface of the earth.

Of course, if oxygen was missing from the primitive atmosphere, the ultraviolet radiation would have readily penetrated the atmosphere and reached the surface of the young oceans. These oceans contained the gases methane

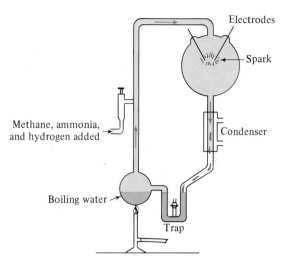

FIGURE 2–7
The apparatus used by Stanley L. Miller in the 1952 experiment that resulted in the synthesis of the basic components of life, amino acids. A mixture of water vapor, methane, ammonia, and hydrogen was subjected to an electrical spark that provided the energy for synthesis. This mixture is thought to resemble the composition of the atmosphere and oceans that existed when "organic" molecules first formed on Earth.

(CH_4), ammonia (NH_3), nitrogen (N_2), and carbon dioxide (CO_2).

Laboratory experiments have shown that exposing a mixture of hydrogen, carbon dioxide, methane, ammonia, and water to ultraviolet light and an electrical spark will produce a large assortment of organic molecules (figure 2–7). The exposure of the mixture to ultraviolet radiation causes photosynthesis, carried on in the absence of chlorophyll. *Photosynthesis* is a chemical reaction in which energy from the sun is stored in the products of the photosynthetic reaction—the organic molecules. The production of organic substance in the shallow oceans that developed early in the earth's history must have resulted in a vast amount of "organic" material, as do the laboratory experiments, but this material did not yet represent life on the planet earth. There is also evidence that some organic molecules reached the earth from outer space via meteorites.

Carbon—The Organizer of Life

Let us now leave these general considerations for a moment and consider more specifically the element that seems to be the most important in allowing the buildup of the organic compounds we have just mentioned. This element is carbon. We even refer to the study of the chemistry of carbon compounds as organic chemistry. In studying the chemistry of carbon, we are constantly concerned with molecules which are several times larger than those with which the inorganic chemist is commonly concerned. The greater percentage of inorganic molecules contain fewer

than 10 atoms, while in the study of organic chemistry, molecules which may exceed 100 atoms per molecule are not uncommon.

This fact arises out of the "combining power" of the carbon atom. The combining power of an atom is indicated by its *valence*. Carbon has a valence of four (the highest normal valence), which means it can combine with four atoms of hydrogen.

To demonstrate how this ability gives carbon an unusually high combining power, let us compare the carbon atom with other common atoms with valences less than four. If we consider chlorine, which occurs as a gas and has a valence of one, we can see that a single atom of chlorine combines with a single atom of hydrogen to produce a molecule of hydrogen chloride, HCl.

Cl—H

Oxygen, which also occurs as a gas, has a valence of two. An atom of oxygen combines with two molecules of hydrogen in the formation of water, H_2O.

Another gas common in our atmosphere, nitrogen, is trivalent. One atom of nitrogen will combine with three atoms of hydrogen to produce the gas ammonia, NH_3.

We return to the carbon atom, with a valence of four, and find that one carbon atom will combine with four atoms of hydrogen to produce the gas methane, CH_4.

We can readily see that the higher the valence of an atom of an element, the greater the complexity of the molecules it can form. If we consider that carbon has a valence of four compared to the valence of one for chlorine, then carbon must have a much greater combining power than chlorine or any other element with a valence of one. Consider chlorine as it combines with the metal sodium (Na), which also has a valence of one.

NaCl

We can see that these two elements can combine to form only one compound, sodium chloride (common table salt).

By contrast, methane, an important constituent of the earth's early atmosphere, represents only the simplest of the group of carbon compounds called *hydrocarbons*—those containing only carbon and hydrogen atoms. Somewhat more complex are the *carbohydrates,* which are composed of carbon and water molecules. Belonging to this group are the sugars and starches. Even more complex carbon compounds that are known as *fats* and *proteins* may or may not have been present in those early oceans as products of inorganic photosynthesis. Aiding carbon further in forming large molecules is the ability of carbon atoms to attach to one another and produce long chains and rings.

We do not know precisely how the organic material took on the characteristics that allowed it to become living substance. Through some process, however, the organic material became chemically self-reproductive, developed the ability to actively seek light, and began to metabolize and grow toward a characteristic size and shape dictated by forces from within. These living entities were composed mostly of a substance with a chemical composition not greatly unlike that of the water in which they originated. We call this material *protoplasm.*

Plants and Animals Evolve

These very earliest forms of life must have been *heterotrophic* organisms, those that depend on an external food supply. That food supply was certainly abundant in the form of molecules of the nonliving organic material from which the first living single-celled organisms originated. The *autotrophs,* those organisms that do not depend on an external food supply but manufacture their own, eventually evolved. These first autotrophic forms may well have been organisms similar to our present-day anaerobic bacteria, forms that can exist without free oxygen. They may have been able to oxidize various inorganic compounds that would release energy that could be used to produce their own food internally. This process is called *chemosynthesis.* At some later date, the more complex single-celled plants probably evolved. Having developed a green pigment called *chlorophyll,* plants could capture the sun's energy and, through photosynthesis, produce their own food supply from the carbon dioxide and water that surrounded them.

The autotrophs had developed security in the form of a built-in mechanism to assure them of a food supply. Heterotrophs, not so endowed, were at a considerable disadvantage because they had to search for this necessity of life. To meet this need, animals eventually emerged with their inherent characteristic of mobility and began to develop an awareness or a consciousness that gave them a much higher place in the biological community than that possessed by the plants.

Plants and animals developed in a beautifully balanced environment where the waste products of one filled the vital needs of the other; we can see this expressed chemically by looking at the processes of photosynthesis and respiration. In the photosynthetic process, energy is being captured and stored in the form of organic compounds. This is an *endothermic* chemical reaction. Respiration is just the opposite—an *exothermic* reaction. It is the reaction by which the energy stored in the foods produced by photosynthesis is extracted by the plants and animals to carry on their life processes.

Photosynthesis $\rightarrow$ *energy is stored (endothermic)*

$$6H_2O + 6CO_2 + Energy \longleftrightarrow C_6H_{12}O_6 + 6O_2$$
Water + Carbon + Energy   Sugar + Oxygen
           dioxide

Respiration $\leftarrow$ *energy is released (exothermic)*

Every living organism that inhabits the earth today is the result of an evolutionary process of natural selection that has been going on since these early times and by which various life forms (species) have been able to inhabit increasingly numerous niches within the earth environment. As these diverse life forms developed in an attempt to exist and adapt to various environments, they also modified the environments in which they lived.

As the plants emerged from the oceans to inhabit the terrestrial environment, they changed the appearance of the landscape from the harsh, bleak panorama we may envision as being typical of a lunar surface to the soft green hues that cover much of the land surface of the earth today. Other changes were manifested in the ocean itself as vast quantities of the hard parts of dead organisms began to accumulate on the bottom. These accumulations of calcium carbonate ($CaCO_3$) and silica (SiO_2) formed the limestone ($CaCO_3$) and diatomaceous deposits (SiO_2 deposits composed mostly of the remains of single-celled plants called diatoms) that we now see exposed on the continents as evidence that these rocks formed from sediments that accumulated on the bottom of the seas. Since over half of the rocks exposed at the surface of continents originally formed on the ocean floor, we can learn much about the oceans of the past by studying these rocks.

Probably the most important environmental modification was that which involved the atmosphere. It was this change that made it possible for most present-day animals to develop. A byproduct of photosynthesis in the early oceans was the free oxygen that makes up 21 percent of our present atmosphere. Carbon dioxide, which at one time must have made up a large portion of our atmosphere, was removed by photosynthesis to produce a concentration of 0.03 percent in the earth's present atmosphere. These changes in the atmosphere must have had a very significant effect on the animal populations and cli-

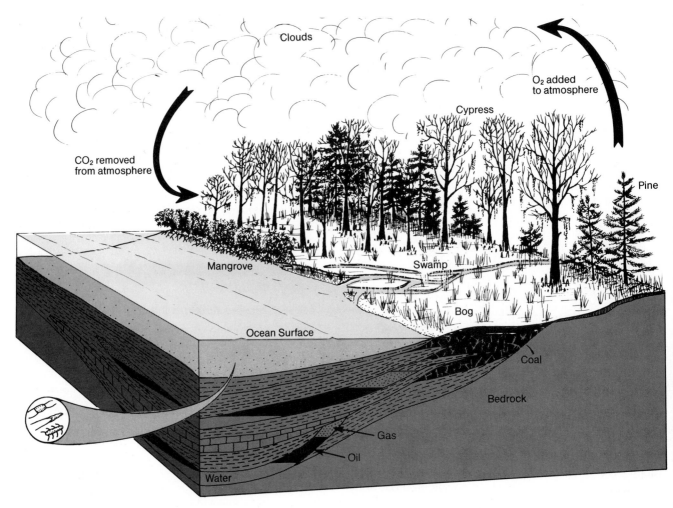

FIGURE 2–8

The Effect of the Evolution of Plants on the Earth Environment. Subsequent to the origin of single-celled plants in the ocean, the earth's atmosphere began to be enriched with O_2 and its CO_2 content was decreased as a result of photosynthesis. As these plants and the microscopic animals that fed off them died and sank to the ocean floor, these remains were incorporated into the marine sediments. Some of the remains were buried before they could be completely decomposed and were ultimately converted to petroleum (oil and gas). When plants moved onto the continents, they evolved into the various forms that support the animal forms that subsequently evolved there. In the low-lying, highly productive wetlands, plant remains accumulated in the sediment to be converted to peat and ultimately coal.

mates that have evolved throughout the earth's history (figure 2–8).

We also see, in petroleum and coal deposits, the remains of plant and animal life that were not completely oxidized but were buried in a reduced (oxygen-free) environment that allowed their energy to remain stored. These deposits provide us with over 90 percent of the energy used for domestic and industrial purposes today. So we, as animals, depend not only upon the present productivity of plants to supply the energy required by our life processes, but we also depend very heavily on the energy stored by plants during the geologic past.

SUMMARY

Our **solar system**, consisting of the sun and nine planets, belongs to the galaxy of stars we call the **Milky Way**. All the stars we see with the unaided eye belong to this galaxy; yet there are some 100 million galaxies thought to exist in the universe. Most seem to be moving away from us as though all were set in motion by a single large explosion.

The galaxies are thought to be accumulations of the debris from this explosion in large eddies, inside of which are smaller eddies where concentration of matter has produced stars and their associated planets. **Stars** result from accumulation of masses large enough to produce sufficient internal temperatures and pressures to set off processes causing them to "shine" as they give off energy at high rates. The **planets** were smaller, so they did not reach this state.

Protoearth was composed mostly of hydrogen and helium, but as it condensed and heated up, these elements were driven off into space. There are indications the earth became molten and developed a layered structure that included a core, mantle, and a crust composed of a thin basaltic unit which ultimately underlaid the oceans and the thicker granite crust of the continents. During this period the earth also developed an atmosphere rich in **water vapor** and **carbon dioxide** produced by volcanic activity. **Methane** and **ammonia** were also present in significant quantities. As the earth's surface cooled sufficiently, the water vapor condensed and accumulated in depressions on the surface to give the earth its first oceans.

It is in these oceans life is thought to have begun. As **ultraviolet radiation** fell on the oceans with their dissolved carbon dioxide, methane, and ammonia, inorganic molecules may have combined to produce carbon-containing molecules that are now formed naturally on earth only by organisms. Chance combinations of these molecules eventually produced **heterotrophic organisms** that were probably similar to present-day **anaerobic bacteria**. Eventually, **chemosynthetic autotrophs** evolved, and later some cells included **chlorophyll** into their makeup. **Plants** were the result. The **photosynthesis** of plants extracted carbon dioxide from the atmosphere and produced the first free oxygen at the earth's surface. This produced an oxygen-rich atmosphere in which animals as we know them could survive. Eventually, both plants and animals evolved into forms that could survive on the stark environment of the continents, producing the lush continental environments we enjoy today.

QUESTIONS AND EXERCISES

1. Why is it theorized that the observed motions of galaxies were initiated by an explosion?

2. In what three components of the earth's makeup must we find the total mass of crystalline rock components that were eroded from the primary crystalline rocks of the earth's surface?

3. List the five most abundant excess volatiles. Why are they called excess volatiles?

4. Why does the fact that new water is continually being released to the atmosphere by volcanic activity not necessarily mean the oceans will progressively cover an increasing percentage of the earth's surface?

5. Why do we believe that free oxygen (O_2) was not released into the atmosphere by volcanic activity?

6. How does the presence of oxygen (O_2) in our atmosphere help reduce the amount of ultraviolet radiation that reaches the earth's surface?

7. How did the photosynthesis that produced the first organic molecules in the early oceans differ from plant photosynthesis? How does chemosynthesis differ from photosynthesis?

8. When we say carbon (C) has a greater combining power than oxygen (O), we are referring to their chemical valence. How is valence related to combining power?

9. Describe some basic characteristics of living things.

10. Discuss photosynthesis and respiration and explain their relationship to the chemical processes of endothermic and exothermic reactions.

11. As plants evolved on the earth, great changes in the earth's environment were produced. Describe some of these major changes caused by plants.

REFERENCES

Glaessner, M. F. 1984. *The dawn of animal life: a biohistorical study.* Cambridge: Cambridge University Press.

Gregor, B. C., Garrels, R. M., Mackenzie, F. T., and Maynard, J. B., eds. 1988. *Chemical cycles in the evolution of the earth.* New York: John Wiley.

Holland, J. D. 1984. *The chemical evolution of the atmosphere and oceans.* Princeton, N.J., Princeton University Press.

Rubey, W. W. 1951. Geologic history of seawater—an attempt to state the problem. *Geological Society of America Bulletin.* 62:1110–19.

Strom, K. M., and Strom, S. E. 1982. Galactic evolution: a survey of recent progress. *Science* 216:4546, 571–80.

SUGGESTED READING

Sea Frontiers

Feazel, C. T. 1986. Asteroid impacts, sea-floor sediments, and extinction of the dinosaurs. 32:3, 169–178.
A discussion of the evidence in marine sediments that may support the theory that the demise of the dinosaurs was caused by the impact of a large meteor.

Scientific American

Barrow, J. D., and Silk, J. 1980. The structure of the early universe. 242:4, 118–28.
Although the universe is inhomogeneous on the small scale of the solar system or a galaxy, it is very homogeneous on the scale of the universe as a whole.

Frieden, E. 1972. Chemical elements of life. 227:1, 52–64.
The roles of the 24 elements known to be essential to life are discussed, including some background on how they may have been selected from the physical environment in which life evolved.

Gott, J. R., and Gunn, J. E. 1976. Will the universe expand forever? 234:3, 62–79.

Based on data regarding the recession of galaxies, average density of matter, and chemical elements, the authors suggest expansion will be reversed.

McMenamin, M. A. S. 1987. The emergence of animals. 256:4, 94-103.
A discussion of how the explosive diversification of animal forms 570 million years ago may have been related to the breakup of a single large continental landmass.

Stebbins, G. Ledyard, and Ayala, Francisco J. 1985. The evolution of Darwinism. 253:1, 72–85.
New advances in molecular biology and new interpretations of the fossil record add to the knowledge of evolution.

Vidal, Gonaalo, 1984. The oldest eukaryotic cells. 250:2, 48–57.
Cells with a nucleus appear to have evolved as marine plankters 1.4 billion years ago.

Wilson, Allan C. 1985. The molecular basis of evolution. 253:4, 164–175.
Mutations within the genes of organisms play an important role in evolution at the organismal level.

3
MARINE PROVINCES

An understanding of the general bathymetric features of the ocean is important because these features will be related ultimately to the origin of the ocean basins as well as physical and biological phenomena to be discussed in following chapters. *Bathymetry* in oceanography is analogous to *topography* in geography. As investigations into the depth of the ocean have proceeded, it has become apparent that a broad shelf-like feature that steepens and slopes off into deep basins has generally developed around the continents. Quite commonly, linear mountain ranges run through the basins. In all oceans, but especially around the margin of the Pacific Ocean, deep linear trenches up to 11 km (6.8 mi) in depth separate the slopes from the deep ocean basin (figure 3–1). The underlying structures and their origin will be discussed in chapter 4.

The hypsographic curve showing the distribution of the earth's solid surface at elevations above and below sea level indicates that 71 percent of the earth's surface is to be found beneath the oceans (figure 3–2). The mean elevation of the continents above the sea (840 m; 2755 ft) and mean depth of the oceans (3865 m; 12,677 ft) is the result of different densities of the continental and ocean crustal rocks. The ocean crust is denser.

ISOSTASY

The crust can be divided into two distinct types. The basaltic oceanic crust has an average density of about 3.0 g/cm^3. The continents are composed of a lighter granitic crustal material ranging in density from 2.67 g/cm^3 near the surface to 2.8 g/cm^3 deep beneath the continental mountain ranges. The continental crust averages about 35 km (22 mi) in thickness but may reach a maximum thickness of 60 km (37 mi) beneath the highest mountain ranges.

The granitic continental blocks and basaltic oceanic crust "float" on the denser mantle beneath. This concept of crustal flotation, which may be compared with the flotation of ice on water, is called isostasy (fig. 3–3).

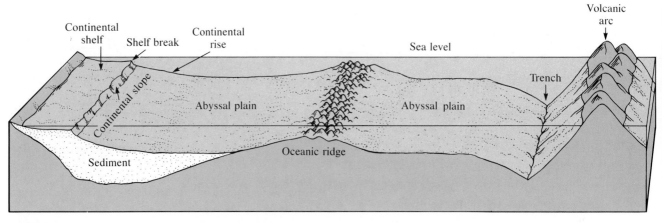

FIGURE 3–1
Marine Physiographic Provinces. A longitudinal profile showing submarine physiographic provinces. To the left of the oceanic ridge is a continental margin typical of the Atlantic Ocean, with a well-developed continental shelf, slope, and rise at the surface of a thick sedimentary deposit. The features to the right are more typical of the Pacific Ocean, where trenches and associated volcanic arcs are common.

FIGURE 3–2
Hypsographic Curve. This curve indicates the percentage of the earth's total area, land area, and ocean area at depths and elevations indicated along the left margin of the figure. (After Sverdrup, Johnson, and Fleming, 1942.)

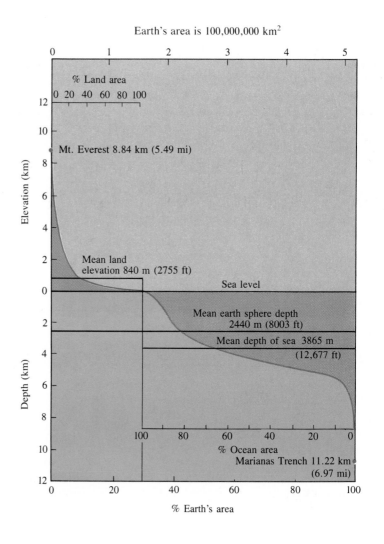

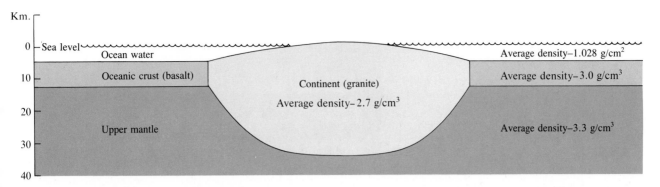

FIGURE 3–3

Isostasy. Isostasy is a state of equilibrium reached by different components of the earth's crust and the mantle. So that the lithosphere (the rock sphere of the earth's structure), which contains the continental and oceanic crustal units as well as the upper mantle, may maintain a uniform average density, continental crustal units that extend up to 14 km (8.7 mi) above the floor of ocean basins must send deep roots into the mantle.

If we take a regularly shaped prism of ice with a density of 0.91 and float it on water with a density of l.0, the ice will sink into the water until it displaces water equal in mass to the ice. When this displacement called *buoyancy* occurs, 91 percent of the ice mass will be submerged. Therefore, 91 percent of the length of the prism of ice will be beneath the surface of the water. Similarly, about 55 percent of the mass of the continents is found submerged in the mantle. Actually, we would expect about 91 percent of the continents to be submerged because they are, on the average, about 91 percent of the density of the upper mantle. However, where high heat flow in the mantle heats and reduces the density of the mantle beneath the continents, this low-density mantle provides *isostatic compensation.* In such situations as exist beneath the southwestern United States and eastern Africa, the low-density mantle material provides the buoyancy that would normally be provided by the deep root of continental crust.

TECHNIQUES OF DETERMINING BATHYMETRY

Needless to say, our greatest understanding of the ocean floor relates to the shallow shelf areas at the margins of the continents, over which the most important commercial fisheries have been located. They are presently being explored extensively because of their high potential for petroleum accumulation. Because of their economic importance and the fact that we have a greater knowledge of this region than any other ocean region, we will begin our description of the ocean floor with these marginal shelves.

However, before we discuss the various submarine provinces, we should discuss briefly the technique that has been most used to obtain data regarding submarine geological processes, *seismic surveying.* Such surveys are made by emitting a sound signal that will travel through the wa-

ter, bounce off an object, and return to be picked up by a receiver. If the velocity at which sound will travel in water has been determined, the distance to the object is equal to the velocity times one-half the time required for the sound signal to make the trip from the surface to the object or ocean floor and back to the receiver (figure 3–4).

If one wishes to know only the distance to an object in the water or the ocean floor, relatively weak high-frequency sound signals can be used. Marine geologists, however, need to know the thickness of various rock units beneath the ocean floor, and for this purpose they use stronger low-frequency signals generated by explosions. These sound signals can penetrate the rocks beneath the ocean floor, reflect off the contacts between rock units, and produce seismic reflection profiles such as those shown in chapters 3 and 5.

A variation of this technique is *side-scan sonar,* which allows a survey ship to map the topography of the bottom along a strip of ocean floor up to 5 km (3 mi) wide. The map is developed with the aid of computers and sound emitters directed away from both sides of the ship.

Due to the gravitational effects on the ocean surface, sea level rises or falls 4 m (13 ft) with each decrease or increase in ocean depth of 1000 m (3280 ft). Altimeters aboard *Seasat A* detected these differences and provided data to produce a map of the ocean surface. The maps in Plate 14 show how these changes in ocean surface elevation reflect major bathymetric features of the ocean floor.

CONTINENTAL MARGIN

Continental Shelf

Extending from the shoreline is a shelflike feature called the *continental shelf,* which is geologically part of the continent. During the geologic past much of it was exposed

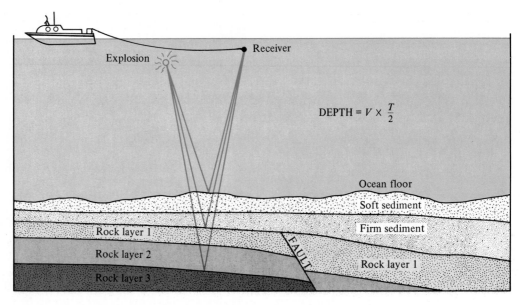

$$\text{DEPTH} = V \times \frac{T}{2}$$

FIGURE 3–4

Seismic Profiling. Low-frequency sound that can penetrate bottom sediments is emitted by an explosion. It reflects off the boundaries between rock layers and returns to the receiver. The depth of each reflecting layer is equal to the velocity of sound travel (V) times one-half the time (T) required for the sound to travel from the source to the reflecting layer and back to the receiver.

above the shoreline. Since the shoreline has moved back and forth across the shelf, its general bathymetry can usually be predicted by looking closely at the topography of the adjacent coastal region. With few exceptions, this coastal topography can be expected to extend beyond the shore and onto the continental shelf. The continental shelf may be defined as a shelflike zone extending from the shore beneath the ocean surface to a point at which a marked increase in slope occurs. This point where an increase in the rate of slope occurs is referred to as the *shelf break,* and the steeper slope beyond the break is known as the *continental slope* (shown in figure 3–1).

The width of continental shelves may vary from a few tens of meters to about 1300 km (800 mi). The broadest shelf developments occur off the northern coasts of Siberia and North America in the Arctic Ocean and in the North and West Pacific from Alaska to Australia. The average width of the continental shelf is about 70 km (43 mi), and the average depth at which the break occurs is about 135 m (443 ft). Around the continent of Antarctica, however, the continental break occurs at a depth of 350 m (2200 ft). The mean slope of the continental shelf is 0°07′ or 1.9 m/km (10 ft/mi).

Sediment is transported across the continental shelf by current flow and mass wasting (submarine sediment slides), which are often generated by earthquakes. Evidence of earthquake-induced sediment failure producing a slide on the gently sloping (0.25°) Klamath River delta off the northern California coast was recorded in 1980. On No-

vember 8, 1980, an earthquake of magnitude 6.5 on the Richter scale occurred 60 km (37 mi) off the coast. During a December survey of the area, using high-resolution seismic reflection and deep-towed side-scan sonar, a slide zone 1 km (0.62 mi) wide and 20 km (12.4 mi) long was discovered. The failure seems to have occurred along the sediment boundary between fine to medium sand and a seaward deposit of mud (clayey silt), shown in figure 3–5.

Figure 3–6 shows seismic profiles of the slide area before and after the earthquake occurred. A continuous gentle slope was recorded in October 1979 prior to the quake. A slide feature characterized by flat terraces and 1 to 1.5m (3.3–5 ft) high scarps at the seaward edge of the terraces appears on the profiles made after the quake.

Continental Slope

The continental slopes beyond the shelf break are features similar in relief to mountain ranges we find on the continents. The break at the top of the slope may be from 1 to 5 km (0.62–3 mi) above the deep ocean basin at its base. In areas where the slope descends into submarine trenches, even greater vertical relief is measured. Off the west coast of South America the total relief from the top of the Andes Mountains to the bottom of the Peru-Chile Trench is about 15 km (9.3 mi).

Continental slopes vary in steepness from 1° to 25° and average about 4°. Around the margin of the Pacific Ocean, where the slope is associated with the processes forming

FIGURE 3–5

Location of Klamath River Delta Failure Zone Resulting from November 8, 1980, Earthquake. The slide zone is 1 km (0.62 mi) wide and about 20 km (12.4 mi) long at an average depth of 60 m (197 ft). It seems to be associated with a sediment contact where nearshore sandy sediment changes seaward to mud. The location of the seismic profiles and side-scan images shown in figure 3.6 are indicated. (Courtesy of Michael Field, USGS.)

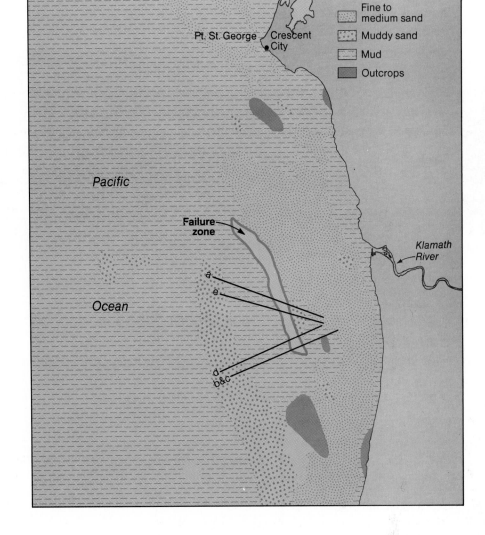

coastal mountain ranges and submarine trenches, the continental slopes are steeper than in the Atlantic and Indian oceans. Slopes in the Pacific Ocean average more than 5°, while those in the Atlantic and Indian oceans are about 3°.

Submarine Canyons

The continental slope and, less commonly, the continental shelf are cut by large submarine canyons that resemble the largest of canyons cut on land by rivers. Like the canyons cut by rivers, the submarine canyons have tributaries and steep V-shaped walls that expose a wide range of rock types varying greatly in geologic age. Exposed in the walls of the canyons are rocks ranging from soft shales to quartzite and granite.

The canyons are obviously erosional features; the problem is to explain how the erosion occurred. If the first person to observe a submarine canyon had observed those off the west coast of the French island of Corsica in the Mediterranean, he or she would have concluded that, without doubt, submarine canyons were related to land river systems, since the canyons lead right into the mouths of the rivers that drain the western flank of the island. Sediments recovered during Deep Sea Drilling Project cruises in the Mediterranean Sea indicate this body of water dried up at least once in its history. Therefore, the submarine canyons off the island of Corsica may well have been created primarily or entirely by stream erosion. The majority of submarine canyons, however, do not tie so nicely with land drainage systems, and many are confined exclusively to the continental slope. The most obvious objection to explaining them as drowned river valleys is the fact that they continue to the base of the continental slope, which averages some 3500 m (11,500 ft) below sea level. Since rivers lose their ability to erode shortly after reaching the ocean, it seems impossible that rivers could have cut canyons this far below sea level.

FIGURE 3–6

Comparison of Pre- and Post-earthquake High-resolution Seismic Reflection Records from Klamath River Delta. Vertical scale bars = 10 m (33 ft). Horizontal scale bars = 250 m (820 ft). Solid triangles mark approximate water depth of 60 m (197 ft). Profile locations are shown on figure 3–5. Lines *c* and *d* are uniboom records made with a strong sound source and show the subsurface structure, whereas lines *a*, *b*, and *e* show primarily the condition of the ocean floor. There is clearly no evidence of sediment failure on the pre-earthquake records (*d* and *e*), while the postearthquake records (*a*, *b* and *c*) show scarps and terraces that have resulted from sediment movement. The unconformity labeled in *d* is an old erosional surface upon which younger sediment has been deposited. (Courtesy of Michael Field, USGS.)

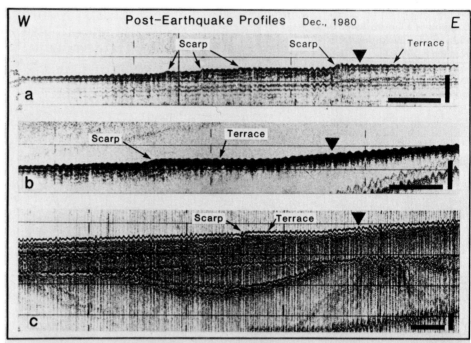

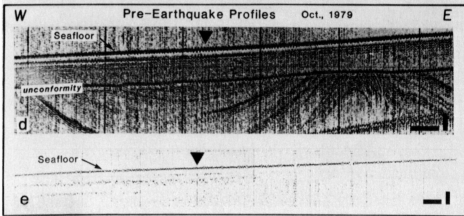

Recent work done with side-scan sonar indicates that the continental slope is dominated by submarine canyon topography along the Atlantic coast from Hudson Canyon to Baltimore Canyon (figure 3–7A). Smooth intercanyon areas are particularly rare on the middle and upper surfaces of the slope. There is also a correlation between the spacing of canyons and the continental slope gradient. No canyons are present where the slope is less than 3°. Canyons are from 2 to 10 km (1.2–6 mi) apart where the slope is between 3° and 5°, and they are 1.5 to 4 km (0.9–2.5 mi) apart where the slope gradient exceeds 6°.

Canyons confined to the continental slope are straighter and have steeper canyon floor gradients than those that extend into the continental shelf. This condition suggests that the canyons are initiated on the continental slope by some marine process and cut their way headward into the continental shelf as they age.

Turbidity Currents

Probably the most widely supported theory concerning the origin of submarine canyons is that which explains the erosion by *turbidity currents,* flows of sediment-laden water that periodically move down the canyon. Robert Dill has observed turbidity flows resulting from river input near the heads of some Pacific-coast submarine canyons that extend well onto the continental shelf. However, turbidity currents in many canyons confined to the continental slope apparently are initiated after sediment moves across the continental shelf into the head of the canyon and accumulates there. The actual initiation of the current may be produced by an earth tremor or other disturbance that makes the mass collapse and start to move down the slope under the force of gravity. As it moves, the sediment mixes with the water, producing a high density mass that moves down the canyon, spreads out, and is deposited at the base of the continental slope as part of a deep-sea fan structure commonly found at the foot of submarine canyons. The flow moves out across the fans somewhat as a river flow moves across the surface of a delta. A pattern of distributaries develops. They are bordered by levees that are deposited as the turbid mass overflows the banks of the relatively shallow distributary channels (figure 3–7B).

It is not surprising that such events have never been directly observed. However, in 1971, measurement of water properties at a depth of 5200 m (17,000 ft) immediately above the Sohm Abyssal Plain in the North Atlantic (figure 3–8) indicated a turbidity current may have occurred just

prior to the time the samples were taken. In these samples, the values for temperature, salinity, and dissolved oxygen were all higher than normally found near the deep ocean floor. These properties resembled those typical of water at 4200 m (13,800 ft). In addition to these anomalies, the water was many times more turbid than is normal for this location.

A similar condition was recorded during a continuous seismic survey southwest of the Canary Islands on June 26, 1981. A dense cloud 2 km (1.2 mi) wide and extending 215 m (700 ft) above the 4900-m (16,000-ft) deep Canary Abyssal Plain floor was recorded (figure 3–8). Up-slope from the cloud, which shrank in a few hours, the water was clear

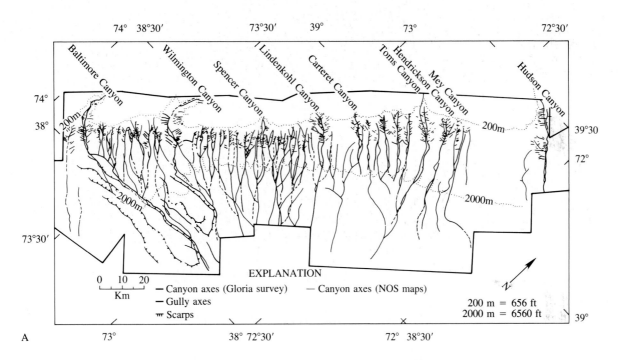

A

FIGURE 3–7

Submarine Canyons. *A*, submarine canyons along the Atlantic coast. There are no submarine canyons between Hudson Canyon and Mey Canyon where the slope gradient is less than 3°. The distance between adjacent submarine canyons decreases as the slope gradient increases. Between Lindenkohl Canyon and Baltimore Canyon, the gradient exceeds 6° for most of the continental slope. Where the slope is steepest, the submarine canyons are spaced as little as 1.5 km (0.9 mi) apart. *B*, erosional and depositional features associated with the formation of submarine canyons.

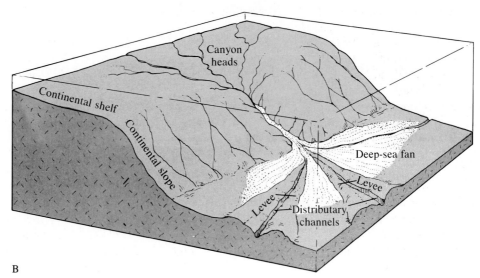

B

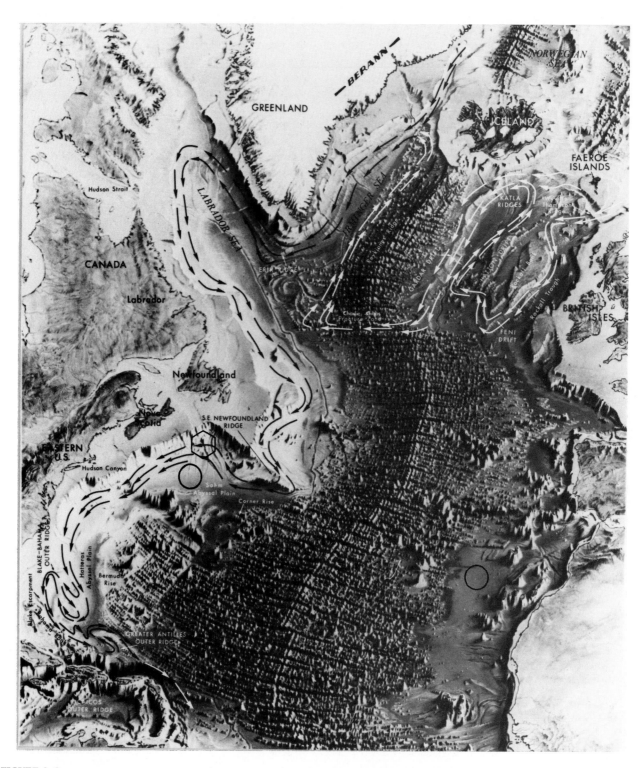

FIGURE 3–8
Floor of the North Atlantic Ocean. Circles indicate sites where turbidity currents may have been accidentally recorded. Arrows indicate the path of North Atlantic Deep Water or western boundary undercurrent. Hexagon indicates cable breaks associated with the 1929 Grand Banks earthquake. (North Atlantic Panorama excerpted from *The world ocean floor* by Bruce C. Heezen and Marie Tharp, published by the Lamont-Doherty Geological Observatory of Columbia University and sponsored by the U.S. Navy through the Office of Naval Research. Copyright © 1977 by Marie Tharp.)

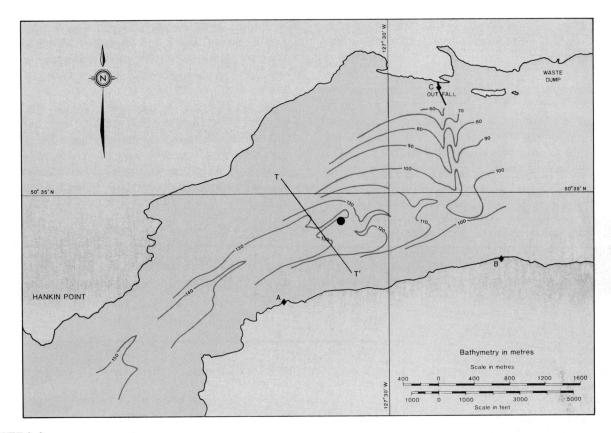

FIGURE 3–9
Rupert Inlet, British Columbia. Bathymetry of submarine channel. Line *T-T'* is path followed by the Canadian Survey Ship *Vector* on August 26, 1976, when it recorded the occurrence of a turbidity current. The black circle denotes the approximate location of a bottom-moored current meter, and the diamonds labeled A, B, and C are locations of positioning transponders. (Reprinted by permission from *Science,* © 1982. Photo courtesy of Alex E. Hay.)

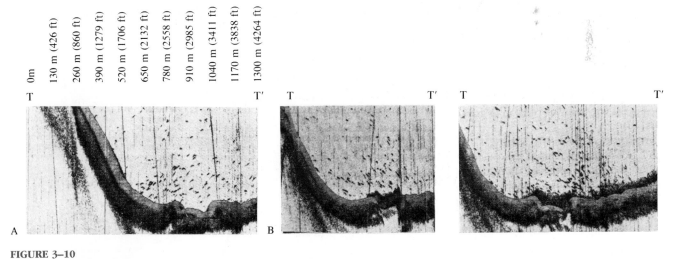

FIGURE 3–10
Acoustic Sounding Profiles along Line *T-T'*. In *A,* the ship crossed the channel at about 12:53 P.M., and no turbidity current was flowing in the channel. Note the levees on either side of the main channel. The discrete spots in the water column above the bottom are thought to be fish. In *B,* the ship was over the channel at 1:03 P.M., and the turbidity current had begun to flow in the main channel. By 1:28 P.M. in *C,* the turbidity current had overspilled the levees. The turbidity current was not detectable 1.5 h after its onset. The steep wall is on the north side of the channel. The inclined vertical lines are the result of transmitted pulses from another sounder. (Reprinted by permission of *Science* 217:4562, 27 Aug., © 1982. Photo courtesy of Alex E. Hay.)

enough to record bottom features with precision. A channel down which the turbidity current could have traveled was clearly present.

The remote sonar recording of a turbidity current generated on August 26, 1976, by the discharge of mine tailings into Rupert Inlet, British Columbia (figure 3–9), may be the only real-time observation of a marine turbidity current moving through its channel. Figure 3–10 shows acoustic sounding profiles across the submarine channel in Rupert Inlet before and during the event that lasted for about one and one-half hours.

Another line of evidence that indicates that turbidity currents move across the ocean bottom is associated with trans-Atlantic cable breaks that occurred on the continental shelf and slope south of Newfoundland after an earthquake on November 18, 1929 (figure 3–8). These cables were broken in a pattern that indicated that the cables closest to

the earthquake were broken first and those that crossed the shelf at greater depths and distances from the epicenter (point on ocean floor directly above earthquake) were broken later. This phenomenon could be explained by a dense flowing mass, moving away from the epicenter down the slope and snapping the cables as it passed. The velocity required in this particular case would be approximately 27 km/h (16.8 mi/h), which seems quite high. The possibility that such a mass may move with this velocity makes it easier to understand how the submarine canyons may have been formed on the continental slope and continental shelf.

Continental Rise

Deep-sea fans accumulate as deposits at the mouths of submarine canyons (figure 3–7B). The merging of these deep-

FIGURE 3–11
The Amazon Cone. The Amazon Canyon branches into two major distributary systems, the Western Levee Complex and the Eastern Levee Complex. They are composed of meandering channels bounded by levees. Note the large debris flows originating on the steep upper slope of the fan cone. (Map courtesy of John E. Damuth and Roger D. Flood. Permission granted by the Geological Society of America.)

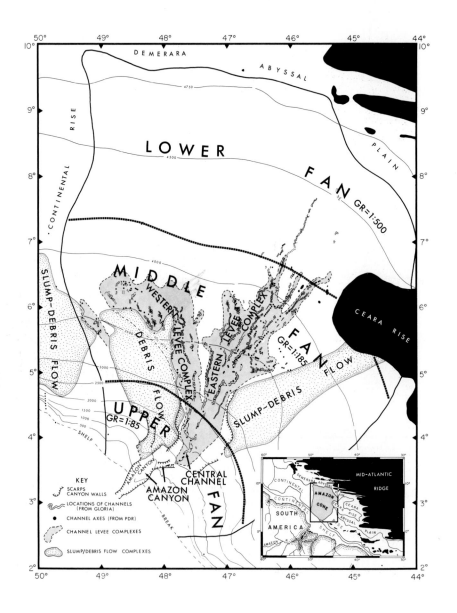

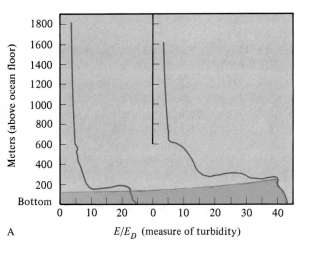

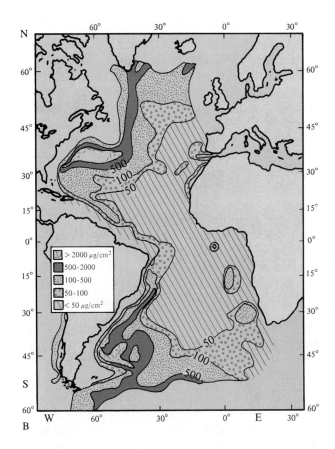

FIGURE 3–12

Benthic Nepheloid Layer. *A,* two light-scattering curves showing the nepheloid layer on the Blake-Bahama Outer Ridge. (See figure 3–8.) *B,* distribution of suspended material in the nepheloid layer in the Atlantic Ocean. Note the high concentrations found only in the western portions of the ocean overlying the continental rise. Compare the concentration pattern in the North Atlantic with the location of the western boundary undercurrent in figure 3–8. (After Biscaye and Eittreim, 1977.)

sea fans along the base of the continental slope is partly responsible for the development of the continental rise.

The Amazon Cone is one of the largest deep-sea fans. It extends 700 km (434 mi) northeast of Brazil (figure 3–11). Examination of the fan's distributary system using GLORIA long-range side-scan sonar revealed the following features. The main channel, which extends seaward from the Amazon Submarine Canyon, begins at a depth of 1400 m (4600 ft). The channel itself is 250 m (820 ft) deep and up to 3 km (1.9 mi) wide, while the entire channel-levee system has a total relief of 300 m (975 ft) and a width of up to 50 km (31 mi). At a depth of about 2200 m (7200 ft), the channel splits into western and eastern systems that extend across the middle of the fan to a depth of about 4200 m (13,800 ft). Additional branching and development of meanders occur across the middle of the fan. The branching seems to result from channel flow breaching the levees, especially on the outside curves of meanders.

The onset of meandering of the distributary channels across the middle fan corresponds well to decreased gradient of the middle fan surface as compared to that of the steeper upper fan. The meander patterns, as well as other features—abandoned meander loops, cutoffs, etc.—closely resemble in form and scale those found on the flood plains of mature river systems on land such as the lower Mississippi River.

It is also clear that mass wasting in the form of slump-debris flows continues to be an important process in distributing sediment across the fan—especially on the steeper upper fan slope.

A major factor in shaping the continental rise is the strong *western boundary undercurrent* (WBUC) or *slope current* that flows toward the equator at the base of the continental slope along the western boundaries of most ocean basins. This deep boundary current and the continental rise over which it flows have been extensively studied along the east coast of North America using submersibles, surface ships, and side-scan sonar.

The Coriolis effect (described in chapter 7) forces the WBUC, which originates in the Norwegian Sea and flows into the North Atlantic through deep troughs between Greenland and Scotland, to veer to the right and flow snugly against the base of the continental slope at all times. Picking up volcanic debris from Iceland and sediment from periodic turbidity flows, and scouring sediment from some portions of the continental slope and rise, the western boundary undercurrent, flowing at velocities that reach 40 cm/s (1 mi/h), creates a *benthic nepheloid layer* (BNL) of suspended sediment above the ocean floor (figure 3–12). As the current is forced around bends in the continental slope, its velocity decreases and the rate of deposition increases, producing the deposits called *drifts* or *ridges* shown in figure 3–8.

Increased knowledge of these continental-margin features and the processes that create them may be of importance in selecting ocean sites for disposal of toxic waste and for effective exploitation of oil, gas, and mineral resources.

DEEP-OCEAN BASIN

Abyssal Plains

Extending from the base of the continental rise are the deep-ocean basins, which include flat surfaces with slopes of less than 1:1000 that cover extensive portions of the basins (figure 3–13). These flat *abyssal plains* are particularly extensive and flat in the Atlantic Ocean, although they are occasionally interrupted by volcanic peaks. Most abyssal plains lie at depths of between 4500 (14,800 ft) and 6000 m (19,700 ft).

The benthic nepheloid layer that is so well developed over the continental slope is also found (although not so well developed) throughout the deep ocean where bottom current speeds average at least 8 cm/s (approx. 0.15 mi/h). Such currents may play an important role in distributing fine sediment on the deep abyssal plains. Just north of the equator in the eastern Pacific, average bottom current velocities measured 30 m (100 ft) above the ocean floor reached 18.5 cm/s (0.4 mi/h).

Trenches

Although the continental rise is a feature commonly found at the base of the continental slope where it meets the abyssal plain, the slope sometimes descends into long, narrow steep-sided trenches. The deepest portions of the

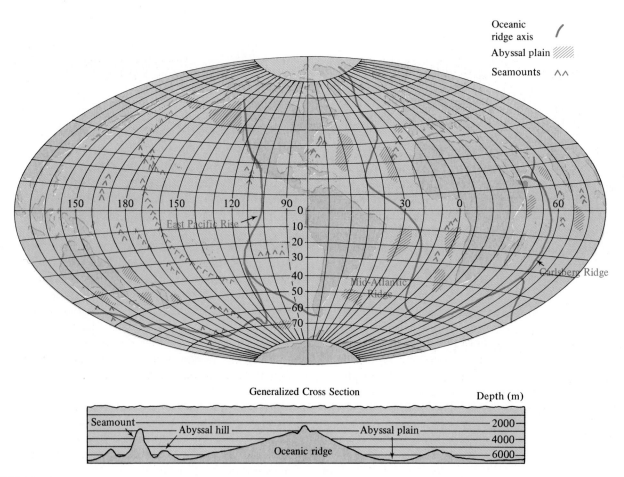

FIGURE 3–13
Oceanic Ridges, Abyssal Plains, and Volcanic Peaks. 65,000 km (40,400 mi) of oceanic mountain ranges extend across the floors of the world ocean. Flat-floored basins with sediment cover produce abyssal plains at the base of many oceanic mountain ranges. Volcanic peaks rising less than and more than 1 km (0.62 mi) above the ocean floor are named abyssal hills and seamounts, respectively. (Base map courtesy of National Ocean Survey.)

FIGURE 3–14
Trenches.

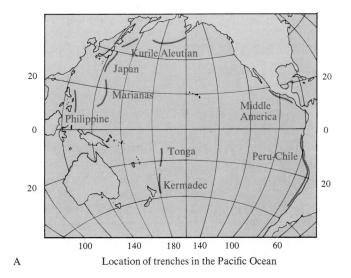

A   Location of trenches in the Pacific Ocean

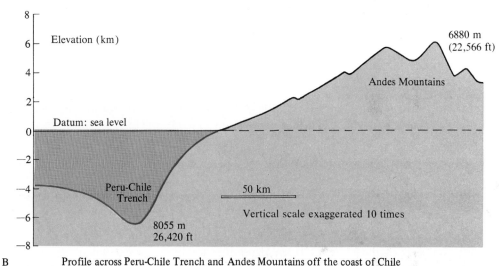

B   Profile across Peru-Chile Trench and Andes Mountains off the coast of Chile

world's oceans are found in these trenches. Such features are characteristic of the margins of the Pacific Ocean along the coast of South America, Central America, and the Aleutian Islands and are also characteristic of the western margin of the Pacific Ocean (figure 3–14). A depth of 11,022 m has been recorded in the Challenger Deep of the Marianas Trench. Table 3–1 presents some of the dimensional characteristics of trenches. The landward side of the trench rises as a volcanic arc that may produce islands (Japan) or a volcanic mountain range at the margin of a continent (the Andes).

Oceanic Ridges and Rises

Extending across some 65,000 km (40,400 mi) of the deep ocean basin are mountainous features that are referred to as *ridges* if they have steep and irregular slopes, and *rises* if the slopes are more gentle. The best explored of these features are the Mid-Atlantic Ridge and the East Pacific Rise.

This system of mountains is entirely volcanic and composed of lavas with a basaltic composition characteristic of the ocean basin crust. The Mid-Atlantic Ridge was a focal point during the development of the theory of global plate tectonics because of its position, separating the Atlantic Ocean into equal halves. Recently, the East Pacific Rise has been studied because of the hydrothermal vents and associated biological communities located in its central rift valley.

Oceanic ridges and rises extend an average of 2.5 km (1.5 mi) above the abyssal plains on either side. Oceanic ridges practically surround Antarctica. A major ridge extending from the southern Indian Ocean to the Red Sea is the Carlsberg Ridge.

Volcanic Features

The evidence of volcanic activity on the bottom of the Pacific Ocean is particularly widespread—more than 20,000

TABLE 3–1
Dimensions of Trenches.

Trench	Depth (m)	Mean width (km)	Length (km)
Peru-Chile	8,100	100	5,900
Aleutian	7,700	50	3,700
Middle America	6,700	40	2,800
Marianas	11,022	70	2,550
Kurile	10,500	120	2,200
Kermadec	10,000	40	1,500
Tonga	10,000	55	1,400
Phillippine	10,500	60	1,400
Japan	8,400	100	800

volcanic features are found there. Some marine volcanic activity produces widespread gently sloping surfaces where large volumes of lava flowed out in broad sheets, solidified, and became *archipelagic aprons.* Such features commonly surround volcanic islands.

Volcanic peaks that extend more than 1 km (0.6 mi) above the abyssal floor are termed *seamounts,* unless they have flattened tops (figure 3–13), in which case they are classified as *table mounts* or *guyots.* Guyots are less common than seamounts and are found in linear trends in the Pacific Ocean. The tops of guyots in the Pacific Ocean are between 1800 and 3000 m (5900 and 9850 ft) below the present ocean surface. Since there is evidence to indicate that the flattened surface of the guyots was produced by wave action when they were exposed above the ocean surface, a significant amount of subsidence of the ocean basin in these regions must have occurred since the guyots were at the surface. An explanation of this subsidence is presented in chapter 4.

Volcanic features on the ocean bottom that are less than 1 km (0.6 mi) above the ocean floor are called *abyssal hills* (figure 3–13). They cover a large percentage of the entire ocean basin floor and have an average relief of about 200 m (650 ft), while many such features are found buried beneath the sediments of abyssal plains.

Fracture Zones

The oceanic ridges are offset by rugged fault scars called *fracture zones* (figure 3–13). These linear bands of mountains and troughs intersect all oceanic ridges at intervals at near right angles. In the Pacific Ocean, where the scars are less rapidly covered by sediment than in other ocean basins, fracture zones extend for thousands of kilometers and have widths of up to 200 km (120 mi).

SUMMARY

Seventy-one percent of the earth is covered by oceans. The continents stand high and bound the oceans because they are composed of granite, which is less dense than the basaltic crust that underlies the oceans. Both of these crustal units "float" on the underlying mantle according to the principles of a concept of crustal flotation called **isostasy.**

Much of the knowledge we have gained about the nature of the ocean floor has been obtained using **seismic surveys** and **side-scan sonar.** Seismic surveys can reveal the topography of the ocean floor and the location of rock unit boundaries beneath it. Side-scan sonar is used to make strip maps of ocean floor topography along the course of the survey ship.

Extending from the shoreline of continents are gently sloping **continental shelves** that extend out to sea for an average distance of 70 km (43 mi). Sediment is transported across the continental shelves by currents and mass wasting (submarine sediment slides). When the shelves reach an average depth of 135 m they steepen into a feature called the **continental slope.** The slopes continue to the deep ocean floor with steepness ranging from 1° to 25° and averaging 4°. Cutting deep into the slopes and reaching well onto the continental shelves, in some cases, are **submarine canyons.** The origin of the canyons is unknown, but they may be caused in part by periodic flows that are a mixture of sediment and water called **turbidity currents.** Turbidity currents deposit their sediment load at the base of the continental slope to produce **deep-sea fans** or **cones** through a series of meandering **distributary channels,** which branch out from the mouth of the submarine canyon. The merging of these fans produces a gently sloping **continental rise** at the base of some continental slopes. A major factor in distributing sediment along the continental slope is the **western boundary undercurrent (WBUC)** or slope current that flows toward the equator along the western boundary of ocean basins. The sediment suspended and carried above the bottom by these currents creates a **benthic nepheloid layer.** When the current slows to make turns along the base of the continental slope, it deposits sediment onto features called **drifts** or **ridges** that become prominent features of the continental rise.

The continental rises gradually become flat, extensive, depositional features of the deep ocean basin called **abyssal plains.** These plains are often penetrated by volcanic peaks. In some places, particularly around the margin of the Pacific Ocean, the continental slope does not merge into a continental rise but continues down into deep linear **trenches**. Trenches are usually bounded on their landward side by **volcanic island arcs** or **continental volcanic mountain ranges** if the trench lies near the continent. **Oceanic ridges** and **rises** are volcanic mountain ranges running through the deep-ocean basins and rising about 2.5 km (1.5 mi) above the abyssal plains on either side. The volcanic peaks protruding from the abyssal plains are called **sea-mounts** if they extend over 1 km (0.6 mi) above the deep-ocean floor and **abyssal hills** if they are smaller. Some peaks once stood above sea level and had their tops flattened by wave action. They are called **guyots** or **table mounts**. Rugged fault scars called **fracture zones** cut across vast distances of ocean floor and offset the axes of oceanic ridges and rises.

QUESTIONS AND EXERCISES

1. Compare the concept of isostasy with that of buoyancy, which explains the flotation of ice in water.

2. Describe the major features of the continental margin: continental shelf, continental slope, continental rise, submarine canyon, and deep-sea fans or cones.

3. Why is the suspended sediment layer best developed along the western margin of the ocean basins?

4. In which ocean basin are most ocean trenches found?

5. Describe the following: seamount, abyssal hill, guyot, oceanic ridge, fracture zone.

REFERENCES

Anderson, R. N., 1986. *Marine-geology—A planet earth perspective*. New York: John Wiley.

Biscaye, P. E., and Eittreim, S. L. 1977. Suspended particulate loads and transports in the nepheloid layers of the abyssal Atlantic Ocean. *Marine Geology* 23:155–72.

Damuth, J. E., Kolla, V., Flood, R. D., Kowsmann, R. O., Monteiro, M. C., Gorini, M. A., Palma, J. J., and Belderson, R. H. 1983. Distributary channel meandering and bifurcation patterns on the Amazon deep-sea fan as revealed by long-range side-scan sonar (GLORIA). *Geology* 11:2, 94–98.

Fairbridge, R. W., ed. 1966. *Encyclopedia of oceanography*. New York: Van Nostrand Reinhold.

Field, M. E., Gardner, J. V., Jennings, A. E., and Edwards, D. E. 1982. Earthquake-induced sediment failures on a 0.25° slope, Klamath River delta, California. *Geology* 10:10, 542–545.

Hay, A. E., Burling, R. W., and Murray, J. W. 1982. Remote acoustic detection of a turbidity current surge. *Science* 217:4562, 833–35.

Hollister, C. D., Flood, R. D., and McCave, I. N. 1978. Plastering and decorating the North Atlantic. *Oceanus* 21:1, 5–13.

Schilling, J. G., Thompson, T., Kingsley, R., and Humphris, S. 1985. Hotspot–migrating ridge interaction in the South Atlantic. *Nature* 313:5999, 187–91.

Shepard. 1973. *Submarine geology*. New York: Harper & Row.

Sverdrup, H. U., Johnson, M. W., and Fleming, R. H. 1942. Reprinted 1970. *The oceans: their physics, chemistry, and general biology*. Englewood Cliffs, N.J.: Prentice-Hall.

Twichell, D. C., and Roberts, D. G. 1982. Morphology, distribution and development of submarine canyons on the United States Atlantic continental slope between Hudson and Baltimore canyons. *Geology* 10:8, 408–12.

Weirich, F. H. 1984. Turbidity currents: monitoring their occurrence and movement with a three-dimensional sensor network. *Science* 224:4647, 384–87.

SUGGESTED READING

Sea Frontiers

Mark, K. 1976. Coral reefs, seamounts, and guyots. 22:3, 143–49.
A discussion of the role of global plate tectonics in explaining the distribution of seamounts, guyots, and the evolution of coral reefs.

Schafer, C., and Carter, L. 1986. Ocean-bottom mapping in the 1980s. 32:2, 122–30.
The use of SeaMARK (Seafloor Mapping and Remote Characterization), a side-scan sonar device, in mapping the continental margin off the coast of Labrador is described.

Shepard, F. P. 1975. Submarine canyons of the Pacific. 21:1, 2–13.
A description of canyons around the Pacific margin and discussion of their possible origin.

Scientific American

Emery, K. O. 1969. The continental shelves. 221:3, 106–25.
The nature of the continental shelves and the effect of the advance and retreat of the shoreline across them as a result of glaciation are discussed.

Heezen, B. C. 1956. The origin of submarine canyons. 195:2, 36–41.
Theories explaining the origin of submarine canyons are presented along with data on the location and nature of such features.

Menard, H. W. 1969. The deep-ocean floor. 221:3, 126–45.
A summary of the dynamic effects of sea-floor spreading is presented with a description of related sea-floor features.

4

THE ORIGIN OF OCEAN BASINS: GLOBAL PLATE TECTONICS

The fact that we live on a dynamic earth on which movement is the rule rather than the exception has long been accepted by geologists. Geologic processes responsible for the ever-changing landscape of the continents are believed to require long periods of geologic time to build such features as mountains. However, many geologists were not prepared until recent years to accept as a process associated with the building of mountain ranges a dynamic characteristic of the earth that was of a much broader scope—the horizontal movement of continents. When geologists speak of the movement of continental masses across the earth's surface, they sometimes refer to this phenomenon as *continental drift.*

Marine geologists who have studied the ocean floor and developed theories concerning why the continents are moving relative to one another may use the term *sea-floor spreading.* The name for this theory is derived from evidence indicating that new oceanic crust and rigid upper mantle material is being formed along the axes of a series of oceanic ridges and rises. This newly formed material then moves at right angles away from the ridge and rise axes. The continents float on this denser material and are carried along by the moving layer of denser rock.

A third term used to encompass the totality of the process is *global plate tectonics.* Study of the process has indicated that there are a number of major plates into which this rock sphere, or *lithosphere,* can be divided. The interaction of these plates as they move builds the structural features of the earth's crust that can be observed by geologists. Thus, *tectonics* refers to the building of the earth's crustal structure and is derived from the Greek tektonikos, which means "to construct."

The possibility that the continents may be moving across the surface of the earth is not new. Such a possibility was suggested early in the 19th century, and the first theory attempting to explain the movement was presented in 1912. With such a long history of awareness of the possibility of such movement, why

Photograph by Rod Catanach, courtesy of Woods Hole Oceanographic Institution.

has the acceptance of theories related to the process been so long in coming?

DEVELOPMENT OF THE THEORY

Alfred Wegener is considered by most scientists to be the pioneer of the modern continental drift theory. This German scientist originally was drawn to the concept in an attempt to explain the ancient climates that were recorded in the rocks deposited in ocean basins and on landmasses of the past. Wegener's theory was published in 1912 and met a great deal of resistance from the scientific community.

Wegener postulated that about 200 million years ago all the continental mass of the earth was one large continent, *Pangaea.* About 180 million years ago, the continent began to break up, and the various continental masses we know today started to drift toward their present positions. Since it was impossible at that time to present evidence that would support a mechanism for such transportation, the theory did not receive wide acceptance.

Although many Southern Hemisphere geologists had accepted continental drift as a geologic reality, it was not until the 1950s that geologists of the Northern Hemisphere began to give it serious attention. The impetus for the renewed attention arose from the study of the earth's ancient magnetism. The British geophysicist S. Keith Runcorn explained his observations of the magnetic properties of the rocks of Europe and North America in terms of continental movements. The details of these considerations will be discussed later in this chapter.

On the basis of this study of the earth's ancient magnetic fields, convincing arguments could be made for the fact that the continents had drifted relative to one another. As study continued, more evidence was gathered to support this movement, and additional data suggested the mechanism by which the movement might have taken place. In the next section, we will outline in some detail the early observations that led to the initial interest in the theory of continental drift and those more recent findings that led to the development of the theory of global plate tectonics.

Continental Jigsaw

Before we look at the bulk of scientific evidence that has been gathered in relation to plate tectonics, we should consider how well the continents do fit together. Attempts were made by many of the early investigators to arrange the continents in a manner that would achieve a reasonable fit and support their data. Most of these attempts used the existing shoreline as the margin of the continent. Based on our previous discussion of the geomorphic provinces of the ocean basin, we considered the continental shelf and

FIGURE 4–1

Computer Fit of Continents by Sir Edward Bullard. Sir Edward Bullard's fit of the continents was attempted in 1965 after a convincing fit pattern had been achieved in 1958 by the Australian geologist S. Warren Carey without the aid of computers. Bullard's fit was in complete agreement with that of Carey. (From *Continental drift* by Don and Maureen Tarling, © 1971 by G. Bell & Sons, Ltd. Reprinted by permission of Doubleday & Co., Inc.)

continental slope to be part of the continental mass. Studies of the magnetic intensity of the earth's crust support such a contention.

There is a distinct increase in the magnetic intensity of oceanic crustal rocks compared to that of the continental crustal rocks. The increased intensity of the magnetism of the rocks is caused by the higher iron content of oceanic crustal rocks.

Sir Edward Bullard, an English geophysicist, constructed a computer fit of all the continents in 1965. The arrangement that produced minimum overlaps and gaps was found to be formed by continents outlined by the 2000-m (6560-ft) depth contour. This contour represents a depth that is approximately halfway down the continental slope (figure 4–1).

CONTINENTAL GEOLOGY

To test the fit of the continents, we may compare the rocks along the margins of the two continents that are thought to have been united. We need to identify rocks of the same type and age on the continents along their common margin. Identification is not easy in some areas, since during the millions of years since their separation, younger rocks may have been deposited and covered those rocks that might hold the key to the past history of the continents. However, there are many areas where such rocks are available for observation, and in these areas we can compare the ages by the use of (1) *fossils,* the remains of ancient organisms preserved in rocks, and (2) *radioactive dating* of the rocks.

The Fossil Record

The use of fossils became an important dating device in the early 19th century when it was realized that a particular layer of sedimentary rock would contain the remains of organisms that, as a group, were unique in time. Rocks laid down earlier and those laid down later would have a distinctly different assemblage of plant and animal fossils. Once these assemblages are recognized, a geologist can tell if rocks exposed in one area are younger or older than in another. The unique character of these assemblages is the result of evolution throughout geologic time. Appendix IV shows the general pattern of this evolution and the geologic time scale.

Dating by such a method is referred to as *relative dating.* One can tell only whether or not an assemblage is younger or older than another, but not its actual age in years. Such assemblages have been found abundantly in sedimentary rocks of the last 600 million years. Note that this is but a little over one-eighth of the time represented by the earth's existence, which is thought to exceed 4.5 billion years. Therefore, this method will not be effective in comparing the ages of the continental margins composed of rocks in excess of 600 million years of age.

Radioactive Dating

Most of the rocks we find on the continents contain small amounts of radioactive elements such as uranium, thorium, and potassium, which break down into atoms of other elements. Each radioactive *isotope* (atom of an element with an atomic weight different than that of other atoms of the element) has a specific *half-life,* the time it takes for one-half of the atoms present in a sample to decay to atoms of some other element. By comparing the quantities of the radioactive isotope with the quantities of their decay products in given continental rocks, the age of the rocks may be determined within 2 to 3 percent of their actual age. Although in very old rocks this error can be quite significant, this method of age determination is a very powerful tool and represents the first possibility that we have had of giving the actual age of rocks in years. We refer to such dating

as *absolute dating.* As compared to the relative dating possible with the use of fossils, absolute dating tells us not only which rocks are younger or older, but how much younger or older (figure 4–2).

Ancient Life and Climates

The fossil record of plants and animals contained in the sedimentary rocks can tell us much about the environments of the past. For instance, it is relatively simple to determine whether an organism lived in the ocean or on land. This can normally be told also by the characteristics of the sedimentary rock itself. However, finer distinctions in environmental conditions may also be determined.

Upon examining the distribution of some fossil assemblages and other characteristics of rocks that give clues to the climate under which they were formed, we find some that could not have formed under the climatic conditions present in the areas where we find them today. These anomalous locations of fossil assemblages and rock types may be explained by assuming that either (1) the factors that control climatic belts today are different from the factors that controlled climates in the past, or (2) the factors controlling climate throughout geologic time have not changed. The second assumption, which seems more logical, leaves us with the conclusion that these assemblages and rock types in the sediments of the past must have been carried to their present position by the movement of the continental masses.

The dominant factor controlling climatic distribution is latitudinal position on the rotating earth. Assuming that the earth's axis of rotation has not changed significantly throughout its history, we may conclude that given latitudinal belts have possessed climatic characteristics that have not changed greatly during the evolution of life on earth.

Laurasia and Gondwanaland

Modern reef-building corals are known to exist in an environment where the water is clear and shallow and its temperature does not fall below 18°C. Although we cannot be sure such conditions have been required throughout the long history of coral evolution, we may assume that the conditions required were similar.

Exactly the same species of coral are found in rocks 350 million years old in western Europe and eastern North America, as well as throughout the Alps and Himalayas. Other fossil evidence indicates that throughout much of the 600-million-year fossil record, a major ocean existed that separated a large continent composed of North America, Europe, and Asia to the north from a large continent composed of South America, Africa, India, Australia, and Antarctica to the south. This ocean mass has been given the name *Tethys Sea,* and the supercontinents were *Laurasia* to the

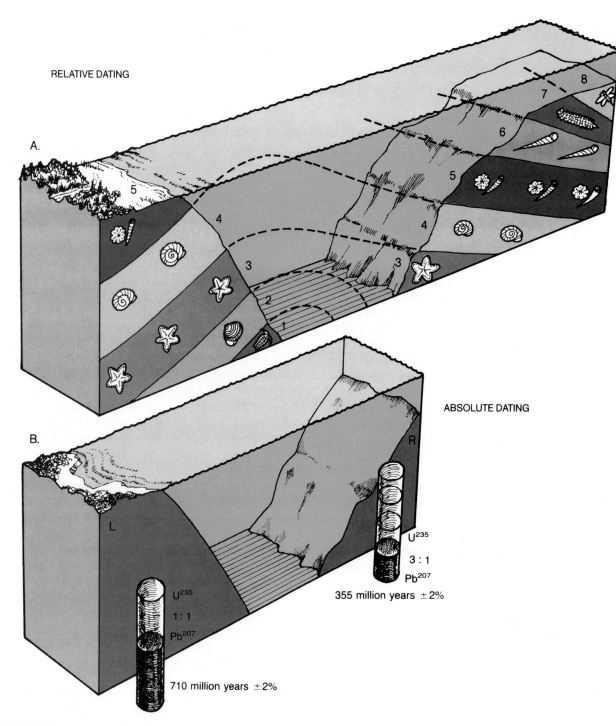

RELATIVE DATING

A.

ABSOLUTE DATING

B.

L

R

U^{235}
3 : 1
Pb^{207}

355 million years ±2%

U^{235}
1 : 1
Pb^{207}

710 million years ±2%

FIGURE 4–2

Relative and Absolute Dating. *A*, fossil assemblages used in relative dating can be found in outcrops L and R. Assemblages 3, 4, and 5 may be matched, telling us the rocks in segments of the outcrops formed at the same time. We cannot, however, tell how many years ago this formation occurred. The concept of superposition states that younger sediments are laid down on older deposits. *B*, Outcrops L and R contain the radioisotope, uranium 235. It has a half-life of 710 million years and decays to lead 207. Dating with radioisotopes is possible for many types of rocks but is more broadly possible with igneous rocks like the basalt and granite that make up most of the oceanic and continental components, respectively, of the earth's crust. At outcrop L the ratio of ^{235}U to ^{207}Pb of 1:1 means that half of the ^{235}U atoms have decayed to ^{207}Pb, so the rocks are one half-life of ^{235}U old, or 710 million years. In outcrop R, the ratio of ^{235}U to ^{207}Pb is 3:1. This means that one-fourth of the original ^{235}U atoms have decayed to ^{207}Pb, so the rock is one-half of a half-life old, or 355 million years.

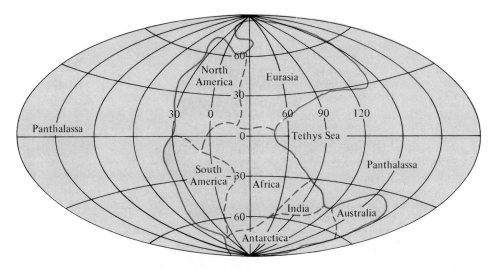

FIGURE 4–3
Pangaea. About 200 million years ago, near the beginning of the Mesozoic era, all the present continents were joined together into one continent—Pangaea. The ocean that surrounded Pangaea was called Panthalassa.

north and *Gondwanaland* to the south. The record also indicates that the two supercontinents periodically came into direct or close contact across the Tethys Sea. It appears that both continents together formed one large landmass, *Pangaea,* (one earth) about 200 million years ago. Pangaea was surrounded by the ocean *Panthalassa* (one sea) (figure 4–3).

Fossils from sediments that were laid down within a land environment also support the existence of the Tethys Sea. From 350 to 285 million years ago there existed two distinct floral assemblages on the two supercontinents. The Laurasian floral assemblage included many species of tropical plants that were incorporated into the sediment to form the extensive coal beds mined throughout the eastern United States and Europe. Throughout Gondwanaland existed an assemblage represented by a few species of plants that were thought to have grown in a cold climate. Supporting this belief are indications of glaciation in South America, Africa, India, and Australia, which at that time must have been very near the southern polar region (figure 4–4A and B).

Continental Magnetism

Although we cannot reconstruct the relative position of continents prior to 200 million years ago with great accuracy, there is enough evidence from that time to the present to give us some idea of the paths followed by the continents after the breakup of Pangaea. The first clues to such movements came from the study of the magnetism of continental rocks.

All igneous rocks contain some particles of magnetite that are natural magnets and align themselves with the earth's magnetic field at the time of the rocks' formation. Volcanic lavas such as basalt are high in magnetite content and solidify from molten material with temperatures in excess of 1000°C (1800°F). As they cool below 600°C

(1100°F), the magnetite particles become oriented in the direction of the earth's magnetic field, recording that field relative to the rock permanently. If the earth's magnetic field changes subsequent to the formation of the igneous rock, the alignment of these particles will not be affected.

Magnetite is also deposited in sediments. While the deposit is in the form of a sediment surrounded by water, magnetite particles have an opportunity to align themselves once more with the earth's magnetic field. This alignment is preserved when the sediment is turned into solid rock.

Although a number of rock types may be used for the study of the earth's paleomagnetism, the basaltic lavas and other igneous rocks high in magnetite content are best for such studies. The magnetite particles act as small compass needles as shown in figure 4–5. They point not only in a north-south direction but also into the earth at an angle relative to the earth's surface called the *magnetic dip* (inclination) that is proportional to latitude. At the equator, the needle does not dip at all but lies horizontal. It points straight into the earth at the magnetic pole. At points between the equator and the pole the angle of dip increases with increasing latitude.

Apparent Polar Wandering

Since the present magnetic poles do not coincide with geographical poles related to latitude, it might be felt that determining latitude by magnetic dip would give some incorrect determinations. However, for the last few thousand years the *average position of the magnetic pole* has coincided with that of the *geographic pole.* If we assume that this was true for the past, we can determine an average position of the magnetic pole during a time interval and consider it to represent the geographic pole.

As these average positions for the magnetic pole are determined for rocks on the continents, it is found that their position changed with time. It *appears* the magnetic pole

FIGURE 4-4

Fossil and Glacial Evidence Supporting Continental Drift. *A*, a fossil coral assemblage that flourished in the Tethys Sea and continental floral assemblages *A* and *B* that developed to the north and south of the Tethys suggest Pangaea was divided into Laurasia to the north and Gondwanaland to the south during the time span covering 280 to 350 million years ago. The *B* flora are thought to be cold-climate forms. This belief is supported by evidence of glaciation. *B*, Laurasia and Gondwanaland may have been in the relative positions shown here during the interval 250 to 350 million years ago. The Tethys coral assemblage could have been deposited in the narrow Tethys Sea while the tropical Laurasian floral assemblage and the polar Gondwanaland floral assemblage were developing on the continents.

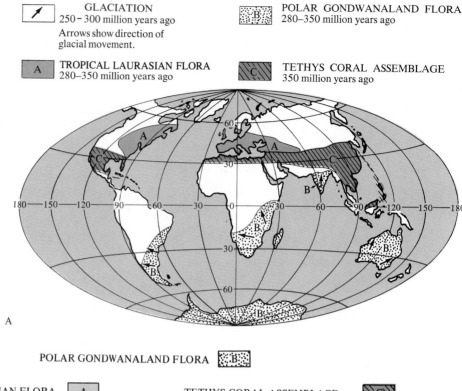

GLACIATION
250–300 million years ago
Arrows show direction of glacial movement.

POLAR GONDWANALAND FLORA
280–350 million years ago

A TROPICAL LAURASIAN FLORA
280–350 million years ago

C TETHYS CORAL ASSEMBLAGE
350 million years ago

A

POLAR GONDWANALAND FLORA B

TROPICAL LAURASIAN FLORA A

TETHYS CORAL ASSEMBLAGE C

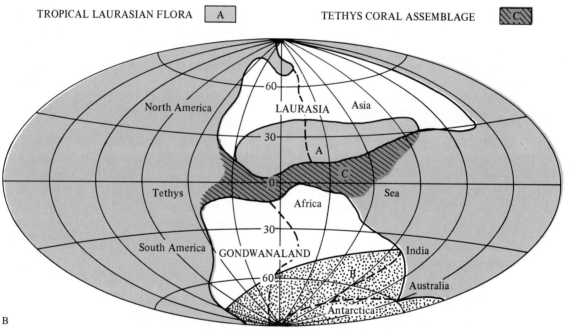

B

was wandering. The wandering curve of the pole for North America shows an interesting relationship with that determined for Europe. Both curves have a similar shape, but for all rocks older than 70 million years, the pole determined from North American rocks lies to the west of that determined by the study of European rocks. In fact, for the interval between 100 million and 300 million years ago, the two curves are separated by an angle approximately equal to the angular width of the Atlantic Ocean. Adjusting the position of the continents so that their wandering polar curves coincide, we can determine with some degree of assurance the relative positions of the continents at various times in the past (figure 4–6).

Estimating the latitude of formation for rocks from many areas studied throughout the continents on the basis of paleoclimatic and paleomagnetic evidence produces similar results. The most logical explanation for the anomalous condition of both climate and magnetic dip recorded in some rocks is that the continents moved after the rocks formed.

Magnetic Polarity Reversals

Not only does it seem that the magnetic poles as determined from continental rocks have wandered throughout geologic time, but the polarity, the direction of the magnetic field, seems to have reversed itself periodically as the poles wandered. Reversals in polarity are not apparent—they are real—and may be described in the following way. A compass needle that points to the magnetic north today would, during a period of reversal, point south (figure 4–7). It is not known why these reversals occur, but they

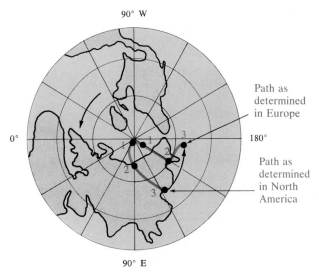

Numbers indicate position of poles hundreds of millions of years ago

FIGURE 4–6

Apparent Wandering Curves of North Magnetic Pole Determined from Rocks of North America and Europe. Apparent wandering curves of the north magnetic pole as determined from North America and Europe going back in time from the present to 300 million years ago follow divergent paths. If one moves the continent and the apparent position of the pole as determined from North America to a position that makes it coincide with the pole as determined from Europe, the direction and distance equal that required to close the Atlantic Ocean between the two continents. This suggests that the poles have not actually wandered. They appear to have done so because of continental movements.

FIGURE 4–5

Nature of Earth's Magnetic Field. Over long periods of time, the earth's average magnetic field can be considered to be that of a large magnet with its poles aligned along the earth's axis of rotation. As are the compass needles in the illustration, magnetite particles in newly forming rocks of the earth's crust are aligned with this field. Once the rocks solidify, the magnetite particles are frozen into position and become fossil compass needles that tell today's investigators about the strength and alignment of the earth's magnetic field when the rocks of which they are a part formed. Note that the magnetic lines of force are parallel to the earth's surface at the equator and strike the earth's surface at increasingly greater angles moving from the equator toward the poles. Thus, the magnetite particles frozen into rocks record, by the angle of their dip relative to the earth's surface, the approximate latitude at which they originally formed (no dip—0° at or near the equator; 90° depending on the earth's magnetic polarity at the time, at or near the north or south pole).

have for the last 76 million years occurred once or twice each million years. The period of time during which a reversal occurs lasts for a few thousand years. It is identified by a gradual decrease in the intensity of the magnetic field of one polarity until it disappears, to be followed by the gradual increase in the intensity of the magnetic field with the opposite polarity. The time during which a particular paleomagnetic condition was in existence can be determined by the radioactive dating of the igneous rock from which the paleomagnetic measurements were taken.

Since the earth's magnetic field serves as a shield to prevent the entry of many harmful forms of radiation, it is thought that periods during which reversals in polarity occurred may have been associated with the dying out of many species of life that are most sensitive to radiation. Research into this relationship indicates that there may be a correlation between the extinction of large numbers of fossil species and the weakening of the magnetic field during polarity changes. The earth's present magnetic field has been weakening for the last 150 years, and some investigators think that the present polarity will disappear in another 2000 years.

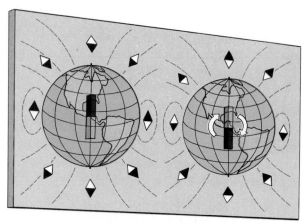

FIGURE 4–7
Reversal of Polarity by Earth's Magnetic Field. *A,* the black end of each of the compass needles above the earth's surface aligns with geomagnetic lines of force and point toward the north magnetic pole when the earth's magnetic polarity is as it is today. *B,* during a period of reversal of the polarity of the earth's magnetic field, the black end of the same compass needles will point toward the south magnetic pole.

MARINE GEOLOGY

We have so far considered data obtained from the conti nents that would bear on the theory of drifting continents For many years such data were all that was available, since extensive sampling of the deep-ocean bottom has only quite recently become technologically feasible. In the following paragraphs we will discuss evidence taken directly from the ocean floor supplementing that which we have discussed from the continents.

Paleomagnetism

A detailed study of the magnetism of the Pacific Ocean floor by Scripps Institution of Oceanography identified narrow strips of magnetic anomalies that run parallel to the East Pacific Rise. Each strip represents a period when the polarization of the earth's magnetic field was either normal (as it is today) or reversed. The boundaries between strips are marked by a zone of low magnetic intensity developed during the change of polarity.

FIGURE 4–8

Geomagnetic Record Contained in the Ocean Floor. The polarity of the earth's magnetic field is recorded at the time of the ocean floor's formation. Like a tape recorder, the ocean floor stores a record of changes in polarity. The pattern recorded on one side of an oceanic ridge is the mirror image of that recorded on the opposite side.

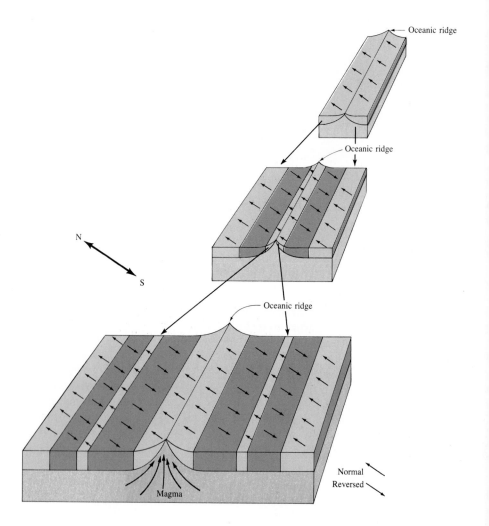

scribed. The abrupt change in the age of ocean floor along
lines running perpendicular to the oceanic mountain range
in the eastern and southern Pacific is the result of transform faults offsetting the axis of the mountain range during
the early stages of plate formation.

PLATE 14

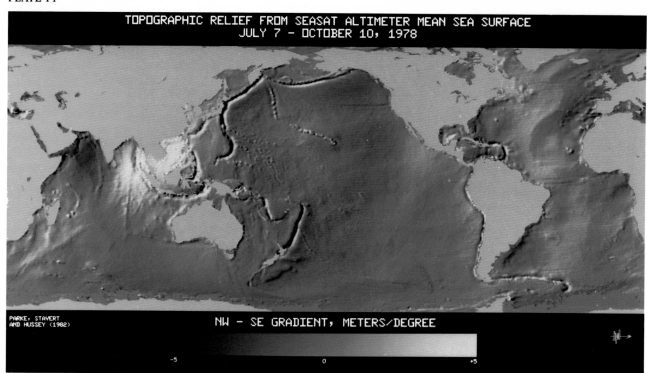

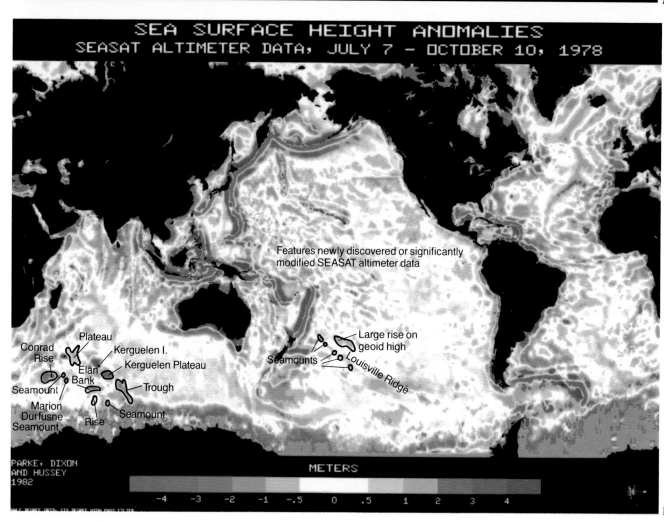

Due to gravitational effects, the ocean surface rises over high volcanic peaks and drops over depressions such as trenches on the ocean floor. An altimeter on the SEASAT satellite that operated in 1978 was able to measure these changes in the ocean surface with sufficient precision to map major bathymetric features on the ocean floor. In the southern Indian Ocean and Pacific Ocean, this remote sensing of the ocean surface was able to identify some new bottom features and provide information that significantly altered the shape or location of previously identified features (circled).

Additional study showed the sequence of polarity changes on one side of an oceanic ridge to be identical to the sequence of reversals on the opposite side of the ridge. While dating the reversal points on both flanks of the ridge, it became apparent that the rocks became older with increased distance from the ridge axis. The evidence indicated new ocean crust was being formed at the oceanic ridges and moving away down the flanks on both sides of the ridge. Having determined from continental studies the date at which many of the more recent reversals occurred, it was possible to measure the width of the ocean floor represented by each polarity strip. Dividing the width by the number of years that polarity lasted produced the rate at which the ocean floor appeared to be moving away from the ridge (figure 4–8).

Confirmation of the spreading of the ocean crust away from the oceanic ridges required obtaining actual samples of the crust at various locations for radioactive dating. In addition, since sediment could not be laid down on crustal material until it formed at the ocean ridges, it would be expected that the fossil assemblages observed in sediments immediately overlying the oceanic crust should be representative of organisms that existed at the time of the crustal formation. A significant aspect of the Deep Sea Drilling Project was to check the age of the ocean bottom by drilling through the sedimentary section into the oceanic crust. Although all attempts to do this did not meet with success, enough data based on fossil assemblages immediately overlying the crust and radioactive age determinations of the crust itself confirmed that the ocean floor is moving in the manner suggested by paleomagnetic data (figure 4–9).

Sea-Floor Spreading

A Mechanism for Spreading An absolute answer to the mechanism that might cause the formation of *lithosphere* (rock sphere) at the oceanic ridges and its movement away from those ridges at right angles has not been determined. However, the lithosphere, which includes the oceanic and continental crustal units and the upper mantle, thickens from a few kilometers near the ridge axis to over 200 km (120 mi) beneath some continental regions. The increasing thickness of the lithosphere as one moves away from the oceanic ridges and rises *(spreading centers)* correlates well with increasing age of the ocean floor, depth of the ocean, and decreasing heat flow (see figure 4–10).

Deep within the earth, radioactive atoms are breaking down and releasing energy that must find its way to the surface of the earth as heat. It is conceivable that this heat moves to the surface through convection cells that carry the heat up in the regions of the oceanic ridges. If this is the case, there must be some regions on the earth where the cooler portions of the mantle descend to complete the convection cell.

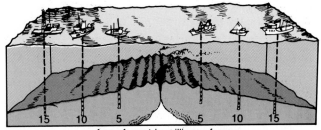

Age of crust in millions of years

FIGURE 4–9
Confirmation of Sea Floor Spreading. If new ocean crust is being added at ocean ridges and spreads away from these structures, the ages of fossil assemblages in sediments immediately above the crust and radiometrically determined ages of the basal sediments and crust should increase as one moves away from the oceanic ridges. Data of this type recovered from the ocean floor support the theory that sea-floor spreading originates at the ridges.

Heat flow measurements taken throughout the earth's crust show that the quantity of heat flowing to the surface along the oceanic ridges is as much as eight times in excess of the average for the earth's crust (1.25 μcal/cm^2/s). In addition, areas where ocean trenches are known to exist have heat flows that are as little as one-tenth the average.

The lateral movement that occurs between the regions of creation (oceanic ridges and rises) and resorption (trenches) of the lithosphere originates within a low-velocity plastic portion of the mantle where temperatures approach the melting point of the mantle. This zone is the *asthenosphere* (weak sphere). These movements in the asthenosphere could carry the more rigid lithosphere, which contains the crust and upper mantle (figure 4–11). It has been observed that the trenches do not fit the isostasy patterns characteristic of the continents, and it appears that there may be some force other than gravity pulling the lithosphere down in these regions. The descending convection cells might well provide that force.

The observed relationship of increasing lithospheric plate thickness, age of ocean floor, and ocean depth, along with a decrease in the rate of heat flow through the ocean floor with increasing distance from the ridge-and-rise spreading centers can be explained by the convection process. The lithosphere at the spreading centers is thin because of the proximity of high temperatures immediately beneath the ocean floor. This heat also causes expansion of the upper mantle rocks, raising the oceanic ridges and rises to elevations well above those of the ocean floor on either side of the zones of upwelling heat. As the lithosphere moves away from the spreading centers toward zones of convection downwelling beneath the trenches, it encounters decreasing rates of heat flow through the ocean floor from the underlying mantle. This results in increasing thickness of the lithosphere as cooling asthenosphere at the base of the lithosphere is converted to rigid litho-

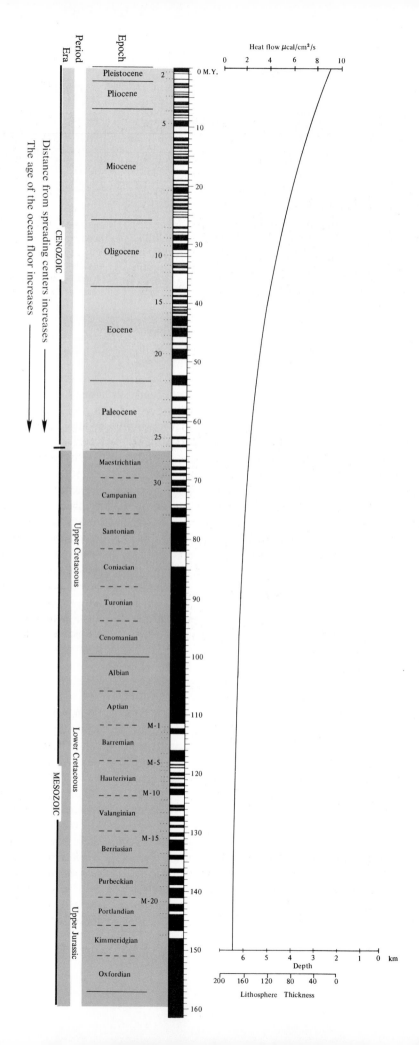

FIGURE 4–10
Magnetic Time Scale, Heat Flow, Ocean Depth, and Lithosphere Thickness. Black bands represent periods of time when the magnetic polarity of the earth was the same as at present, and white bands represent periods when it was reversed. The curve shows that the ocean gets deeper and the lithosphere thicker with increasing age, while the rate of heat flow to the earth's surface through the lithosphere decreases as the lithosphere ages. Although this curve could not be accurately applied to any particular region of the ocean basins, it indicates that, in general, heat flow decreases, while ocean depth, thickness of the lithosphere, and age of the ocean floor increase with increasing distance from the spreading centers. The geologic time scale is shown (see Appendix IV).

66

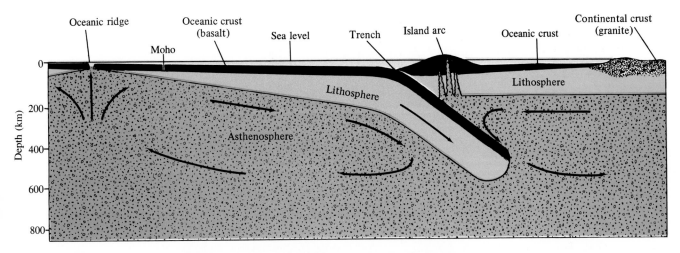

FIGURE 4–11

Movement in the Asthenosphere. Lateral flow within the asthenosphere is a possible mechanism for the lateral movement of the lithosphere, including the crust. The asthenosphere is a low-strength plastic zone within the mantle where pressure is low enough and temperature sufficiently near the melting point of the material to allow it to flow slowly.

sphere. The thermal contraction of hot, fluffy lithosphere into cold, brittle lithosphere would account for the increasing depth of the ocean away from the spreading centers.

The relationship of plate characteristics and ocean depth to convection-cell motion represents only one possibility. Many investigators do not believe convection is an important factor in plate motion. Some believe the plates are pulled down into the mantle by the weight of their cooler, dense edges. At present there seems to be no clearly defined process that can be identified as driving the lithospheric plates.

PLATE BOUNDARIES

The process of sea-floor spreading sets up stresses within the lithosphere as it moves across the spherical surface of the earth. These stresses have broken the present lithosphere into about a half-dozen major and another half-dozen or more minor plates, some of which are shown in figure 4–12. The plate boundaries where new lithosphere is being added along oceanic ridges are called *constructive* or *divergent plate boundaries;* boundaries where plates collide and one subducts beneath the other are called *destructive* or *convergent boundaries.* Lithospheric plates also move past one another along *transform* or *shear boundaries.*

Constructive Boundaries

Oceanic Ridges and Rises Project FAMOUS, the French-American Mid-Ocean Undersea Study, conducted in 1974, made observations in the rift valley running along the axis of the Mid-Atlantic Ridge. These observations provide a clue to the spreading process (figure 4–13).

Fissures oriented parallel to the ridge axis range from hairline size near the center of the rift valley to more than 10 m (33 ft) in width near the margins of the valley. This may indicate the plates are being continuously pulled apart rather than being pushed apart by upwelling of material beneath the ridges. Possibly, the upwelling of magma beneath the oceanic ridges is more that of filling in the void left by the separating plates of lithosphere. Whatever the mechanism of emplacement, a large mass of magma or viscous igneous material must exist near the surface beneath the ridges. The effect of this source of heat so close to the surface is manifested in the thermal expansion of rocks of the oceanic ridges and rises and a significant thinning of the lithosphere beneath them. The expansion causes the ridges and rises to be elevated to mountainlike proportions above the deep-ocean basins.

Destructive Boundaries

Ocean Trench System Oceanic ridges and trench systems both represent significant discontinuities in the earth's crustal structure and are characterized by earthquakes and volcanic activity. There is a significant difference in the earthquake activity of the two regions, however. Shallow quakes usually less than 10 km (6 mi) in depth are associated with the constructive ridge boundaries where the lithosphere is thin, and movements that cause earthquakes in the trench regions may occur at depths down to 700 km (435 mi).

The earthquake *focus,* the point at which the movement that causes the quake occurs, is usually shallow in the immediate area of the trench. As earthquake activity moves toward the continent, the depth of the focus becomes

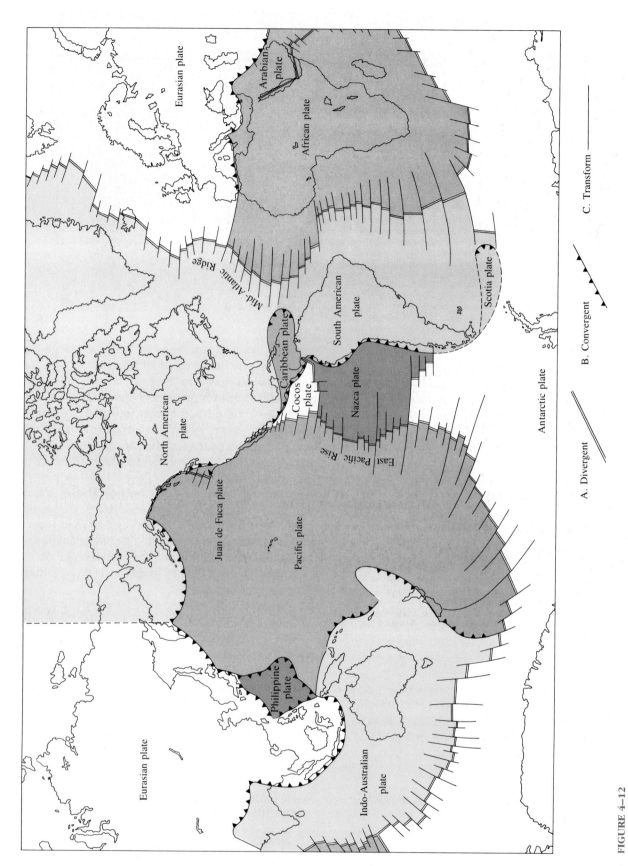

FIGURE 4–12

Crustal Plates and Associated Features. (A colored version of this figure is presented in Plate 10.)

Courtesy U.S. Geological Survey.

A. Divergent B. Convergent C. Transform

FIGURE 4–13
Rift Valley Fissures. *A,* a fissure in the rift valley of the Mid-Atlantic Ridge photographed from the submersible *Alvin* during Project FAMOUS, 1974. (Courtesy of Woods Hole Oceanographic Institution.) *B,* a more readily observable fissure above sea level in the rift valley of Iceland. Here the Mid-Atlantic Ridge rises above the ocean surface as an island because of the high rate of volcanic activity in the Iceland hot spot. (Courtesy of Br. Robert McDermott, S.J.)

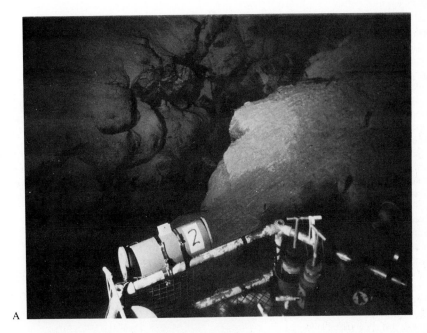

A

B

greater and reaches a maximum value of 700 km (435 mi). There appears to be a band 20 km (12.5 mi) thick that dips from the trench region under the continents within which all the earthquake foci are located. The angle of dip is approximately 45°, becoming steeper at greater depths (figure 4–14A).

This band within which the earthquakes originate must lie within the slab of lithosphere that is descending. The fact that the oceanic crust generated at the oceanic rises and ridges is subducted at the trenches is supported by the fact that no oceanic crustal material more than 200 million years old has been recovered from the ocean bottom.

As the lithosphere is carried into the asthenosphere, it is heated by being subducted to greater depths. Water and other volatiles are freed, producing a low-density mixture that rises to the surface on the continental side of the trenches to produce basaltic volcanoes that may become island arcs.

Continental Arc–Trench System Should an oceanic plate subduct beneath a plate with a continent at its leading edge, the melting of the subducted oceanic plate occurs beneath the continent. Consequently, the rising basalt melt passes through and mixes with the granite of the continen-

FIGURE 4–14

Effects of Plate Collisions. *A,* classic oceanic trench systems develop where oceanic lithosphere subducts beneath a plate containing a continent some distance behind its leading edge. A basaltic island arc–trench system develops. The sloping belt in which earthquake foci are located is the Benioff seismic zone. *B,* when an oceanic lithospheric plate subducts beneath a plate with continental crust at its leading edge, an andesitic continental arc develops. The volcanic rock andesite is named for the Andes Mountains, where a continental arc exists. Andesite has a composition between that of granite and basalt. It is thought to form as melted basalt from the descending oceanic plate rises through the continental granite to produce the volcanic continental arc. *C,* as two plates carrying continents near their leading edge converge, trenches do not develop because sediments that accumulated in the sea between the continents are too light to be subducted. The shortening of the earth's crust in such regions is accommodated by the folding of the sediments into mountain ranges such as the Appalachians, Alps, Himalayas, and Urals. The lower lithosphere and asthenosphere may be involved in low rates of subduction beneath the mountains.

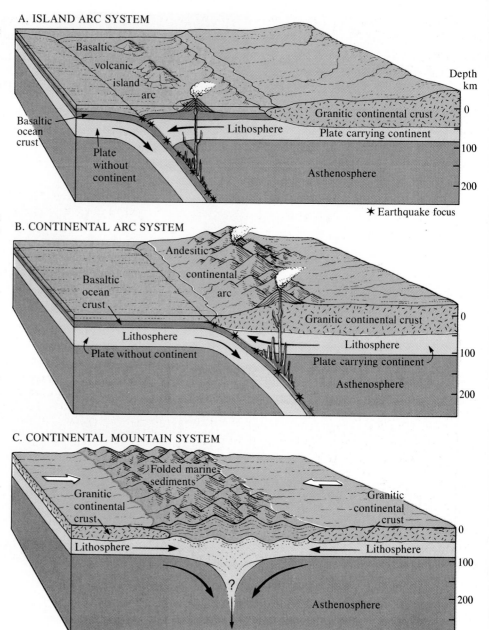

tal crust. This results in a continental arc of volcanoes along the edge of the continent (figure 4–14B). The volcanoes are composed of a rock called andesite, which is of a composition intermediate between that of basalt and granite and is thought to form as a result of the mixing of the two basic crustal rock types.

If the spreading center producing the subducting plate is far enough from the subduction zone, an oceanic trench is well developed along the margin of the continent. The Peru-Chile Trench is an example. It is associated with the Andes Mountains continental arc of volcanoes from which the rock andesite takes its name. The Cascade Mountains continental arc extending from northern California into Canada has no visible trench associated with it. This is probably primarily the result of the nearness of the Juan de Fuca spreading center to the edge of the North American Plate.

Continental Mountain Systems If two lithospheric plates collide and both contain continental crust near their leading edge, the surface expression will be a mountain range composed of the folded sedimentary rocks derived from sediments deposited in the sea that previously separated the continental blocks. These sedimentary rocks do not subside in a zone of subduction because of their relatively low density, having been derived from the continents. The oceanic crust itself may, however, subside beneath such mountains. The exact nature of the subsidence of oceanic crustal material in these regions has not been determined (figure 4–14C).

Some degree of folded mountain formation is always associated with plate collisions, but ocean trenches are absent in some areas—probably for a number of reasons. One possible explanation for their absence has to do with the rate of plate convergence. Where the rate of convergence is less than 6 cm (2.4 in.) per year, the crust may be able to absorb the compression by folding itself into mountain ranges. Higher rates of convergence cause one plate to break free and move past the other while sinking back into the mantle, thus producing the ocean trench.

Shear Boundaries

Transform Faults   With new ocean crust moving away from oceanic ridges at various rates and in various directions toward an ultimate collision with crust that originates along other oceanic ridges, head-on collisions between plates produce the island arc–trench systems and the con-

tinental mountain ranges such as the Alps and Himalayas. Since the earth is spherical, all collisions between the various plates of the earth's crust indicated in figure 4–12 cannot be head-on; there should be evidence of plates sliding past one another. Close examination of the margins of the various plates shows such contacts do exist. A classic example of such a contact is the San Andreas Fault, which runs across California from the head of the Gulf of California to the San Francisco area. This shear plate boundary and others like it are called *transform faults* (figure 4–15).

Fracture Zones   Close observation of the axes of oceanic ridges and rises shows that they do not move in a continuous meandering path through the world's oceans but are offset periodically by transform faults that run perpendicular to the axes. Continuing along the alignment of the transform fault beyond the offset axes are *fracture zones*. The transform faults represent active displacements of the axes, while the fracture zones are evidence of past transform fault activity. On opposite sides of the transform fault, the lithospheric plates are moving in opposite directions, while there is no relative plate motion occurring along a fracture

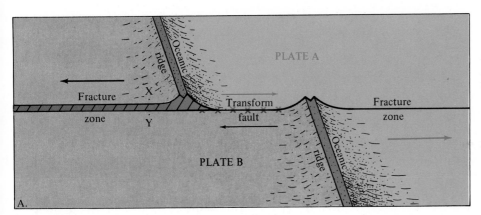

X – Shallow focus
earthquakes

→ Arrows indicate
the direction of
plate movement.

FIGURE 4–15
Transform Faults—Fracture Zones. *A,* the axes of oceanic ridges are offset by transform faults due to the stresses resulting from the motion of rigid plates moving across the spherical surface of the earth. Earthquakes are common and of relatively great magnitude along the transform faults. They are called transform faults because they cut across the larger physiographic features of the ocean floor, oceanic ridges and rises. Extending beyond the offset ridges are fracture zones, where earthquakes are rare because there is no relative movement on the opposite sides of the fracture zone. Fracture zones are scars of old transform fault activity and are maintained as significant topographic features because the ocean floor on one side of the fracture zone is older than the ocean floor immediately opposite it on the other side of the fracture zone. Point *X* represents younger ocean floor than point *Y.* Therefore, point *Y* has subsided more because of cooling and thermal contraction than the ocean floor at point *X.* Thus, there is an escarpment along the fracture zone in Plate B that faces the older ocean floor on the side of the fracture zone where point *Y* is located. There is also an escarpment on Plate A, but it faces the opposite direction. *B,* the San Andreas Fault is a classic example of a transform fault that offsets the oceanic ridge or rise and forms a boundary between plates that are sliding past each other. Along this contact, the Pacific Plate is moving north relative to the North American Plate.

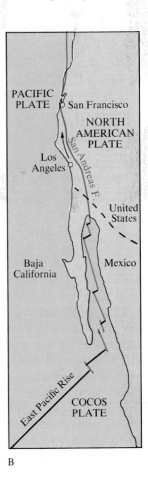

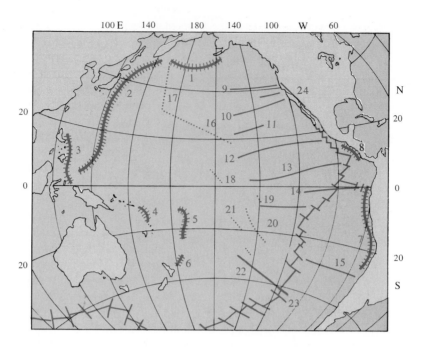

1. Aleutian Trench	9. Mendocino Fracture Zone	17. Emperor Seamounts
2. Kuril-Japan-Mariana Trench System	10. Murray Fracture Zone	18. Line Islands
3. Mindanao Trench	11. Molokai Fracture Zone	19 Marquesas Islands
4. New Hebrides Trench	12. Clarion Fracture Zone	20. Tuamotu Islands
5. Tonga Trench	13. Clipperton Fracture Zone	21. Society Islands
6. Kermadec Trench	14. Galapagos Fracture Zone	22. Eltanin Fracture Zone
7. Peru-Chile Trench	15. Easter Fracture Zone	23. East Pacific Rise
8. Mid-America Trench	16. Hawaiian Islands	24. San Andreas Fault

FIGURE 4–16

Physiographic Features of Pacific Ocean Basin. (See plate 12 for a detailed view of these features. Base map courtesy of Open University Press.)

zone. In other words, the transform faults serve as plate boundaries, and the fracture zones are embedded in a single plate (figure 4–15). Earthquakes with foci above a depth of 10 km (6 mi) are common along the transform faults, while the fracture zones are aseismic.

Major fracture zones extend into the Pacific Plate for thousands of kilometers from the Mendocino Fracture Zone, which intersects the San Andreas Fault in northern California, to the Easter Fracture Zone in the South Pacific. The alignment of these zones is shown in figure 4–16. The physiographic relief associated with the fracture zones may extend through a width of up to 200 km (124 mi). The ocean floor on either side of a fracture zone may have depths differing by as much as 1500 m (4900 ft). For example, the ocean bottom on the southern side of the Mendocino Fracture Zone is more than 1000 m (3300 ft) deeper than that to the north. Therefore, this feature is sometimes referred to as the *Mendocino Escarpment*. Not

only did significant vertical displacement take place along the Mendocino Fracture Zone, but offsets of magnetic anomalies indicate a 1170-km (725-mi) left-lateral displacement. This means that a person standing on the south side of the fracture zone looking north would have to travel 1170 km (725 mi) to the left (west) on the north side of the fracture zone to encounter crust formed at the same time as that on which he or she stands.

INTRAPLATE FEATURES

Seamounts and Tablemounts Highly characteristic of the floor of the Pacific Plate are seamounts and table-mounts. A theory related to the formation of such features in association with the oceanic ridge systems could account for the existence of many seamounts and tablemounts (figure 4–17). Considering the oceanic ridge system on a worldwide basis, active oceanic volcanoes are characteristic

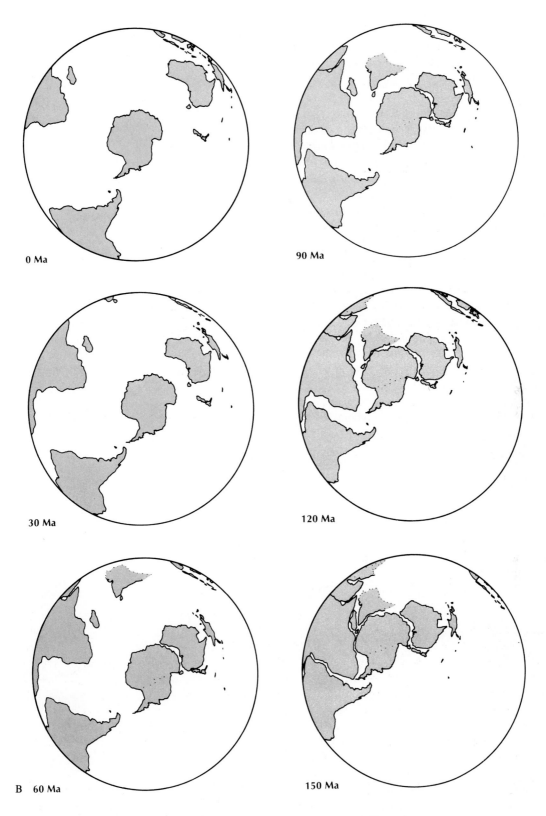

0 Ma

90 Ma

30 Ma

120 Ma

B   60 Ma

150 Ma

span of time, the present-day coastlines have been maintained throughout. Ages are quoted in Ma (millions of years ago) at intervals of 30 million years. The projections are based on data taken from the first 60 legs of the Deep Sea Drilling Project, conducted during a decade of deep-ocean drilling from August 1968 to August 1978. (Courtesy of P. L. Firstbrook, B. M. Funnell, A. M. Hurley, and A. G. Smith in association with Scripps Institution of Oceanography, University of California, San Diego.)

Continuing northwest from Midway Island, submerged seamounts have the same alignment as the island chain from Midway to Hawaii. At about 188°W longitude and 30°N latitude, an abrupt change in the alignment of the seamounts occurs. From this point on, the alignment is slightly west of true north. The age of the seamounts at this juncture of the two trends is thought to be between 30 and 40 million years, while the oldest seamounts in the northward-trending Emperor Seamount Chain formed some 70 million years ago. If the Emperor Seamount Chain and the Hawaii seamounts and islands were generated by the same hot spot, there must have been an abrupt change in the direction of the movement of the Pacific Plate between 30 and 40 million years ago (figure 4–16).

The Line Islands to the south of the Hawaiian Islands do not show as distinct a pattern of progressive formation as the Hawaiian Islands and the Emperor Seamount Chain. Beginning with Christmas Island near the southern end of the Line Islands, just north of the equator, an age of approximately 75 million years seems to be the minimum that could be determined. At the northern end of the chain, the age appears to be no more than 85 million years. Although the islands at the southern end are younger, the difference in age is not so great as might be expected had the hot spot theory accounted for this alignment.

South of the equator are other northwest-southeast-trending volcanic island chains—the Marquesas Islands, Tuamotu Islands, Society Islands, and Cook Islands. It appears that groups of islands in a given chain may have

formed more or less contemporaneously. Although there are indications these chains of islands may have a hot spot origin, the evidence is not so clearly developed as for the Hawaiian Islands.

Paleoceanic Reconstruction Figure 4–22 shows the changing positions of the continents as a result of the global plate tectonics process during the last 150 million years of the earth's history.

Metallic Ores and Plate Tectonics

The origin of metallic ores that contain rich deposits of copper, lead, zinc, and silver associated with the destructive margins of lithospheric plates is beginning to be understood. Their origin is related to the process of plate destruction (figure 4–23). Copper sulfide ores associated with sections of oceanic crust that have been pushed up onto continental plates, *ophiolites,* are part of this process. A good example of this kind of ore is the Troodos Massif along the southern coast of Cyprus, which has been mined since early in the development of the Mediterranean civilizations.

The constructive margins of plates associated with oceanic ridges may be the source of this metallic enrichment. The metals found here are predominantly iron and manganese, with significant amounts of copper, nickel, cobalt, zinc, and barium. Analysis of such sediments in a 200-

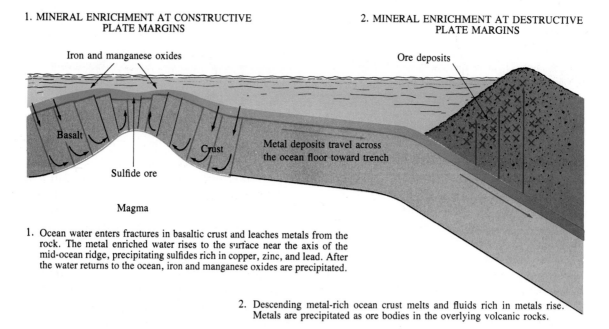

1. MINERAL ENRICHMENT AT CONSTRUCTIVE PLATE MARGINS

2. MINERAL ENRICHMENT AT DESTRUCTIVE PLATE MARGINS

Iron and manganese oxides

Ore deposits

Basalt

Crust

Metal deposits travel across the ocean floor toward trench

Sulfide ore

Magma

1. Ocean water enters fractures in basaltic crust and leaches metals from the rock. The metal enriched water rises to the surface near the axis of the mid-ocean ridge, precipitating sulfides rich in copper, zinc, and lead. After the water returns to the ocean, iron and manganese oxides are precipitated.

2. Descending metal-rich ocean crust melts and fluids rich in metals rise. Metals are precipitated as ore bodies in the overlying volcanic rocks.

FIGURE 4–23
Theory of Metallic Ore Production.

km^2 (77-mi^2) area of the Atlantis Deep in the Red Sea have shown that they contain more than 3 million tons of zinc, 1 million tons of copper, almost 1 million tons of lead, and 5000 tons of silver. Similar iron- and manganese-rich deposits were recovered by submersibles diving along the Mid-Atlantic Ridge during Project FAMOUS in 1974.

SUMMARY

Scientific investigation of evidence suggesting continents are moving relative to one another began with **Alfred Wegener** at the start of the 20th century. After investigations into the way continents might be fit together and the stronger magnetic field associated with the ocean crust compared to the continental crust, many geologists and geophysicists who had resisted such beliefs throughout the first half of the 20th century began to believe the continents were indeed moving. The evidence gathered by the late 1960s indicated all the present continents were combined into one large continent, **Pangaea**, about 200 million years ago. Just prior to the formation and possibly also after the breakup of Pangaea, two large continents appear to have existed, **Laurasia** in the north and **Gondwanaland** to the south.

The discovery that the **earth's magnetic field** has changed **polarity** throughout the earth's history and that the record of these changes was permanently recorded in the rocks of the ocean crust made it possible to develop a hypothesis of **sea-floor spreading** as the process by which the continents could be moved. According to the hypothesis, new rock is added to rigid **lithospheric plates** by intrusion and extrusion of molten material along oceanic ridges and rises. These plates are composed primarily of dense mantle rocks with a thin crust of **basaltic lava** beneath the ocean basins. As material is added to the plates near the axis of the oceanic ridges and rises, it moves away down the slopes of the mountain ranges to make room for new material surfacing at the axis. Thus, the crustal rocks beneath the oceans should increase in age as one moves from the axis of the mountain ranges down their slopes into the deep-ocean basins. Continents located on a lithospheric plate are carried passively along as the plate moves away from the oceanic ridge or rise. The sea-floor spreading hypothesis was confirmed when the **Glomar Challenger** began the Deep Sea Drilling Project to sample the sediments and crust beneath the oceans in 1968. The drilling proved the ocean crustal rocks were indeed older as one moved away from the oceanic ridges and rises into the deep-ocean basins. This movement of the lithosphere appears to be made possible by the existence of a plastic region of mantle material with temperatures near the melting point of the mantle, the **asthenosphere**, immediately beneath the lithosphere.

As the lithospheric plates move away from the high-heat-flow oceanic ridges and rises toward low-heat-flow trenches, they cool and thicken. The age and depth of the ocean floor increase with increasing distance from the **spreading centers**.

As new mass is added to the lithosphere at the oceanic rises and ridges—**constructive** or **divergent** plate boundaries—the leading edge of the plates is subducted into the mantle at **ocean trench systems** or beneath **continental mountain ranges** such as the Himalayas—**destructive** or **convergent** plate boundaries. Additionally, there are plate boundaries called **transform faults** along which oceanic ridges and rises are offset and plates slide past one another—**slip-slide** or **shear** plate boundaries. Although the earth is over 4.5 billion years old, no rocks have been found on the ocean floor that are as much as 200 million years old. We do not know exactly how the convectionlike movement is driven, but it is obvious that the process produces recycling of material between the lithosphere and the deeper mantle.

Additional intraplate features are **seamounts** and **tablemounts** (guyots) that formed as volcanic peaks and islands along the spreading centers. As they moved away from the spreading centers on the cooling lithospheric plates, the volcanoes became inactive within 30 million years and submerged to their present positions. Where average monthly temperatures exceed 18°C and deposition of sediment is minimal, **coral reefs** develop around volcanic peaks. With continued subsidence, they will pass from **fringing reefs** to **barrier reefs** and **atolls**. There is evidence that reefs can survive subsidence rates of over 10 m per 1000 years. Another feature that appears to have its classic development in the Pacific Ocean is the **hot spot**, a region of local upwelling of molten material. The volcanic activity presently producing the island of Hawaii also produced all the volcanic peaks of the **Hawaiian Island chain** and the **Emperor Seamounts** during the past 70 million years. Very long **fracture zones** such as the **Mendocino Fracture Zone** and others in the North Pacific display extreme horizontal and vertical displacements. There is evidence that **metallic ore deposits** associated with continental mountain ranges originated as part of the global plate tectonics process.

QUESTIONS AND EXERCISES

1. Why couldn't Alfred Wegener convince most scientists that the continents were indeed moving across the earth's surface?

2. What evidence was developed in the mid-20th century that supported the possibility of continental movement?

3. How do the magnetic properties of the continents and ocean basins differ? Why?

4. Describe how fossil evidence was used in relative dating of rock units found on the earth.

5. What discovery made absolute dating of rock units possible? Why was this technique a significant improvement over relative dating?

6. What is the age of a rock that has a ratio of ^{235}U to ^{207}Pb of 1 to 3?

7. Describe the evidence found on the present continents that suggests Pangaea broke into two continents, Laurasia and Gondwanaland, between 250 and 350 million years ago.

8. How does the magnetic dip of magnetite particles found in igneous rocks tell us at what latitude they formed?

9. How does continental movement account for the two apparent wandering curves of the north magnetic pole as determined from Europe and North America?

10. What property of the magnetic record found in oceanic crustal rocks gave rise to the idea of sea floor spreading?

11. Describe a possible mechanism responsible for moving the continents over the earth's surface. Include a discussion of the lithosphere and asthenosphere.

12. Describe the general relationships that exist among distance from the spreading centers of the oceanic ridges, heat flow, age of the ocean crustal rock, and ocean depth.

13. Discuss the three types of plate boundaries, spreading centers of the oceanic ridges, trenches, and transform faults. Explain why earthquake activity is usually confined to depths less than 10 km (6 mi) along the spreading centers and transform faults while it may occur as deep as 700 km (435 mi) near the trenches. Construct a plan view and cross section showing each of the types of boundary and direction of plate movement.

14. What evidence indicates that the plates are more likely being pulled down in the trench areas than being pushed away from the oceanic ridge spreading centers?

15. What conditions of plate collision are likely to create a folded mountain range such as the Himalayas rather than a trench–island arc system?

16. Describe the relationship between fracture zones and transform faults, including a comparison of the earthquake activity along each.

17. Explain why seamounts and guyots increase in age and depth with increased distance from the oceanic ridges. Discuss the length of time they probably were active volcanoes.

18. Describe how the hot spot theory explains the age, alignment, and degree of submergence characteristic of volcanoes found along the Hawaiian Island and Emperor Seamount chains.

19. Describe the relationship of metallic ore deposits in the lithosphere to the plate tectonics process.

REFERENCES

Anderson, D. L. 1975. Accelerated plate tectonics. *Science* 187:1077–79.

Bullard, Sir Edward, 1969. The origin of the oceans. *Scientific American* 221:66–75.

Carrigan, C. R., and Gubbins, D. 1979. The source of the earth's magnetic field. *Scientific American* 240:2, 118–33.

Davies, P. J., Symonds, P. A., Feary, D. A., and Pigram, C. J. 1987. Horizontal plate motion: A key allocyclic factor in the evolution of the Great Barrier Reef. *Science* 238:1697–1700.

Dietz, R. S., and Holden, J. C. 1970. The breakup of Pangaea. *Scientific American* 223:30–41.

Haggerty, J. A., Schlanger, S. O., and Silva, I. P. 1982. Late Cretaceous and Eocene volcanism in the southern Line Islands and implications for hot spot theory. *Geology* 10:8, 433–36.

Heezen, B. C., and MacGregor, I. D. 1973. The evolution of the Pacific. *Scientific American* 229:102–12.

Heezen, B. C., and Rawson, M. 1977. Visual observations of the sea-floor subduction line in the Middle-America Trench. *Science* 196:423–26.

McKenzie, D. P., and Sclater, J. G. 1973. Evolution of the Indian Ocean. *Scientific American* 228:62–72.

Menard, H. W. 1986. The ocean of truth. Princeton, N. J.: Princeton University Press.

Shepard, F. P. 1977. *Geological oceanography.* New York: Crane, Russak.

Smith, R. V., and Kinsey, D. W. 1976. Calcium carbonate production, coral reef growth, and sea level change. *Science* 194:937–38.

Tarling, D., and Tarling, M. 1971. *Continental drift: A study of the earth's moving surface.* Garden City, N.Y.: Doubleday.

SUGGESTED READING

Sea Frontier

Burton, R. 1974. Instant islands. 19:3, 130–36.
A description of the 1973 eruption on the island of Hilmaey south of Iceland and its relationship to plate tectonics processes.

Dietz, R. S. 1976. Iceland—where the midocean ridge bares its back. 22:1, 9–15.
A description of the rift zone of Iceland and its relationship to the Mid-Atlantic Ridge and sea-floor spreading.

Dietz, R. S. 1977. San Andreas: an oceanic fault that came ashore. 23:5, 258–66.
A discussion of the San Andreas Fault and its relationship to global plate tectonics.

———. 1971. Those shifty continents. 17:4, 204–12.
A very readable and informative presentation of the crustal features associated with and the possible mechanism of plate tectonics.

Emiliani, C. 1972. A magnificent revolution. 18:6, 357–72.
A discussion of advances in studying earth science from the sea, including climate cycles, plate tectonics, and the economic potential of resources lying within marine sediments.

Kenyon J. 1987. Lighthouse Reef: A Caribbean atoll. 33:6, 427–37.
A description of this atoll, a type of reef that is not common in the Caribbean Sea.

Mark, K. 1976. Coral reefs, seamounts, and guyots. 22:3, 143–49.
A well-written discussion of the formation of coral reefs, including the theory of Charles Darwin and the significance of plate tectonics.

———. 1972. Ocean fossils on land. 18:2, 95–106.
A discussion of the significance of marine fossils found in rocks now many miles inland is centered around the work of an early American paleontologist, James Hall.

Rona, P. 1984. Perpetual sea-floor metal factory. 30:3, 132–141.
A discussion of how metallic mineral deposits form in association with hydrothermal vents located on oceanic ridges and rises.

Smith, F. 1981. Baja: Yesterday, today, and tomorrow. 27:4, 194–205.
Global plate tectonics and its effect on Baja California are discussed.

Scientific American

Bonatti, Enrico, and Crane, Kathleen. 1984. Oceanic fracture zones. 250:5, 40–51.
The role of oceanic fracture zones in the plate tectonics process is related to the spherical shape of the earth.

Bonatti, E. 1987. The rifting of continents. 256:3, 96–103.
From studying the Red Sea, scientists develop a model of how the process that rifts continents apart proceeds.

Courtillot, V., and Vink, G. E. 1983. How continents break up. 249:1, 43–49.
The process of continental rifting and deformation is considered.

Dietz, R. S., and Holden, J. C. 1970. The breakup of Pangaea. 229:4, 30–41.
A description of the breakup of the hypothetical supercontinent and the movement of continents to their present positions. The possible mechanism of continental drift is explored.

Edmond, John M., and Von Damm, Karen. 1983. Hot springs on the ocean floor. 248:4, 78–93.
Hydrothermal vents along oceanic ridge rift valleys produce metallic "ores" and interesting biological communities.

Francis, Peter, and Self, Stephen. 1983. The eruption of Krakatau. 249:5, 172–87.
An analysis of the 1883 event based on the volcanic deposits and the timing of air and sea waves created by the eruption.

Hallam, A. 1975. Alfred Wegener and continental drift. 232:2, 88–97.
A history of Wegener's development of his theory of continental drift presented in 1912 and the reasons for its general rejection until the 1960s.

Hékinian, Roger. 1984. Undersea volcanoes. 251:1, 46–55.
The volcanism associated with spreading center rift valleys adds new rock to the lithosphere.

Schlater, J. G., and Tapscott, C. 1979. The history of the Atlantic, 240:6, 156–75.
The history of the formation of the Atlantic Ocean basin is described based on data gathered from measurements of heat flow and magnetism as well as rock samples recovered by the Deep Sea Drilling Project.

Vink, E. Gregory, Morgan, W. Jason, and Vogt, Peter R. 1985. The earth's hot spots. 252:4, 50–57.
A hot spot, such as that underlying the island of Hawaii, leaves distinguishing marks on the earth's dynamic surface.

5

MARINE SEDIMENTS

Over one-half of the rocks exposed at the surface of the continents above the shoreline were formed by the lithification of sediment laid down in past ocean environments. From this observation, it appears that there is a significant relationship between the formation of the continental masses and the surrounding ocean basins, but the exact nature of this relationship has for years eluded the inquiring geologist. Geologists once believed that if they could examine the entire sedimentary column in the deep-ocean basin, which they assumed to have been permanently in existence since the initial formation of the earth's oceans, a very great portion of the history of the earth might well be recorded in these sediments.

Because of recent advances in technology, it has been possible to sufficiently examine the sediments in the deep-ocean basins to make it seem certain such is not the case. It seems unlikely that sediment representing more than 200 million years of the earth's 4.6-billion-year history will be found in the deep-ocean basins. This, however, does not dim the marine geologist's interest in the study of these sediments, since they will help determine the details of this shorter history of the present ocean basins. Clues to past climates, movements of the ocean floor, ocean circulation patterns, and nutrient supply for the marine plant population are embedded in the sedimentary deposits throughout the ocean basins.

Sediment Texture

Sediment texture, which is determined primarily by grain size, is indicative of the energy condition under which a deposit is laid down. The abbreviated Wentworth scale presented in table 5–1 classifies particles in categories ranging from boulders to colloids. Between these extremes are cobbles, pebbles, granules, sand, silt, and clay-sized particles. Deposits that are laid down in areas where wave action is strong (areas of high energy) may be composed primarily of the larger particles—cobbles and boulders. The deposition of

Silica skeletons of radiolarians, single-celled marine animals, are shown here magnified 280 times life size. Geologists study the evolution of these microfossils as a basis for assigning ages to ocean sediments. Picture was taken using camera attached to Cambridge S4 scanning electron microscope at Scripp's Analytical Facility. Photo courtesy of Scripps Institution of Oceanography, University of California, San Diego.

TABLE 5–1
Wentworth Scale of Grain Size for Sediments.

Particle	Min. Size (mm)
Boulder	256
Cobble	64
Pebble	4
Granule	2
Sand	0.062(1/16)
Silt	0.004(1/256)
Clay	0.002(1/4096)
Colloid	<0.002

Source: Wentworth (1922) after Udden (1898).

Maturity increases
Degree of sorting increases
Clay content decreases
Rounding of sand particles increases

—— Clay particle

FIGURE 5–1
Sediment Maturity.

clay-sized particles occurs in areas where the energy level is low and the current speed is minimal.

Sediments composed of particles that are primarily within the same size classification are *well sorted.* A beach sand is usually a well-sorted deposit. Glacial till that may be found on the continental shelf is an example of a poorly sorted deposit because it contains particles ranging in size from the colloid to boulders that have been carried by glaciers and dropped out as the glacier melted.

As particles are carried from the source to their point of deposition, they increase in *maturity.* This results from their association with moving water, which has the capacity to carry particles of a certain size away from a deposit and leave behind other larger particles. As the time in the transporting medium increases, particles of sand size or larger become more rounded through chemical weathering and abrasion. Increasing *sediment maturity* is indicated by (1) decreasing clay content, (2) increasing degree of sorting, and (3) increasing rounding of the grains within the deposit.

The poorly sorted glacial till, which contains relatively large quantities of clay-sized and larger particles that have not been well rounded, is an example of an immature sedimentary deposit. The beach sand we used as an example of a well-sorted sediment contains very little clay-sized material and is usually composed of well-rounded particles that have undergone considerable transportation. The beach sand is an example of a mature sedimentary deposit. Figure 5–1 illustrates the nature of sediments of varying degrees of maturity based on these criteria.

Sediment Transport

The lithogenous sediment carried to the margins of the ocean by the rivers and glaciers flowing from the continents either settles out to fill estuaries, becomes a part of a delta, is spread across the continental shelf by currents, or is carried beyond the shelf to the deep ocean basin. The

greatest volume of sediment transport in the ocean is achieved by the longshore currents near the margins of the continent. Lower-energy currents distribute the finer components of the sediment along the margin of the continental shelf and even into the deep-ocean basin.

Figure 5–2 shows the relationship between average horizontal current velocity and the erosion, transportation, and deposition of particles ranging in size from 0.001 to 100 mm (0.00004–4 in.). As might be expected, the curve that

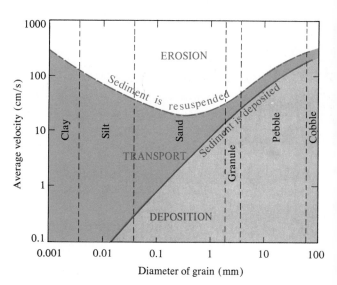

FIGURE 5–2
Hjulstrom's Curves. Horizontal current velocity vs. erosion and deposition by particle size. Observing the curve that shows the average velocity of current at which particles settle out—deposition—we see that larger particles settle out at higher current velocities than do smaller particles. This is not surprising. What is a bit surprising is that it requires a much higher velocity current to pick up—erode from the ocean floor—a tiny clay particle than a larger sand particle. The clay particles require a high-velocity current to erode them because of the strong cohesive attraction that exists between deposited clay particles.

separates transportation of particles from their deposition shows that larger particles will settle out at higher current velocities than smaller ones.

Observing the curve that separates the process of erosion from transportation, we can see that it takes higher velocity currents to erode (pick up and carry away) larger particles from the sand-sized particles up. The surprising fact that clay-sized particles require a higher velocity current than that needed to erode the larger sand-sized particles results from the fact that clay-sized particles are flat rather than round like typical sand-sized particles. Thus they have a greater surface area per unit of mass in contact with one another when they are laid down in a deposit. This causes a great cohesive force in clay sediments. To overcome this cohesive attraction and pick up these particles once they are deposited requires a high velocity current.

Particles less than 20 μm (1μm = 1 millionth meter) in diameter can be carried far out over the open ocean by prevailing winds. Such particles either settle out as the velocity of the wind decreases or find their way into the ocean during precipitation. They commonly serve as nuclei around which raindrops and snowflakes form. Some particles that reach high altitudes can be carried very rapidly by the jet stream that exists in the midlatitudes.

COMPOSITION OF MARINE SEDIMENT

Clues to the origin of sediments can be found in their mineral composition. Certain minerals such as quartz and feldspar are characteristic rock-forming minerals that, when found in a marine sediment, indicate it was derived from rocks. Although silica (SiO_2) and calcite ($CaCO_3$) both may be found in other types of deposits, they are the most abundant compounds in sediments derived from organisms. The minerals that precipitate from ocean water are numerous but, as will be discussed later in the chapter, are characteristic of such origin. Within the following paragraphs we will describe the various origins of marine sedimentary particles.

Lithogenous Sediment

Igneous Rocks. *Lithogenous* means "derived from rocks." Most of the rock mass that supplies lithogenous sediment is that which makes up the continents. The volcanic islands found in the open ocean are also important sources of lithogenous sediment. All parts of the earth's crust originally formed from the solidification of molten material into *igneous* rocks (*ignis* is Latin for "fire"). Igneous rocks solidified at temperatures and pressures well above those of atmospheric conditions and in an environment in which free oxygen and water were scarce.

Igneous rocks are composed of discrete crystals of naturally occurring compounds called *minerals*. These minerals may be grouped into two basic categories: *ferromagnesian* and *nonferromagnesian*. Ferromagnesian minerals are relatively high density, dark in color, and contain significant amounts of the elements iron and magnesium. Common varieties are olivine, augite, hornblende, and biotite. The nonferromagnesian minerals quartz, feldspar, and muscovite contain no iron or magnesium and have a lower density and lighter color than the ferromagnesian minerals. All these minerals are—to varying degrees—unstable under the conditions that prevail at the earth's surface and begin to undergo chemical and physical breakdown known as *weathering*. As they break down, the particles are carried to the ocean, where they are deposited. By far the greatest quantity of lithogenous material is found around the margins of the continents, but there are no parts of the ocean basins where traces of this sediment cannot be found.

Figure 5–3 shows the major types of igneous rocks and their general mineral composition and texture. *Intrusive rocks* such as granite cool and solidify slowly beneath the surface of the earth. The individual crystals grow to large size and give the rocks a coarse-grained texture. *Extrusive rocks* cool more rapidly at or near the earth's surface, so the mineral crystals don't have time to grow large. They have a fine-grained texture.

Clays The products of the decomposition of feldspars (aluminum- and silica-rich) and ferromagnesian (iron- and

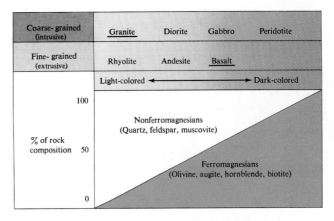

FIGURE 5–3

Igneous Rock Classification. This table shows the most common types of igneous rocks. The coarse-grained ones cool slowly deep beneath the earth's surface, and granite, the typical intrusive rock of the continents, is the most common variety. Peridotite is found in the earth's mantle. The fine-grained rocks, which are associated with volcanic activity, cool rapidly at or near the earth's surface. The most common variety is basalt, which is characteristic of the oceanic crust. The rocks on the left of the table are light colored and contain mostly nonferromagnesian minerals. Moving to the right, ferromagnesian minerals replace nonferromagnesian minerals, and the rocks become darker in color.

magnesium-rich) minerals found in igneous rocks comprise four basic groups of *clays* (hydrated aluminum silicates) identifiable in marine sediment. *Chlorite* is a greenish platy mineral that forms from the decomposition of ferromagnesian minerals and contains magnesium and iron. *Montmorillonite* contains sodium and potassium, as well as calcium, magnesium, and iron. The chemical composition of *illite* is usually restricted to the metallic ions of potassium, iron, and magnesium. *Kaolinite* is a product of the weathering of potassium feldspar and muscovite and contains only the metallic ions aluminum and potassium. In addition, *zeolites* are a group of hydrated aluminum silicates resulting primarily from the weathering of feldspars and possessing varying quantities of potassium, sodium and calcium. Phillipsite is one of the most abundant zeolites found in marine sediments.

Wind Transport A large percentage of the lithogenous particles that find their way into the deep-ocean sediments far from the continents are transported by the prevailing wind systems that remove small particles form the subtropical desert regions of the continents. Figure 5–4 shows the content of small shards of quartz that are present in the surface sediments of the ocean floor. It shows a close relationship between the location of strong prevailing wind systems and high concentrations of quartz shards in ocean sediment. The pattern of clay particle distribution in these sediments would likely be similar.

Biogenous Sediment

Biogenous means "derived from organisms." The insoluble remains of organisms, such as bones and teeth of animals

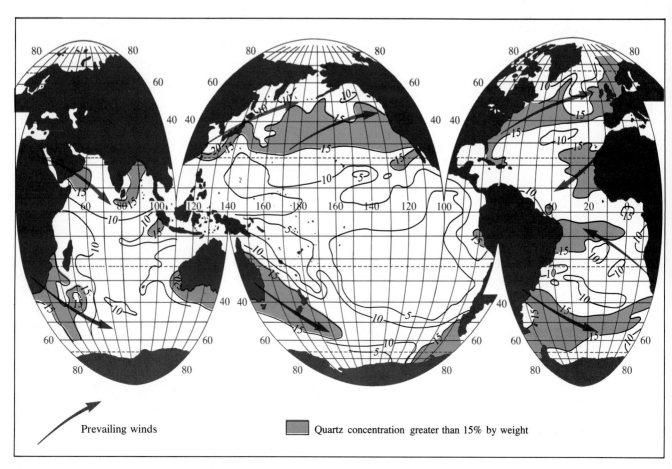

FIGURE 5–4

Lithogenous Quartz (SiO₂) in Surface Sediments of the World's Oceans. Occurring as chips and shards mostly within the 5–10 μm (0.0002–0.0004 in.) range, most quartz found in oceanic sediments is believed to have been transported by winds. Desert areas of Africa, Asia, and Australia are important sources of eolian (windblown) quartz. Turbidity flows have been significant sources of quartz in the equatorial, northwestern, and northern Atlantic Ocean and the Bay of Bengal, where deep-bottom currents have also modified sediment distribution. The distribution of lithogenous clays transported by winds may show a similar pattern. (After Leinen et al. 1986.)

FIGURE 5–5

Microscopic Skeletons from the Deep Sea. Magnified 160 times by scanning electron microscopy, the skeletons above are foraminifera and radiolaria found in sediment cores taken at Site 55 in the Caroline Ridge areas of the west Pacific Ocean by the Deep Sea Drilling Project. They were recovered in water 2843 m deep, with further penetration of 130 m into the ocean bottom. Rounded globose forms are foraminifera, whose skeletons are calcareous. Forms with reticulate skeletons are radiolaria and are silicous. In life, these tiny organisms lived near the ocean surface, but upon death their skeletons fell to the sea bed to become entombed in the sediments. By carefully studying them, scientists can determine how long ago they lived and can learn much about the history of the ocean. (Courtesy of the Deep Sea Drilling Project, Scripps Institution of Oceanography, University of California, San Diego.)

and the protective coverings of minute plants and animals, are deposited on the ocean bottom. The most common chemical compounds found in biogenous sediment are calcium carbonate ($CaCO_3$) and silica (SiO_2).

Contributing most of the silica component of biogenous particles, which rarely exceed 100 μm in diameter, are microscopic plants *(diatoms)* and animals *(radiolarians)*. Foraminifera, close relatives of radiolarians, are responsible for most of the calcium carbonate component, but other significant calcareous contributors are the larger pteropods and microscopic algae, *coccolithophores* (figure 5–5).

Massive, undisturbed, and detrital reef material is also included within the biogenous classification. Composed primarily of the calcareous skeletal structure secreted by corals and algae, coral reefs have been deposited as sedimentary rock, although by biological rather than physical processes.

Hydrogenous Sediment

Hydrogenous means "derived from water." These deposits are formed by a chemical reaction that occurs within seawater. Manganese deposits, phosphorite, and glauconite are minerals that form by chemical precipitation from water. The rate of accumulation of this type of sediment is very slow.

Manganese Nodules   Among the most important sediments on the ocean floor in terms of economic potential are the manganese* nodules (polymetalic nodules) found on the deep ocean floor (figure 5–6). Such nodules have been known to be relatively abundant in all of the major ocean basins since the voyage of the *Challenger.* The major components of these nodules are MnO_2 and Fe_2O_3, with the average manganese dioxide content ranging around 30 percent and the average iron oxide content around 20 percent by weight. Also occurring in economically significant concentrations in the nodules are copper, cobalt, and nickel. Although these concentrations are usually less than 1 percent, they can exceed 2 percent by weight.

Since iron and manganese both occur in hydrogenous deposits in concentrations greater than those they reach in igneous rocks, there exists a problem of explaining the concentration of these elements in the marine deposits. This problem has not yet been successfully solved, but it has been concluded that there are three possible sources for manganese and iron. It is thought that they are derived from (1) the weathering of volcanic material produced by volcanic activity on the ocean floor; (2) concentration in hydrothermal springwater found at the axes of oceanic ridges and rises (plate 3); and (3) runoff from the continents that carries the iron and manganese minerals as soluble compounds from the continents. There seems to be little variation in the chemistry of deep ocean water throughout the world ocean. Studies on the relationship between manganese nodules and organisms suggest that the presence of microorganisms, primarily bacteria, may be a determining factor in the formation of nodules. The Manganese Nodule Project (MANOP), which began in 1977, uses sediment traps to catch particles falling to the ocean floor and a Bottom Lander to carry out seabed experiments designed to explain how nodules form and why they have the observed range of composition.

Figure 5–7 shows the general distribution of rich deposits of manganese nodules in the world ocean. It appears that the area with the greatest economic potential is in the north equatorial Pacific from Mexico to south of the Hawaiian Islands.

---

*The element manganese (Mn) has an atomic weight (54.94) that is more than twice the atomic weight of magnesium (Mg) (24.31). Manganese falls next to iron (Fe), which has an atomic weight of 55.85, on the periodic table of elements (Appendix III).

A   B

FIGURE 5–6

Hydrogenous Sediment. *A,* photograph of the surface of a manganese nodule, magnified 156 times by a
scanning electron microscope. The small tube-and-dome structures were built by organisms for shelter, according to Dr. Jimmy Greenslate at Scripps Institution of Oceanography. He has studied 71 nodules from the Pacific
Ocean and found biologically derived structures both buried inside them and covering large portions of their
surfaces. Nodules may owe their existence to organisms that participate in their formation. *B,* manganese nodules on the Pacific Ocean sea floor are formed as chemical precipitates. They contain quantities of manganese,
iron, copper, nickel, and cobalt. Although the manganese is inferior to manganese ores mined on land, it is
still a valuable commodity, making harvesting of the lumps a probability in the future. (Courtesy of Scripps
Institution of Oceanography, University of California, San Diego.)

Cobalt-Rich Ocean Crusts Recent investigations have
revealed cobalt-rich crust on the upper slopes of islands
and seamounts between the depths of 1000 and 2500 m
(3280 and 8200 ft). These crusts are up to twice as rich in
cobalt as manganese nodules. If the deposits are found to
be sufficiently extensive, they would provide a potential
source of cobalt for the United States and other nations.
Many of the potential deposits are on the flanks of islands
within the limits of the 200-mi-wide Exclusive Economic
Zone.

Metallic Sulfide Deposits Deposits of iron, copper, and
zinc sulfides are directly associated with the hydrothermal
vents found along the axes of oceanic ridges and rises.
These deposits are created by the interaction of these metals with H_2S (hydrogen sulfide) gas at temperatures of
about 350°C (662°F) (figure 5–8).

Phosphorite Phosphorus in the form of P_2O_5 is found
abundantly as a precipitate in nodules, as a thin crust on
the continental shelf, and on banks at depths above 1000
m (3280 ft). Concentrations of phosphates in such deposits
commonly reach 30 percent by weight. Again, the exact
process of formation is not determined, but the deposits
seem to be associated with areas of upwelling water rich in
phosphorus.

Deposits of phosphorite in nodules or bedded crusts
may be found in water close to the edge of the continental
shelf (figure 5–7). These deposits are not likely to be
mined in the near future because of the extensive deposits
of phosphate on land.

Glauconite This generally greenish hydrous silicate has
a complex and variable composition which includes ions
of potassium, magnesium, and iron. Glauconite may form
by submarine weathering of ferromagnesian minerals, although the exact mode of formation is not known. Glauconite, which is found to depths of 2500 m (8200 ft) but
more commonly on topographically high areas near the
coastline, is considered to form only in the marine environment. Forming as casts in organic remains and as encrus-
tations, glauconite is more commonly found in grains of silt
or sand size in mud and sand deposits. Deposits in which
the glauconite is sufficiently abundant are frequently referred to as *green sands* or *green muds* due to the colora-
tion that results from the presence of glauconite.

Carbonates The two most important carbonates in marine sediment are *aragonite* ($CaCO_3$) and *calcite* ($CaCO_3$).
Aragonite and calcite have the same chemical composition
but different crystalline structures. Aragonite is the least stable form and over a period of time changes to calcite. Car-

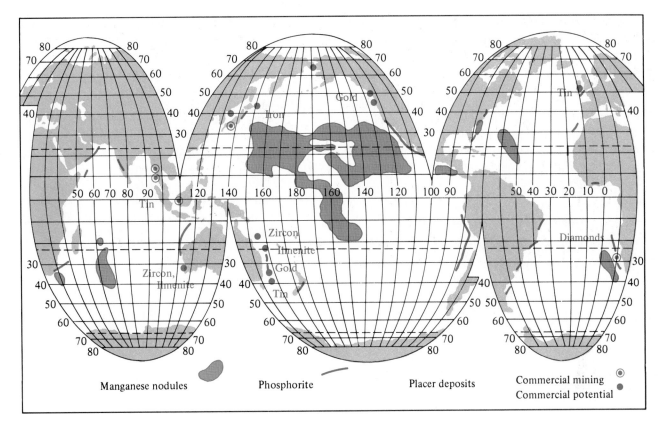

FIGURE 5–7
Manganese Nodule, Phosphorite, and Placer Deposits. (Base map courtesy of *National Ocean Survey*).

bonate precipitation may occur directly from ocean water through physiochemical precipitation when water conditions are right. The most common forms of precipitated carbonate are aragonite crystals less than 2 mm (0.08 in.) in length and oolites, onionlike spheres that precipitate

FIGURE 5–8
Coating of Metallic Sulfides. This photo of metallic sulfide deposits containing copper, iron, and zinc was photographed from *Alvin* by Dr. Robert D. Ballard during the exploration of hydrothermal vents located at 21°N on the East Pacific Rise. (Photo courtesy of WHOI.)

around a nucleus and reach diameters of less than 2 mm (0.08 in.).

Since ocean water is essentially saturated with calcium (Ca^{2+}), the only factor preventing precipitation of calcium carbonate (the most common carbonate) is the absence of carbonate ions (CO_3^{2-}). Carbon dioxide (CO_2) is important in carbonate chemistry. To see how it influences the precipitation of carbonates, we need to look at the various forms in which CO_2 is stored in ocean water. The following reactions tell us that when carbon dioxide combines with water in the ocean, carbonic acid (H_2CO_3) is formed. Carbonic acid may break down and produce a hydrogen ion (H^+) and a bicarbonate ion (HCO^-_3). On further breaking down of the bicarbonate ion, an additional hydrogen ion may be formed along with carbonate ion (CO_3^{2-}).

$$H_2O + CO_2 \rightleftharpoons H_2CO_3 \rightleftharpoons H^+ + HCO_3 \rightleftharpoons H^+ + H^+ + CO_3^{2-}$$

| Water | Carbon dioxide | Carbonic acid | Hydrogen ion | Bicarbonate ion | Hydrogen ions | Carbonate ion |

Low pH ⟵———————————⟶ High pH
7.5   9.0

The presence of carbon dioxide and the carbonate ion in ocean water is mutually exclusive. If there is a high concentration of CO_2, there will be little or no CO_3^{2-} in the water. Removal of carbon dioxide raises the concentration of CO_3^{2-}, so we might look to areas where photosynthesis is occurring at rates that remove significant amounts of carbon dioxide from the water for areas of carbonate precipitation. Heating water also reduces its ability to dissolve carbon dioxide and decreases carbon dioxide concentration. Both factors are at work in areas where precipitation is known to occur. For instance, in the Bahama Banks, water that is moving north from the deep flow of the Florida Straits moves up over the shallow bank, where it is heated and exposed to high plant productivity. Both changes tend to reduce carbon dioxide content of the water and enhance precipitation of carbonates.

Cosmogenous Sediment

Typical particles of cosmic origin found in marine sediment are microscopic magnetic spherules rich in iron and rocky chondrules (figure 5–9). They had long been thought to have been shed from meteors entering the earth's atmosphere. Recent studies have, however, resulted in a different conclusion. Based on the comparison of their shape, size and composition with that of meteorites, investigators believe they form in the asteroid belt as sparks produced when asteroids collide. They then rain down on the earth as a general component of space dust.

Types of Marine Sedimentary Deposits

As the result of marine sediment deposition, the relative amounts of lithogenous, biogenous, hydrogenous, and cosmogenous particles found in marine sedimentary deposits vary considerably. The two largest categories into which marine sedimentary deposits can be divided are neritic de-

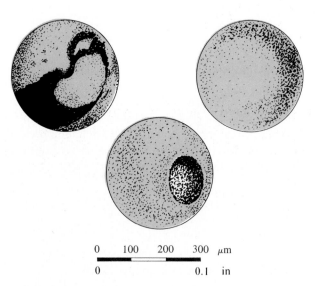

| 0 | 100 | 200 | 300 | μm |

| 0 | | | 0.1 | in |

FIGURE 5–9
Cosmogenous Sediment. Cosmic spherules from the *Challenger* expedition.

posits, found near the continents, and oceanic deposits, characteristic of the deep-ocean basins.

Neritic sediment refers to material with a wide range of particle size that is composed primarily of lithogenous particles derived from the continents. It accumulates rapidly on the continental shelf, slope, and rise. Neritic sediment also contains biogenous, hydrogenous, and cosmogenous particles. These constitute only a minor percentage of the total sediment mass because of the rapid deposition rate of lithogenous particles near the continents. Neritic sediment is distributed along the continental shelf by surface currents and carried down the submarine canyons of the continental slope to the continental rise by turbidity currents. Here the sediment is further distributed by deep boundary currents.

Oceanic sediment is the very fine grained material that accumulates at a slower rate on the deep-ocean-basin floor. It is possible for a greater variety of oceanic sediment types to accumulate because the rate of deposition of lithogenous particles is greatly reduced from that observed on the continental margins. Although the lithogenous component dominates the sediment found in most of the deeper basins of the ocean floor, biogenous and hydrogenous components are abundant in many areas of the ocean bottom.

SEDIMENTS OF THE CONTINENTAL MARGIN (NERITIC SEDIMENTS)

Due to the rise in sea level that occurred with the melting of glaciers at the end of the last ice advance, many rivers of the world today deposit their sediment in estuaries rather than carry it onto the continental shelf as they did at various times during the geologic past. In many areas the sediments that cover the continental shelf, *relict* sediments, were deposited from 3000 to 7000 years ago and have not yet been covered by recent deposits. Such sediments presently cover about 70 percent of the world's continental shelf.

Turbidites

Although it seems unlikely that wave action and ocean current systems could carry coarse material beyond the continental shelf into the deep-ocean basin, there is evidence that much neritic sediment has been deposited at the base of the continental slope forming the continental rise. These accumulations thin gradually toward the abyssal plains. Such deposits are called *turbidites* and are thought to have been deposited by turbidity currents that periodically move down the continental slopes through the submarine canyons, carrying loads of neritic material that spreads out across the continental rise (figure 5–10).

Turbidites are characterized by graded bedding. This means that each deposit that is laid down has coarser material at its base and decreasing particle size toward the top of the deposit. This results from particles of different sizes settling out of a moving water mass as the velocity decreases. The coarser material settles out first, while the finer material stays in suspension longer. The seaward end of a turbidite deposit may be composed mostly of fine clay-sized particles, which represent a contribution to the sediment of the abyssal plains beyond the continental rise.

During the millions of years that sediments accumulated at the margins of continents, the continental shelf, slope, and rise developed as the surface features of a sedimentary wedge more than 10 km (6 mi) thick that contains over 75 percent of the sediment to be found on the ocean floor. Less than 25 percent of marine sediment will be found in the deep-ocean basins that cover in excess of 80 percent of the ocean floor.

Glacial Deposits

Poorly sorted deposits containing particles of all sizes, from boulders to clay, may be found in the high-latitude portions of the continental shelf. These glacial deposits were laid down by melting glaciers that covered the continental shelf area during the Pleistocene epoch, when glaciers were more widespread than they are today and sea level was lower. Glacial deposits are still forming around the continent of Antarctica and around the island of Greenland by ice-rafting, a process by which rock particles trapped in glaciers are carried out to sea by icebergs that break away from glaciers as they push into the coastal ocean. As the icebergs melt, the lithogenous particles are released and settle to the ocean floor.

Carbonate Deposits

During the geologic past, deposits of limestone ($CaCO_3$) in the marine environment appear to have been widespread. Some contain fossil evidence indicative of a biogenous origin, while others appear to have formed as chemical precipitates. There appear to be very few places where nonbiogenous carbonates are presently forming. All these deposits are found in low-latitude shallow waters of continental margins or adjacent to islands. The Bahama Banks is the largest region where significant carbonate deposition is presently occurring, although deposits are also forming on the Great Barrier Reef, in the Persian Gulf, and in other

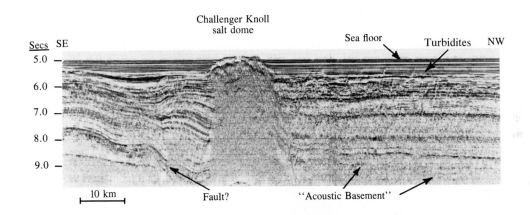

FIGURE 5–10
Seismic Profile Showing Turbidites. This multifold seismic section was made in the area of the Sigsbee Knolls in the Gulf of Mexico. It shows a spectacular salt dome (Challenger Knoll) and Pleistocene turbidites. The latter are closely spaced, uniform horizontal reflectors immediately below the sea floor. The Challenger Knoll was cored during the first voyage of the *Glomar Challenger* in the Deep Sea Drilling Project. (Courtesy J. Lamar Worzel, Geophysical Laboratory of Marine Science Institute, University of Texas.)

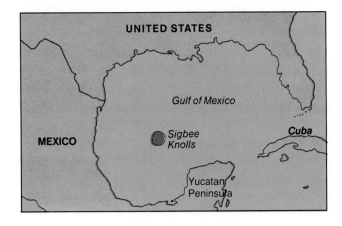

local areas in the low latitudes. Of course, coral reefs are biogenous deposits of calcium carbonate found extensively on the continental margins of low-latitude continents and around oceanic islands.

SEDIMENTS OF THE DEEP-OCEAN BASINS (OCEANIC SEDIMENTS)

As was previously discussed, most of the neritic sediment is deposited on the continental margin as part of the continental shelf, continental slope, or the continental rise. Since the continental rise is a transitional feature between the continental margin and the deep-ocean basin, we may well conclude that neritic material is deposited in significant quantities in the deep-ocean basin as part of the deep-sea fan deposits making up the continental rise. This is true, but when we consider the total surface area represented by the deep-ocean basin, we find that more commonly no neritic material will be found in the sedimentary section. Covering most of the deep-ocean floor are two types of oceanic sediments, abyssal clay and oozes.

Abyssal Clay

Accumulating at a rate of approximately 1 mm (0.04 in.) per 1000 year, *abyssal clays* cover most of the deeper ocean floor. They are commonly red-brown or buff in color due to phillipsite and oxidized iron. Abyssal clay con-

sists predominantly of clay-sized particles derived from the continents and carried by winds or ocean currents to the open-ocean regions. These particles, along with some cosmic and volcanic dust, settle slowly to the ocean floor.

A significant component of the abyssal clay is represented by the minerals montmorillonite and phillipsite, which are thought to form from the interaction of ocean water with volcanic and other lithogenous particles.

Oozes

At somewhat shallower depths, biogenous material makes up a significant portion of the sediment that reaches the ocean bottom. It consists primarily of the minute protective hard coverings of microscopic plants and animals that accumulate on the ocean floor. If sediment contains 30 percent or more skeletal material by weight, that deposit is called an *ooze*. Oozes may be classified as *calcareous* or *siliceous*, depending on the chemical composition of the dominant type of biogenous material making up the deposit.

Oozes are not present on the continental margin. This is not because the skeletal remains of plants and animals do not accumulate there but because the rate of accumulation of lithogenous sediment is so great that organically derived material never makes up 30 percent or more of the sediment. Oozes can form only where the deposition of material other than the remains of plants and animals occurs at a very low rate.

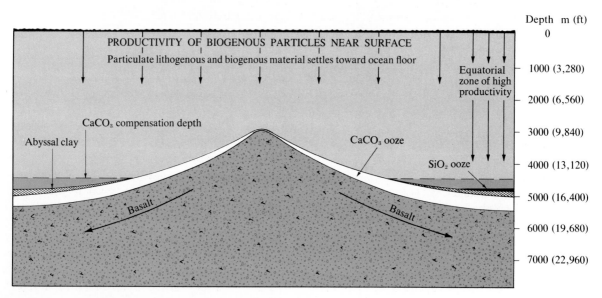

FIGURE 5–11

Sea-Floor Spreading and Sediment Accumulation. New basaltic oceanic crust is forming at ridges that run across the ocean floor and rise to depths of about 3000 m. As the crust moves away from the ridges and sinks to greater depths, the type of sediment that accumulates on it changes. Until it sinks below the calcium carbonate compensation depth, the typical sediment that forms is calcium carbonate ooze. Below this depth calcium carbonate is dissolved, and abyssal clays dominate. If the crust passes beneath a region of high biological productivity, such as the Pacific equatorial region, oozes could again accumulate.

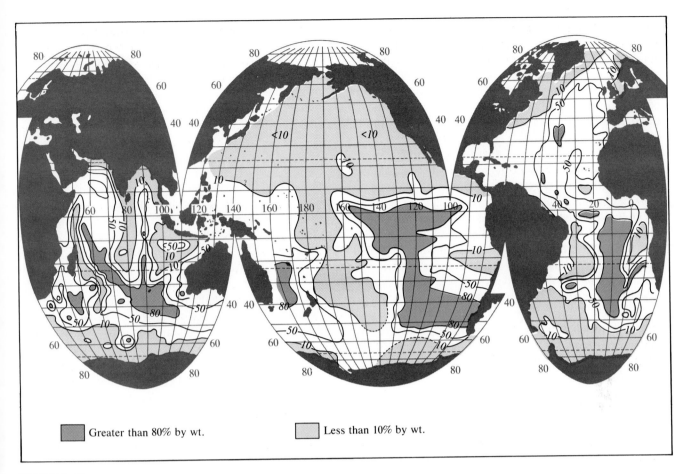

FIGURE 5–12

Calcium Carbonate in Surface Sediments of the World's Oceans. Concentrations of calcium carbonate ($CaCO_3$) in ocean sediments is negatively correlated with ocean depth. The deeper the ocean floor, the less calcium carbonate found in the sediment. It is also low in sediments accumulating beneath cold, high-latitude waters. Extensive areas of concentrations of calcium carbonate in the sediment in excess of 80 percent are associated with relatively shallow ocean floor on the Carlsberg Ridge in the Indian Ocean, the East Pacific Rise, and the Mid-Atlantic Ridge. Note that the deepest ocean basin, the North Pacific, lies for the most part beneath the calcium carbonate compensation depth and has very little calcium carbonate in its accumulating sediment. (After Biscaye et al., 1976; Berger et al., 1976, and Kolla et al., 1976.)

More specific names may be given to oozes that indicate the types of organic remains most abundant in the deposits. *Diatom oozes* are siliceous oozes that contain mostly the remains of diatoms. Other descriptive names commonly used on this same basis are *radiolarian ooze, foraminifera ooze,* and *pteropod ooze.* The predominant type of ooze is *Globigerina ooze,* which is especially widespread in the Atlantic and South Pacific oceans. *Globigerina* is an abundant and widespread genus of foraminifera.

The rate of accumulation of biogenous material on the ocean floor depends upon three fundamental processes—productivity, destruction, and dilution. *Productivity* of the planktonic* forms that contribute most of the biogenous

material in oceanic sediments is greater in the areas of upwelling along the margins of continents and associated with the diverging water masses in equatorial regions.

Destruction of skeletal remains occurs primarily through solution. The ocean is undersaturated with silica at all depths. So for siliceous shells to last until they are incorporated in the sediment, they must be thick, as thin fragments will be dissolved into the ocean water before they can settle out and be covered by the sediment. The solubility of calcium carbonate varies with depth in the ocean. At the surface, where the water temperatures are relatively high, the water is generally saturated with calcium carbonate. At greater depths, the temperature decreases and the carbon dioxide content increases, allowing the calcareous fragments to be readily dissolved. Below 4500 m (14,760 ft) the carbon dioxide content increases to the point that cal-

---

*In this instance, primarily floating, microscopic plants and animals.

cium carbonate dissolves quite readily. Carbonate oozes are generally rare below a depth of 5000 m (16,400 ft). The depth at which calcium carbonate is dissolved as fast as it falls from above is the *calcium carbonate compensation depth*. It may be as deep as 6000 m (19,680 ft) in portions of the Atlantic Ocean, but may be above 3500 m (11,480 ft) in regions of low biological productivity in the Pacific Ocean.

Dilution is illustrated by the fact that calcareous oozes are not found on the continental margins where rapid rates of deposition of lithogenous sediment prevail. Even in areas of upwelling where productivity is high, oozes do not occur on the continental shelf because the concentration of biologically derived silica or calcium carbonate remains well below 30 percent by weight as a result of the high rate of deposition for lithogenous materials in these areas. The rates of deposition for biogenous sediment range between 1 and 15 mm (0.04–0.6 in.) per 1000 years; depending on the net effect of productivity and destruction. Figure 5–11 shows the relationships among calcium carbonate compensation depth, sea-floor spreading, productivity, and destruction in determining what type of sediment will accumulate on the ocean floor.

Figure 5–12 shows the percentage by weight of calcium carbonate accumulation in the ocean basins. It shows that high concentrations (in excess of 80 percent) are found high on the oceanic ridges, whereas little is found in deep-ocean basins like the North Pacific, where the bottom lies beneath the calcium compensation depth.

DISTRIBUTION OF OCEANIC SEDIMENT

Oceanic sediments cover about 75 percent of the ocean bottom. Table 5–2 shows that calcareous oozes cover almost 48 percent of the deep-ocean floor. Abyssal clay covers 38 percent, and siliceous oozes 14 percent of the total area. Calcareous oozes are the dominant oceanic sediment in the Indian and Atlantic oceans, while abyssal clay is the principal oceanic sediment in the Pacific Ocean (figure 5–13). This is undoubtedly related to the fact that the Pacific Ocean is deeper, so that more of its bottom lies beneath the calcium carbonate compensation depth. Siliceous oozes cover a smaller percentage of the ocean bottom in all the oceans because regions of high productivity of diatoms and radiolarians, the major components of these deposits, are

TABLE 5–2

Distribution of Oceanic Sediment. This graph shows that the percentage of an ocean basin floor covered by calcareous ooze decreases with increasing mean depth of the basin. This is probably because the deeper an ocean basin is the greater the percentage of its floor that lies beneath the calcium carbonate compensation depth. The mean depths calculated for this table exclude the shallow adjacent seas where little oceanic sediment accumulates. Notice that the dominant oceanic sediment found in the deepest ocean basin, the Pacific Ocean, is abyssal clay, while calcareous ooze is the most widely deposited abyssal sediment in the shallower Atlantic and Indian oceans.

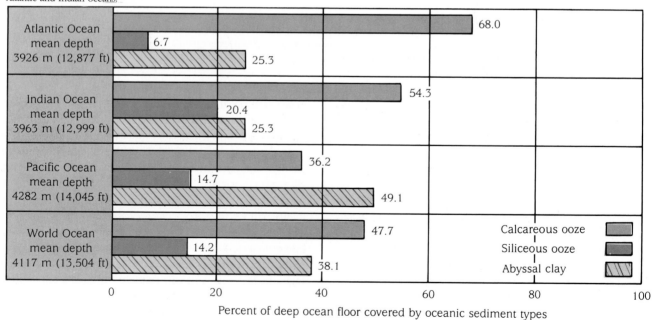

Percent of deep ocean floor covered by oceanic sediment types

Source: After Sverdrup, Johnson, and Fleming, 1942.

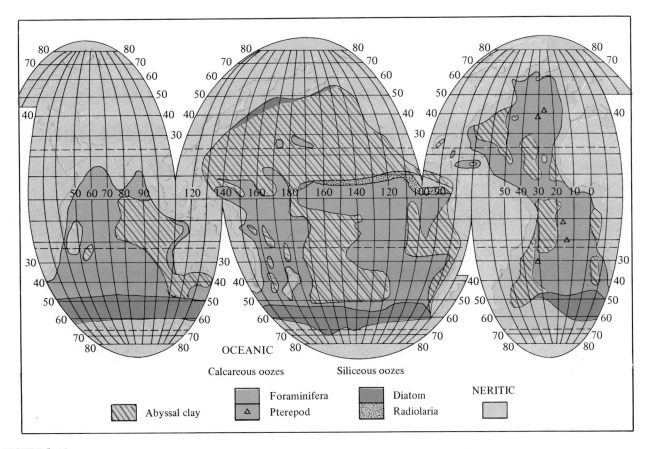

FIGURE 5–13
World Distribution of Neritic and Oceanic Sediments. Beyond the neritic deposits of the continental margins lie the oceanic marine deposits. Abyssal clays are found in the deeper ocean basins, where the bottom lies beneath the calcium carbonate compensation depth. Calcareous oozes are found best developed on the relatively shallow deep-ocean environments of the oceanic ridges and rises. Siliceous oozes are found beneath areas of unusually high biological productivity like the Antarctic and equatorial Pacific Ocean. (Base map courtesy of National Ocean Survey.)

restricted to areas of high biological productivity along the equator and in the Antarctic. Biogenous silica can be differentiated from lithogenous silica because biogenous silica contains molecular water; it is called *opal* ($SiO_2 \cdot nH_2O$). The distribution of opal in the surface sediments of the oceans is shown in figure 5–14.

Fecal Pellets

One major problem that puzzled marine geologists studying oceanic sediment is that sediments on the deep-ocean floor very closely reflect the particle composition of the surface water directly above. This was difficult to understand because it would typically take these very tiny particles from 10 to 50 years to sink from the ocean surface to abyssal depths. During this interval, a horizontal ocean current of only 1 cm/s (about 0.02 mi/h) would carry them from 3000 to 15,000 km (1800–9300 mi) laterally before they reached the deep-ocean floor.

However, study of GEOSECS samples shows that 99 percent of particles that fall to the ocean floor do so as part of fecal pellets produced by tiny animals living in the water column above. These pellets (figure 5–15), though still small (50–100 μm [0.002–0.004 in.]) in their smallest dimension) are large enough to allow the particulate matter to reach the abyssal ocean floor in 10 to 15 days. If this process is important in transporting particles from surface water to abyssal depths, it could readily explain the similarity in particle composition of the surface waters and the sediment immediately below them.

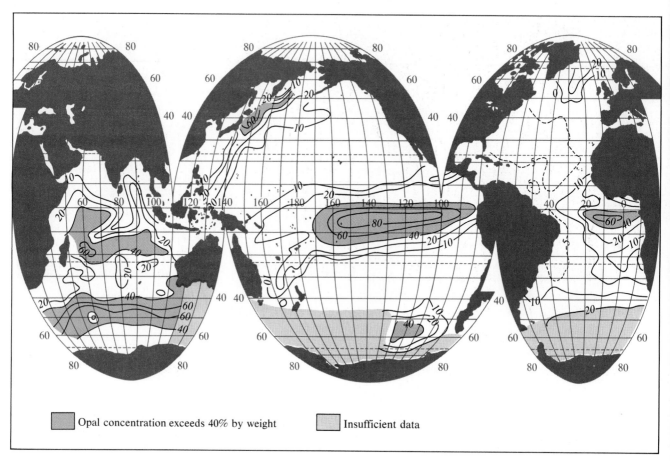

FIGURE 5–14

Biogenous Silica (SiO$_2 \cdot n$H$_2$O—Opal) in Surface Sediments of the World's Oceans. The distribution pattern of opal in surface sediments of the ocean shows maximum concentrations in areas of highest biological productivity. These particles are produced by diatoms and radiolarians in the surface waters. In the equatorial and northwest Pacific, opal is produced predominantly by radiolarians. Elsewhere, diatom remains are more abundant. The highly productive waters south of the Antarctic Convergence show up well in the south Indian Ocean and southeast Pacific Ocean. In the southern Atlantic Ocean and southwest Pacific Ocean there are insufficient data to determine the relationship between highly productive surface waters and opal concentrations in the sediment. (After Leinen et al., 1986.)

SUMMARY

Sediments that accumulate on the ocean floor are classified by origin as **lithogenous** (derived from rock), **biogenous** (derived from organisms), **hydrogenous** (derived from water), and **cosmogenous** (derived from beyond the earth's atmosphere). **Sediment texture**, determined in part by the size and sorting of particles, is affected greatly by the type of transportation that brought it to the deposit (water, wind, or ice) and the energy conditions under which it was deposited.

Lithogenous sediment is composed of fragments of rocks. **Biogenous** sediment is composed primarily of the compounds **calcium carbonate** (CaCO$_3$) from the remains of **foraminifera**, **pteropods**, and **coccolithophores** and sil-

ica (SiO$_2$) from the remains of **diatoms** and **radiolarians**. **Hydrogenous** sediment includes a wide variety of materials that precipitate from the water or form from interaction of substances dissolved in the water with materials on the ocean floor. **Manganese nodules, phosphorite, glauconite, carbonates,** and **zeolites** are examples. **Cosmogenous** sediment is composed of **nickel-iron spherules** and **silicate chondrules** that probably represent fragments resulting from asteroid collision.

Neritic sediment accumulates rapidly along the margins of continents. It is dominated by sediment of lithogenous origin. Due to the recent rise in sea level from melting glaciers, many rivers throughout the world, including those

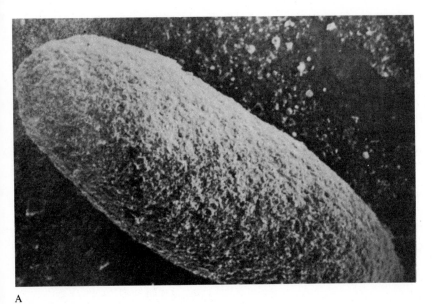

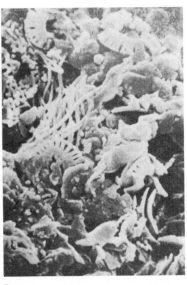

A

B

FIGURE 5–15

A, fecal pellet produced by zooplankton. The pellet is 200 μm (0.008 in.) long. *B,* view of the surface of the fecal pellet showing the remains of small phytoplankton and detritus. (Photos courtesy of Susumu Honjo, WHOI.)

flowing into the Atlantic along the east coast of the United States, are not now depositing sediment on the continental shelves but rather in their estuaries. About 70 percent of the sediment covering the continental shelves is **relict sediment** from 3000 to 7000 years old.

Turbidity currents are thought to transport shelf sediment down submarine canyons and deposit it as **turbidite** on the deep-ocean floor. More than 75 percent of the sediment mass found on the ocean floor is part of the thick sediment wedge underlying the continental shelves, slopes, and rises. In high latitudes this accumulation includes poorly sorted glacial deposits.

Oceanic sediment accumulates at low rates on the floor of the deep-ocean basins, far from the continents. In the deeper basins, where most biogenous sediment is dissolved before reaching bottom, **abyssal clay** deposits predominate. **Oozes** composed of over 30 percent by weight of biogenous sediment are found in shallower basins on

ridges and rises. The rate of biological **productivity** measured against the rates of **destruction** and **dilution** of biogenous sediment determines whether abyssal clay or oozes will form on the ocean floor. Much of the destruction of calcium carbonate occurs because the carbon dioxide content of the ocean water increases with depth. At the **calcium carbonate compensation depth**, ocean water dissolves calcium carbonate at a rate equal to the rate it settles from above, thus preventing accumulation on the ocean floor below that depth.

Although it would take from 10 to 50 years for individual sediment particles in the ocean surface waters to settle to the bottom, most appear to be combined into larger aggregates as **fecal pellets** of small marine animals and reach the ocean floor in 10 to 15 days. This process of particle transport explains the similarity in particle composition of the surface waters and sediment on the ocean floor immediately below them.

QUESTIONS AND EXERCISES

1. What characteristics of marine sediment indicate increasing maturity?

2. Why is a higher velocity current required to erode clay-sized particles than larger, sand-sized particles?

3. List the four basic sources of marine sediment.

4. Referring to figure 5–3, describe how the rocks granite, andesite, and basalt differ in texture, color, mineral composition, and density.

5. List the four common clays produced by weathering of common rock-forming minerals.

6. List the two major chemical compounds of which most biogenous sediment is composed and the organisms that produce them.

7. What is the chemical composition of hydrogenous deposits of manganese nodules, phosphorite, and glauconite? In which environments do they form? What is the relationship between a deposit's formation and its environment?

8. What are the two most common forms in which calcium carbonate precipitates from ocean water? Describe the environment in which such precipitation is most likely to occur. Include a discussion of biological processes and water temperature conditions that enhance precipitation.

9. Describe the most common types of cosmogenous sediment and give the probable source of these particles.

10. Describe the basic differences between neritic sediment and oceanic sediment.

11. Explain why many rivers are not now carrying sediment to the continental shelf, but depositing it in their estuaries.

12. Discuss the processes by which sediments are carried to and distributed across the continental margin.

13. How do oozes differ from abyssal clay? Discuss how productivity, destruction, and dilution combine to determine whether an ooze or abyssal clay will form on the deep-ocean floor.

14. Refer to figure 5–11 and explain why abyssal clay deposits on the floor of deep-ocean basins are commonly underlain by calcareous ooze.

15. How do fecal pellets help explain why the particles found in the ocean surface waters are closely reflected in the particle composition of the sediment directly beneath? Why would one not expect this?

REFERENCES

Berger, W. H., Adelseck, C. G., Jr., and Mayer, L. A. 1976. Distribution of carbonate in surface sediments of the Pacific Ocean. *Journal of Geophysical Research* 81:15, 2617–29.

Biscaye, P. E., Kolla, V., and Turedian, K. K. 1976. Distribution of calcium carbonate in surface sediments of the Atlantic Ocean. *Journal of Geophysical Research* 81:15, 2592–2602.

Heath, G. R. 1982. Manganese nodules: Unanswered questions. *Oceanus* 25:3, 37–41.

Hollister, C. D., Flood, R. D., and McCave, I. N. 1978. Plastering and decorating the North Atlantic. *Oceanus* 21:1, 5–13.

Inter-University Program of Research on Ferromanganese Deposits of the Ocean Floor. Phase 1 Report. Unpublished.

Kolla, V., and Biscaye, P. E., 1976. Distribution of calcium carbonate in surface sediments of the Atlantic Ocean. *Journal of Geophysical Research* 81:15, 2602–16.

Kyte, F. T., Zhou, L., and Wasson, J. T., 1988. New evidence on the size and possible effects of a late Pliocene oceanic asteroid impact. *Science* 241:4861, 63–65.

Leinen, M., Cwienk, D., Heath, R. R., Biscaye, P. E., Kolla, V., Thiede, J., and Dauphin, J. P. 1986. Distribution of biogenic silica and quartz in recent deep-sea sediments. *Geology* 14:3, 199–203.

Shepard, F. P. 1973. *Submarine geology*. New York: Harper and Row.

Spencer, D. W., Honjo, S., and Brewer, P. G. 1978. Particles and particle fluxes in the ocean. *Oceanus* 21:1, 20–26.

Sverdrup, H. U., Johnson, M. W., and Fleming, R. H. 1942. Reprinted 1970. *The oceans: their physics, chemistry, and general biology*. Englewood Cliffs, N.J.: Prentice Hall.

Weaver, P. P. E., and Thompson, J., eds. 1987. *Geology and geochemistry of abyssal plains*. Palo Alto: Blackwell Scientific.

Wentworth, C. K. 1922. A scale of grade and class terms for clastic sediments. *Journal of Geology*. 30:377–92.

SUGGESTED READING

Sea Frontiers

Dudley, W. C. 1982. The secret of the chalk. 28:6, 344–49.
A discussion of T. H. Huxley's erroneous conclusion regarding the significance of coccoliths in marine sediment.

Dugolinsky, B. K. 1979. Mystery of manganese nodules. 25:6, 364–69.
The problems of origin, growth, and the environmental implications of manganese nodule mining.

Feazel, C. T. 1986. Asteroid impacts, sea-floor sediments, and extinction of the dinosaurs. 32:3, 169–78.
A discussion of the possibility that high concentrations of iridium and osmium in a marine clay deposited at the time dinosaurs and many other species died out 65 million years ago may have resulted from the collision of the earth with a meteor 6 mi in diameter.

La Que, F. L. 1979. Nickel from nodules? 25:1, 15–21.

The economic and political problems related to nodule mining are discussed.

Shinn, E. A. 1987. Sand castles from the past: Bahamian stromatolites discovered. 33:5, 334–43.
The formation of stromatolites, sand domes trapped by algal growth, in the Bahamas is discussed.

Victory, J. J. 1973. Metals from the deep sea. 19:1, 28–33.
The formation and economic potential of mining manganese nodules are discussed.

Wood, J. 1987. Shark Bay. 33:5, 324-33.
The history of Shark Bay, Australia, since its discovery by Dirk Hartog in 1616 is summarized. The stromatolites and "tame" dolphins that attract tourists are also discussed.

Scientific American

Nelson H., and Johnson, K. R. 1987. Whales and walruses as tillers of the sea floor. 256:2, 112–18.
The 200,000 walruses and 16,000 gray whales that feed by scooping sediment from the Berling Sea continental shelf suspend large amounts of sediment that is transported by bottom currents.

6

THE NATURE OF WATER

Photograph by Ron Rovtar.

Water accounts for over 85 percent of the mass of most marine organisms. It is the medium in which the chemical reactions that sustain life take place, and it provides support for inhabitants of the open ocean. The nature of ocean water is determined by the modification of the properties of pure water by (1) its interaction with the organisms that live in the ocean and (2) the geologic components of the basins that contain the world ocean.

As the only substance that exists naturally in all three states—solid, liquid, and gas—on the earth's surface, water has some other remarkable properties that we must understand to fully appreciate the ocean environment. Water has the greatest heat capacity of any naturally occurring liquid on the earth, abnormally high melting and boiling points for a substance of its molecular weight, a high surface tension, and a great ability to dissolve other substances. Perhaps its most peculiar property is that of being less dense as a solid than in the liquid state.

All the properties listed above result from the unique shape of the water molecule. It is composed of one atom of oxygen (O) and two atoms of hydrogen (H). The hydrogen atoms are separated by an angle of 105°, as shown in figure 6–1, resulting in the asymmetrical distribution of electrically charged particles within the molecule. The oxygen atom has a nucleus composed of eight positively charged (+) particles called *protons,* with an equal number of negatively charged (−) particles called *electrons* in a cloud surrounding the nucleus. Each hydrogen atom has one proton in its nucleus and one electron orbiting the nucleus. When the hydrogen atoms bond with the oxygen atom to produce a water molecule, the electrons tend to stay close to the oxygen atom with its large positive charge. This makes the oxygen end of the molecule assume a slightly negative character, leaving the positively charged hydrogen nuclei somewhat "exposed" to give that region of the molecule a slightly positive character. The molecule is dipolar with a slight positive charge at one end and a slight negative charge at the other.

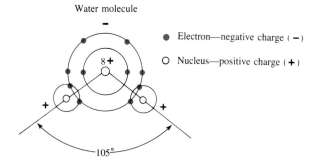

Water molecule

● Electron—negative charge (**–**)

○ Nucleus—positive charge (**+**)

8 **+**

105°

FIGURE 6–1

Dipolar Water Molecule. When two hydrogen atoms combine with an oxygen atom, they produce a water molecule, in which there is a 105° angle between the hydrogen atoms. The electrons continue to orbit their respective nuclei. However, because the oxygen nucleus has a larger positive charge than the hydrogen nuclei, the hydrogen electrons spend more time on the oxygen side, which causes the oxygen end of the molecule to have a slight excess of negative charge and each of the hydrogen nucleii to have a slight excess of positive charge.

SOLVENT PROPERTIES OF WATER

The polarity of the distribution of electrical charge in the water molecule causes it to become oriented in an electrical field with its positively charged hydrogen nuclei toward the negative plate and its negatively charged oxygen end toward the positively charged plate. This orientation of water molecules in an electrical field reduces its strength to $\frac{1}{80}$ of its strength out of water.

Due to their polar nature, water molecules form bonds between one another. These bonds are referred to as *hydrogen bonds*. The positively charged hydrogen areas of the water molecule attract the negatively charged oxygen portion of the neighboring water molecules, bonding together electrostatically. When polar molecules are bound together in complexes, as are the water molecules, their ability to reduce the intensity of an electrical field and, therefore, the electrostatic attraction between *ions* (electrically charged atoms or molecules) of opposite charges is greatly increased. In fact, water can reduce these attractions (bonds) to $\frac{1}{80}$ their strength out of water.

If we consider sodium chloride, a compound containing *ionic bonds,* we can see that simply by placing that substance in water, we have reduced the electrostatic attraction between the sodium and chloride ions to $\frac{1}{80}$ its original strength. This results from the fact the ions have a given amount of energy to exert in attraction to neighboring charged particles. Most of this energy is used up in establishing attractions to the water molecules, and very little remains for the ionic bonding to other ions. This makes it much easier for the ions to be separated. Once they become separated, the sodium ions are attracted to the nega-

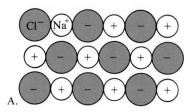

SOLID STATE

A.

Ionic bonds hold sodium ions (Na$^+$) and chloride ions (Cl$^-$) together in the compound sodium chloride (NaCl).

IN SOLUTION

Water molecules

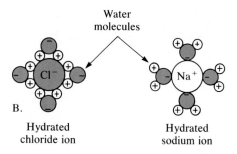

B.

Hydrated chloride ion

Hydrated sodium ion

The positively charged hydrogen end of the dipolar water molecule is attracted to the negatively charged Cl$^-$ ion; the negatively charged oxygen end is attracted to the positively charged Na$^+$ ion. The ions become hydrated when surrounded by water molecules that are attracted to them. In this state they have become solute ions in a water solution.

FIGURE 6–2

Sodium Chloride. Opposite electrical charges of sodium (Na$^+$) ions and chloride (Cl$^-$) ions produces an electrostatic attraction called an *ionic bond.* At top is a representation of the arrangement of sodium and chloride ions in a crystal of solid salt (Na$^+$Cl$^-$). Below, when the salt crystal is put into water, water molecules attach themselves to the Na$^+$ and Cl$^-$ ions in a manner similar to that shown here.

tive pole of the water molecule, and the chloride ions are attracted to the positive pole (figure 6–2).

THERMAL PROPERTIES OF WATER

We must define a few terms before we go further in our discussion. The first is *heat,* which is a form of energy associated with and proportional to the energy level of molecules. It may be generated by combustion, chemical reaction, friction, or radioactivity. So that we can measure the amount of energy that is being added to or removed from the molecules, we must introduce the term *calorie* as a unit for measuring quantity of heat. A calorie is defined as the amount of heat required to raise the temperature of 1 g of water 1°C. So that we can measure the amount of energy that the molecules within the substance have, we will need to understand the meaning of the word *temperature*. Temperature is defined as the direct measure of the *kinetic energy* of the molecules that make up a substance. Kinetic energy is energy of motion, so the higher the temperature, the greater the velocity of the molecules of the substance for which the temperature is being measured.

Heat Capacity

A direct product of the hydrogen bond is the high heat capacity of water. The heat capacity of water compared to that of most other substances is great, and we use this value as a standard against which we compare the heat capacity of other substances. The amount of heat that is required to raise the temperature of 1 g (0.035 oz) of any substance 1°C (1.8°F) is the *heat capacity* of the substance. The heat capacity of water has been chosen as the amount of heat in the unit of heat quantity, the *calorie*. The heat capacity of most other substances is less than 1 cal.

Freezing and Boiling Points of Water

The Celsius (centigrade) temperature scale is based on the characteristics of water, with the freezing point of water (the temperature at which it changes from a liquid to a solid) representing 0° (32°F), and the boiling point of water (the temperature at which it changes from a liquid to a gas) representing 100° (212°F). If we compare some other compounds similar to water in their makeup, we will see that the freezing and boiling points of water are uniquely high. The other compounds occur in nature as gases, while water occurs in the gaseous, liquid, and solid states of matter within the narrow range of atmospheric conditions.

We now consider the nature of the forces that must be dealt with to change the state of a compound. There exists between all molecules of any compound a relatively weak attraction that results from the electrostatic attraction between the positively charged nuclear parts of one molecule and the negatively charged electrons of another. This attraction slightly exceeds the electrostatic repulsion that exists between the nuclear parts of the same molecules and the electrons of the same molecules.

This intermolecular force is known as *van der Waals force,* and it becomes significant only when molecules are very close together, such as in the solid or liquid states. Generally, the heavier the molecule, the greater the van der Waals attraction between individual mole-

cules of the compound. Therefore, with increasing molecular weight, a greater amount of energy is needed to overcome that attraction and allow a change of state, say from a solid to a liquid or a liquid to a gas, to occur. Consequently, the melting and boiling points of compounds generally increase as the molecular weight increases.

We can see that if this attraction is to be broken, energy must be given to the molecules so that they can move more rapidly, in the ways described in figure 6–3, to overcome this force. It is this attraction that must be broken if we are going to produce a change of state in any substance from solid to liquid or liquid to gas.

We find that in a crystalline solid state, although bonds are constantly being broken and reformed, the predominant relationship between molecules is that of a rather firm attachment produced by the nearness of the molecules and the great effect the van der Waals force has upon molecules that are close together. *Vibration* is the dominant type of molecular movement that will be observed in the solid state as the molecules vibrate but remain in a relatively fixed position.

In the liquid state the molecules have gained enough energy to overcome many of the van der Waals forces that bound them together in the solid state. The molecules have enough freedom to move relative to one another. In this state we see all forms of molecular movement—vibration, rotation, and translation. The molecules are free to move relative to one another but

VIBRATION

Movement typical of molecules in a crystalline solid

ROTATION

Movement in liquids and gases

TRANSLATION

All gas molecules move randomly by translation

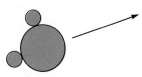

FIGURE 6–3
Motion of Molecules.

are still attracted by one another. Bonds are being formed and broken at a much greater rate than in the solid state.

In the gaseous state, *translation* has become the dominant type of motion. Molecules are now moving at random, and there exists no attraction between individual molecules. The only effect that one molecule will have on another is that which is produced by collision during random movement.

Returning to our comparison of the properties of water and compounds of similar composition containing two hydrogen atoms and an atom of another element, we will use H_2S, H_2Se, and H_2Te. The molecular weights of the four compounds are H_2O (18), H_2S (34), H_2Se (80), and H_2Te (129). Figure 6–4 shows that, as was predicted in the discussion of the van der Waals force, the freezing points and boiling points for H_2S, H_2Se, and

H_2Te increase with increased molecular weight. If we were to insert H_2O into the scale, the freezing point and boiling point should be about $-90°C$ ($-130°F$) and $-68°C$ ($-90.4°F$), respectively. The fact that water freezes at $0°C$ ($32°F$) and boils at $100°C$ ($212°F$) seems to be a violation of natural order. Here we see again the very great significance of the polar nature of the water molecule and the hydrogen bond that this structure produces. The high freezing and boiling points of water are the manifestations of the additional kinetic energy required to overcome not only van der Waals force but also the hydrogen bonds to achieve a change of state.

Latent Heats of Melting and Vaporization

Closely associated with water's unusually high freezing and boiling points are its high latent heat of melting and latent heat of vaporization. The continuous addition of heat to a substance in the solid or liquid state will bring about a change of state in the substance. A liquid converts to a solid at a temperature called its *freezing* (melting) *point,* and a liquid is changed to a gas at a temperature defined as its *boiling* (condensation) *point.*

When a substance changes state, there may be no increase of temperature at that point where a change of state occurs even though heat is continuously being added. The heat energy is being used to break the intermolecular bonds required to complete a change of state. While the bonds are being broken, a mixture of the substance in both states is in equilibrium. Only upon completion of the change of state will the temperature again rise. The heat that is added to 1 g of a substance at the melting point to break the required bonds to complete the change of state from solid to liquid is the *latent heat of melting.* The heat applied to effect a change of state at the boiling point is the *latent heat of vaporization.*

We will now consider the uniquely high latent heats of melting and vaporization of water that are responsible for the great effect water exerts on the earth's surface temperatures. The use of the word *latent* in describing the heats of vaporization and melting is necessary because the heat that must be added to a given mass of ice or water to convert it to a higher energy state, water or water vapor, is held in reserve, or "hidden," by that mass of water or water vapor. When water vapor returns to a liquid, it condenses and the heat is released into the surrounding air. Release of heat also occurs with the freezing of water, which represents the change of state from liquid to solid. During condensation and freezing, the same amount of heat that was required to change the state of that water from a liquid to a gas or from a solid to a liquid is released.

A practical application of these principles of heat transfer can be seen in the use of ice for the purpose of

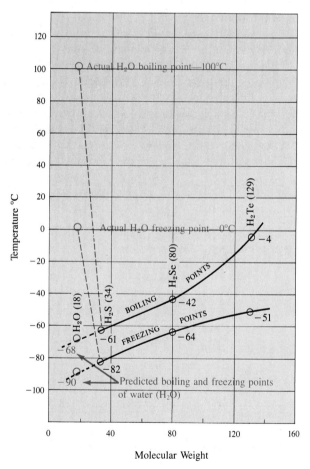

FIGURE 6–4

Molecular Weight—Freezing and Boiling Points. Because of the hydrogen bonds that must be broken in addition to overcoming the van der Waals force in achieving changes of state, the freezing (0°) and boiling (100°) points of water are higher than those that would be expected for a compound of its molecular makeup ($-90°$ and $-68°$).

refrigeration. A block of ice set in a closed container among food articles will lower their temperature because heat energy is drawn from the food articles and added to the molecules of ice to change them to the liquid state as the ice melts. The principle of cooling air with water is similar: In arid climates, the hot dry air that passes through a surface coated with liquid water will lose heat to the water as the water is converted to a vapor. Thus, after passing through or across the water-covered surface, the air is considerably cooler.

To help clarify this phenomenon, we will discuss the heat transfer and the change of state of H_2O from a solid at $-40°C$ ($-84.8°F$). Observe the heat-temperature graph that has a vertical temperature scale ranging from $-40°$ to $120°$ and a horizontal scale that begins at 0 and continues through 800 cal (figure 6–5). For 1 g of ice, we find that the addition of 20 cal of heat raises the temperature $40°$, from $-40°C$ to $0°C$. Thus, the heat capacity of ice is 0.5 cal/g, half that of liquid water.

As the heat is continuously added, we observe that there is no increase of temperature until we have added a total of 100 cal to the gram of ice that we are converting to the liquid state. We do not observe an increase in temperature during the addition of these 80 cal of heat because all the heat energy added is used to break the bonds that are holding the water molecules in the solid state. The temperature will remain unchanged until all the necessary bonds are broken and the mixture of ice and water has been changed to 1 g of water. The amount of heat required to convert 1 g of ice to 1 g of

water, 80 cal, is termed the *latent heat of melting,* and it is higher for water than for any other commonly occurring substance.

The bonds that are broken in converting most substances from a solid to a liquid are the van der Waals bonds. In water, however, not only the van der Waals bonds but also some of the hydrogen bonds must be overcome. It is necessary only to break the ice structure down into numerous small clusters surrounded by individual water molecules to the extent that these remaining ice clusters can move relative to one another and allow the mass to assume a liquid state. Liquid water, particularly at low temperature near the freezing point, could well be described as a pseudocrystalline liquid, as there are still many small clusters of ice crystals contained within it. Once enough hydrogen bonds have been broken to allow freedom of movement among the clusters and individual water molecules, the temperature of the water will again rise with the addition of heat.

As we add heat beyond the 100 cal required to convert the gram of ice to water, we see the temperature of the water increase. As it does, note that it requires 1 cal of heat to raise the temperature of water 1°C. We must then add a total of 200 cal before 1 g of water, starting at $-40°C$ ($-84.8°F$), reaches the boiling point of 100°C (212°F). At this temperature we note the development of another plateau, and it is far more prominent on the graph than that which represents the latent heat of melting, 80 cal/g. This plateau at 100°C represents an addi-

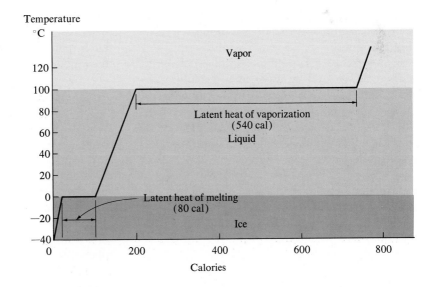

FIGURE 6–5

Change of State Graph—Water.
Starting with 1 g of ice at –40°C, we add 20 cal of heat to raise its temperature to the melting (freezing) temperature (0°C.) This tells us that the heat capacity of ice is only 0.5 cal/g. The addition of another 80 cal of heat (the latent heat of melting) does not increase the temperature because this heat is being used to break enough hydrogen bonds to allow our gram of ice to flow as a liquid. Once the change of state is achieved, the addition of another 100 cal raises the temperature of the gram of water 100°C to the boiling temperature of water. At this temperature, an addition of another 540 cal of heat (the latent heat of vaporization) does not increase the temperature of our gram of water because it is being applied to the breaking of all remaining hydrogen bonds. Once this change of state is completed, the additional heat will raise the temperature of the water vapor.

FIGURE 6–6

Hydrogen Bonds in H₂O. *A,* water in the solid state of ice has molecules firmly fixed into an open crystalline structure by hydrogen bonds that seldom break. *B,* in the liquid state, hydrogen bonds are forming and breaking continuously. Some tiny ice crystals may remain but do not prevent water from flowing. *C,* the gaseous form of water, water vapor, requires that the molecules be free of any bonds so they can move individually by translation. There are no hydrogen bonds.

A. SOLID
Crystalline structure is three dimensional

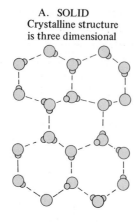

B. LIQUID

C. GAS

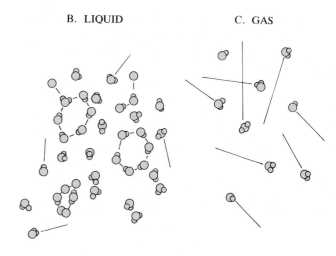

tion of 540 cal, the *latent heat of vaporization,* to the gram of water before complete conversion to the vapor state.

Why does it require so much more heat energy to convert 1 g of water to water vapor than was required to convert 1 g of ice to liquid water? We defined a gas as a substance in which the molecules were moving at random, free from the influence of other molecules, except when they collide. To make the conversion from ice to water, not all the hydrogen bonds had to be broken but only enough to allow freedom of movement along the various ice clusters that remained and the individual molecules that were also present in the system. To convert water to water vapor, every molecule must be freed from the attraction of other water molecules. Therefore, every hydrogen bond must be broken (figure 6–6). To do this requires a great amount of heat energy.

We can now see that large amounts of heat energy are absorbed or released by water as it changes state, particularly if the change of state is from a liquid to a vapor or vice versa. Of course, we do not have temperatures of 100°C (212°F) at the surface of the ocean, where this phenomenon of conversion of water to vapor occurs in nature. There temperatures are about 20°C (68°F) or lower.

The conversion of a liquid to a gas below the boiling point is called *evaporation.* At ocean-surface temperatures, individual molecules that are being converted from the liquid to the gaseous state have a lower amount of energy than do molecules of water at 100°C. Therefore, to gain the additional energy necessary to break free of the surrounding water molecules, an individual molecule must capture energy from its neighbors. This phenomenon explains the cooling effect of evaporation. The molecules left behind have lost heat energy to those that escape. To produce 1 g of water

vapor from the ocean surface at 20°C (68°F) requires more than the 540 cal of heat that are required to make this conversion at the boiling point. The *latent heat of evaporation* at 20°C (68°F) is 585 cal/g. This higher value is due to the fact that more hydrogen bonds must be broken at this lower temperature for a gram of water molecules to enter the gaseous state.

We can readily see the significance of the evaporation-condensation cycle to the earth's surface temperatures. Evaporation removes heat energy provided by the sun and stored in the oceans. This energy is carried into the atmosphere as the water vapor rises and is released there when the vapor condenses and falls as precipitation. This phenomenon and the high heat capacity of water account for the mild winters in humid coastal areas. In such areas abundant rainfall releases heat into the atmosphere, producing winter temperatures that are much higher than those found in the drier regions of the continental interior.

SURFACE TENSION

Next to mercury, water has the highest surface tension of all commonly occurring liquids. We observe a surface tension phenomenon in filling a container with water to the brim and even beyond. You will note that the water can be piled up above the rim of the container forming a convex surface that represents the water interface with the atmosphere. Water drops are also a very common manifestation of surface tension. These phenomena are the product of the tendency for water molecules to attract one another or to cohere at the surface of any accumulation of water. Because of this cohesive tendency, it is possible to float on water objects that are much heavier than water. A razor blade carefully laid on the surface will float, although it is

ter below 4°C, we encounter considerations that must include the hydrogen bond. As the temperature of water is lowered from 4° to 0°C (39.2° to 32°F), we observe that its density decreases.

This anomalous behavior can be explained only by considering the molecular structure of water and the hydrogen bond. As we lower the temperature of 1 g of water from 20°C and reduce the amount of thermal agitation of the water molecules, the unbonded water molecules occupy less volume due to their decreased energy. As we approach the freezing point below 4°C (39.2°F), however, this reduction in volume due to the decreased energy of unbonded molecules is not sufficient to compensate for another phenomenon that is occurring. Ice crystals, open six-sided structures in which water molecules are bound together by hydrogen bonds in a widely spaced pattern, are becoming more abundant. The greater rate of increase in the formation of hydrogen-bonded ice crystals as the temperature approaches the freezing point accounts for the decreased density of water below 4°C (39.2°F) (figure 6–8).

Table 6–1 summarizes the physical and biological significance of the properties of water.

SALINITY OF OCEAN WATER

The properties of ocean water are determined to a great extent by the properties of pure water. In fact, it is the great capacity of water as a solvent that causes the oceans to be filled with a well mixed solution—the "saltwater" we associate with the oceans. The solute that makes the oceans "salty" is composed primarily of ions of elements that have been dissolved from the rocks that make up the earth's crust or released into the oceans from the earth's mantle through volcanic activity.

Though the proportion of water to dissolved salts may vary within the ocean, the major component ions are distributed in ocean water in relatively constant proportions. This *constancy of composition* was first established by Dittmar in his analysis of the water samples from the *Challenger* expedition (discussed in chapter 1). Water makes up, on the average, 96.5 percent of the ocean's mass, so this substance determines most of the physical properties that we observe in ocean water.

It seems probable that when the techniques are developed for making such measurements, all the known elements will be found to be dissolved in ocean water. Merely six elements, however, account for 99 percent of the dissolved solids in ocean water. These major components are chlorine, sodium, magnesium, sulfur (as SO_4), calcium, and potassium (table 6–2 and figure 6–9).

Salinity is defined as the total amount of solid material dissolved in a kilogram of seawater when all the carbonate has been converted to oxide, all bromine and iodine replaced by chlorine, and all organic matter completely oxidized. Given this definition, it may appear that salinity is a very complex property and very difficult to measure. However, since the oceans are so very well mixed, and the relative abundance of the major constituents is essentially constant, we have a condition that makes the measurement of salinity relatively simple.

It is necessary to measure the concentration of only one of the major constituents in order to determine the salinity of a given water sample. The constituent that occurs in the greatest abundance and is the easiest to measure accurately is the chloride ion, Cl^-. The portion of the weight of a given sample of water that is the direct result of the presence of this ion is called *chlorinity* and is usually expressed in the terms of grams per kilogram of ocean water (g/kg) or parts per thousand (‰). The

TABLE 6–2
Ocean Salinity. Considering the average salinity of the World Ocean, 34.7 by weight, a 1000-g (35-oz) sample of this water would contain 34.7 g of dissolved solids. Six ions account for 99.28% of the dissolved solids.

Major Constituents (over 100 parts per million)		Minor Constituents (1–100 parts per million)		Trace Elements (less than 1 part per million)	
Ion	*Percentage*	*Element*	*Percentage*		
Chloride, Cl^-	55.04	Bromine	65.0	Nitrogen	Iodine
Sodium, Na^+	30.61	Carbon	28.0	Lithium	Iron
Sulfate, SO_4^{-2}	7.68	Strontium	8.0	Rubidium	Zinc
Magnesium, Mg^{+2}	3.69	Boron	4.6	Phosphorus	Molybdenum
Calcium, Ca^{+2}	1.16	Silicon	3.0		
Potassium, K^+	1.10	Fluorine	1.0		
	99.28				

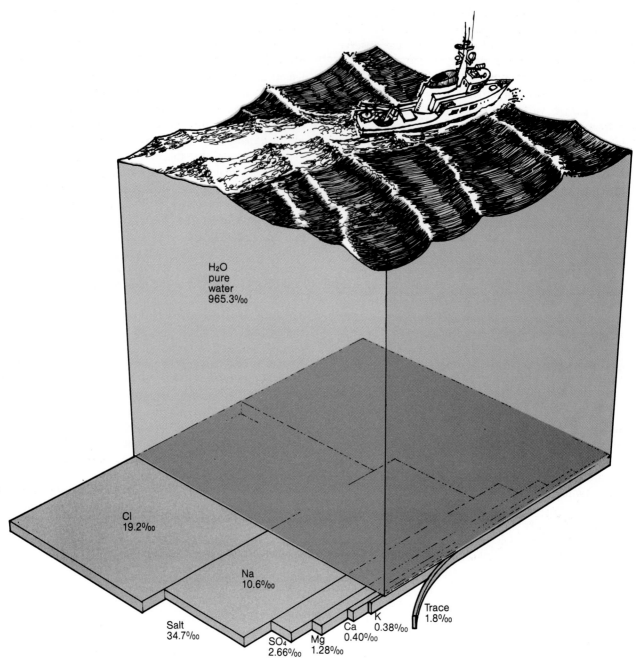

FIGURE 6–9
Constituents of Ocean Salinity.

chloride ion accounts for 55 percent of the dissolved solids in any sample of ocean water. Therefore, by measuring its concentration, we can determine the total salinity in parts per thousand by the following relationship:

$$\text{Salinity (‰)} = 1.80655 \times \text{Chlorinity (‰)}$$

For example, given the average chlorinity of the ocean, 19.2 ‰, the ocean's average salinity can be calculated as 1.80655 × 19.2 ‰, or 34.7 ‰.

From 1902 to 1975, the Hydrographic Laboratory at Copenhagen, Denmark, provided Standard Seawater samples to assure that salinity determinations made throughout the world would be based on the same reference. In 1975, this duty was taken over by the Institute of Oceanographic Services located in Wormly, England. Standard Seawater consists of ocean water analyzed for chloride ion content to the nearest ten-thousandth of a part per thousand. It is then sealed in ampules and labeled to be sent to laboratories throughout the world. The chlorinity of water samples taken and measured in other parts of the world can then be compared if the titration solution or conductivity instrumentation is calibrated with the standard water sample.

The chemical determination of the chloride content in seawater can be made with a high degree of accuracy but requires a great deal of time and care. It is extremely difficult to carry out the chemical titration aboard ship. With the advancements in instrumentation in the field of oceanography, this task has been greatly simplified. Since electrical conductivity of ocean water is known to increase with increased temperature and salinity, salinity is now most commonly determined by measuring the electrical conductivity of ocean water with a modern conductance-measuring instrument, the *salinometer*. Salinity can be determined to better than 0.003 ‰ by this method.

Salinity and Density

Because the components of salinity have a density greater than the density of water, the density of water increases with increased salinity. Physical oceanographers are greatly concerned with the relationships among water density, salinity, and temperature, since density distribution is an important determinant of ocean circulation. Density of ocean water cannot be determined in place, so to determine density distribution within a region of study, oceanographers measure tem-

perature and salinity and then use these values to calculate density.

The temperature of maximum density for fresh water, 4°C (39.2°F), is lowered by increasing pressure or adding solid particles such as salt, both of which will inhibit the formation of the ice clusters. Thus, more energy must be removed to produce crystals equal in volume to those that could be produced at 4°C (39.2°F) in fresh water, causing a greater reduction in temperature. This interference with the formation of ice crystals also produces a progressively lower freezing point for water as more solids are added. The freezing point and temperature of maximum density converge as they drop and coincide when the salinity reaches 24.7 ‰ at a temperature of −1.33°C (29.6°F) (figure 6–10). Above a salinity of 24.7 ‰ the density of water increases with decreasing temperature until it freezes. Thus, average seawater (34.7 ‰) has no maximum density anomaly.

RESIDENCE TIME

The *residence time,* or the average length of time an atom of an element will spend in the ocean, can be calculated by the following equation:

$$\text{Residence time} = \frac{\text{Amount of element in the oceans}}{\text{Rate at which element is added to the oceans}}$$

The oceans are considered to be in a *steady state,* that is, the rate at which an element is removed from the oceans must be equal to the rate at which it is being added. For that reason, we could use either of these values as the denominator in the equation.

FIGURE 6–10
Salinity, Freezing Point, and Temperature of Maximum Density. As the salinity of water increases, the temperature of maximum density and the initial freezing temperature decrease. The temperature of maximum density decreases at a greater rate, so the two converge at a salinity of 24.7 ‰ and both have a value of −1.33°C. Water that is saltier than 24.7 ‰ (ocean water) will freeze before it reaches its theoretical temperature of maximum density, which is lower than its freezing temperature. Therefore, ocean water continues to become more dense as it cools, until it freezes. It is most dense at its freezing temperature.

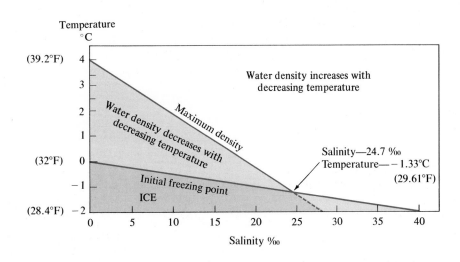

TABLE 6–3
Residence Time of Some Elements in the Oceans.

Element	Amount in ocean (g)	Residence time in years
Sodium, Na	147×10^{20}	2.6×10^{8}
Potassium, K	5.3×10^{20}	1.1×10^{7}
Calcium, Ca	5.6×10^{20}	8.0×10^{6}
Silicon, Si	5.2×10^{18}	1.0×10^{4}
Manganese, Mn	1.4×10^{15}	7.0×10^{3}
Iron, Fe	1.4×10^{16}	1.4×10^{2}

Source: Data from Goldberg and Arrhenius, 1958.

The residence time of an element in the ocean depends on how reactive it is with the marine environment. Those elements that are more reactive have a shorter residence time. The reactive elements aluminum and iron have residence times of 100 and 140 years, re-spectively, while less reactive sodium has a residence time of 260 million years (table 6–3).

Elements are removed from ocean water through biological reactions, or they are incorporated into sediment. Each year a 1-m (3.3-ft) layer of the ocean's surface is evaporated. This water is returned directly by precipitation or indirectly by runoff from the continents. The ocean, with an average depth of 4000 m (13,120 ft), would completely evaporate in 4000 years, but because the water is returned, this is the time required to completely recirculate the ocean. Thus, the residence time of water in the oceans is 4000 years.

The process by which water is recycled among the ocean, atmosphere, and continents is called the *hydrologic cycle* (figure 6–11). Of the total yearly evaporation of water from the earth's surface, 83 percent is removed from the oceans, and 17 percent is removed from the surface of the continents. Seventy-six percent of this wa-

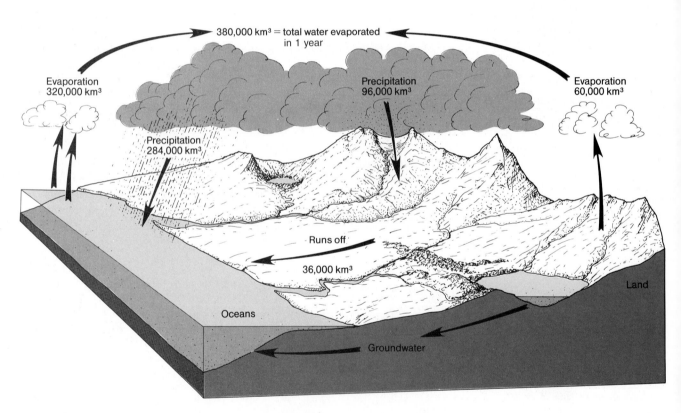

FIGURE 6–11

Hydrologic Cycle. Ninety-seven percent of the earth's water is in the world ocean, 2 percent is in glaciers, 0.6 percent is groundwater, 0.3 percent is in the atmosphere, and about 0.1 percent is in rivers and lakes. Although there is some evidence that indicates the glaciers are entering a stage of melting, we will consider them to be in a steady state. Water is removed from the liquid reservoirs by evaporation and returned to them by precipitation. There is a net atmospheric transfer of water vapor from the oceans to the continents, where it is deposited as precipitation and returns to the oceans by stream flow or as groundwater.

FIGURE 6–12
GEOSECS Sampling Bottles. One of the two rosettes carried aboard Scripps Institution of Oceanography's research vessel *Melville* during GEOSECS Pacific operations shows nonmetallic bottles used in collecting seawater samples. The ends of each bottle can be triggered to close by the scientist operating the shipboard console. (Photo from Scripps Institution of Oceanography, University of California, San Diego.)

ter is removed from the atmosphere by precipitation directly back into the oceans. Twenty-four percent falls as precipitation onto the continent. During a year, the atmosphere transports water vapor equivalent to 8.5 percent of that which is being evaporated from the oceans to the continents. This water is returned to the oceans by stream runoff and as groundwater. At any given time, 97 percent of the earth's water is contained in the ocean basins, 2 percent is in glaciers, 0.6 percent is groundwater, 0.3 percent is in the atmosphere, and about 0.1 percent is in rivers and lakes.

GEOSECS

The GEOSECS project mentioned in chapter 1 is a huge systematic program designed to increase understanding of circulation patterns and mixing processes in the oceans. All the major ocean basins have been sampled throughout the depth of the water column. The sampling system includes electronic sensors that telemeter temperature, salinity, dissolved oxygen, pressure, and particulate content to a console aboard ship. As rosettes of 30-L (9-gal) sampling bottles are lowered into the ocean, oceanographers have the option of deciding to sample at any depth where an interesting value for any of the properties shows on the console (figure 6–12).

The water collected is analyzed for some 23 chemical species, 15 isotope species, and for the amount and types of particulate matter. From the results thus far obtained, oceanographers feel the GEOSECS program will not only provide a much improved understanding of physical processes but also supply much information about biological cycles in the ocean (figure 6- 13).

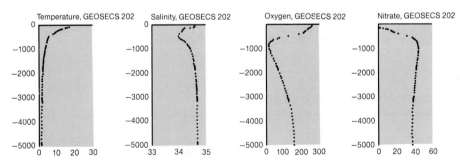
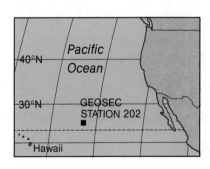

FIGURE 6–13
GEOSECS Data. *A,* the map shows the location of GEOSECS Station 202 located at latitude 26°N, longitude 139°W. *B,* the four curves show the plot of temperature, salinity, dissolved oxygen, and nitrate measured from the surface to a depth of 5,000 m (16,400 ft) at Station 202. They show that ocean water properties may change significantly from the surface to a depth of 1,000 m (3,280 ft). The rate of change below that depth is much less. (Plots courtesy of Woods Hole Oceanographic Institution.)

With this brief introduction to the chemistry and physical properties of water, we will be prepared to discuss considerations regarding the origin of the oceans and life in those oceans in the next chapter. A more comprehensive discussion of chemical factors that relate to the geological, physical, and biological phenomena of the oceans will be undertaken as they are encountered throughout the text.

SUMMARY

Water, which accounts for over 85 percent of the mass of most marine organisms, has some unusual properties resulting from the shape of the water molecule. This shape causes the **water molecule** to have a **dipolar** distribution of electrical charge. The **hydrogen bond**, the bond between water molecules resulting from the dipolar nature of the molecule, plays the major role in giving water its many unusual properties.

Water is a great **solvent** because its dipolar molecules can attach themselves to charged particles, **ions**, that make up many substances (such as **sodium chloride**), hydrate them, and put them into solution.

The presence of the hydrogen bond also accounts for the unusual **thermal properties** of water such as its high **freezing and boiling points**, high **heat capacity**, and **latent heats of melting and vaporization**.

The **surface tension** phenomenon that makes water drops form and allows it to be poured into a container until it stands well above the sides of the container results from the water molecules being strongly attracted by the hydrogen bond to those water molecules beside and beneath them. This **cohesive** force combined with the **adhesive** attraction of water to glass, organic substances, rock, and soil produces the **capillary force** that pulls water to great heights in small tubular openings.

Salinity is the amount of dissolved solids in ocean water, averaging about 34.7 g of dissolved solids per kilogram of ocean water. Salinity is usually expressed in **parts per thousand** (‰). Over 99 percent of the dissolved solids in ocean water are accounted for by six ions (in order of decreasing abundance): chloride, sodium, sulfate, magnesium, calcium, and potassium. In any sample of ocean water these ions will be in the same relative proportions, making it possible to determine salinity by measuring the concentration of only one of them, usually the chloride ion.

The **density** of water increases with decreased temperature, as does the density of most substances, to a temperature of 4°C (39.2°F). Below 4°C, its density decreases with decreased temperature. This is due to the fact that the rate of formation of open ice crystals increases dramatically below 4°C, the **temperature of maximum density** for water. By the time water freezes, the density of the ice is only about 0.9 that of water at 4°C. Density increases with increased salinity and pressure. Due to the fact that the components of salinity have a greater density than water, increased salinity increases the density of the ocean water.

The **residence time** of elements in the ocean water ranges from 100 years for aluminum to 260,000,000 years for sodium. The time required to totally recirculate the water in the oceans, its residence time, is 4000 years. The oceans contain 97 percent of the earth's water. Water is cycled among the ocean, atmosphere and continents by the **hydrologic cycle**, which involves evaporation, precipitation, and surface and groundwater runoff from the continents.

An advanced system of sampling and analyzing the properties and content of ocean water used in conducting the **GEOSECS program** has helped give oceanographers an increased understanding of physical and biological processes of the oceans.

QUESTIONS AND EXERCISES

1. Describe the condition that exists in the water molecule to make it dipolar.

2. Discuss how the dipolar nature of the water molecule makes it such an effective solvent for ionic compounds.

3. Define freezing point and boiling point.

4. Why are the freezing and boiling points of water higher than would be expected for a compound of its molecular makeup?

5. How does the heat capacity of water compare with that of other substances? Describe the effect this characteristic of water produces in climate.

6. Why does the heat energy added as latent heats of melting and vaporization for water not produce an increase in temperature? Why is the latent heat of vaporization so much greater than the latent heat of melting?

7. Describe how excess heat energy absorbed by the earth's low-latitude regions could be transferred to the heat-deficient higher latitudes through a process that makes use of water's latent heat of evaporation.

8. How does hydrogen bonding produce the surface tension phenomenon of water?

9. As water cools, two distinct changes in the behavior of molecules take place. One tends to increase density, while the other decreases density. Describe how the relative rates of their occurrence cause pure water to have a temperature of maximum density at 4°C and make ice less dense than liquid water.

10. What condition of salinity makes it possible to chemically determine the total salinity of ocean water by measuring the concentration of only one constituent, Cl^-?

11. As water becomes more saline, the temperature of maximum density and the freezing temperature of water decrease and converge. At what salinity does water cease to have a temperature of maximum density above its freezing temperature?

12. If there is an estimated 18×10^{20} g of magnesium (Mg) in the ocean, and it is being added (or removed) at the rate of 56.5×10^{12} g per year, what is the residence time of magnesium in the ocean? (see appendix I).

13. List the reservoirs of water on the earth and the percentage of the earth's water each holds. Explain how the hydrologic cycle moves water among these reservoirs.

REFERENCES

Davis, K. S., and Day, J. S. 1961. *Water: The mirror of science.* Garden City, N.Y.: Doubleday.

Hammond, A. L. 1977. Oceanography: Geochemical tracers offer new insight. *Science* 195:164–66.

Harvey, H. W. 1960. *The chemistry and fertility of sea waters.* New York: Cambridge University Press.

Kuenen, P. H. 1963. *Realms of water.* New York: Science Editions.

MacIntyre, F. 1970. Why the sea is salt. *Scientific American* 223:104–15.

Pickard, G. L. 1975. *Descriptive physical oceanography,* 2nd ed. New York: Pergamon Press.

Revelle, R. 1963. Water. *Scientific American* 209:93–108.

SUGGESTED READING

Sea Frontiers

Gabianelli, V. J. 1970. Water—the fluid of life. 16:5, 258–70.
The unique properties of water are lucidly described and explained. Topics covered include hydrogen bond, capillarity, heat capacity, ice, and solvent properties.

Scientific American

Derjaquin, B. V. 1970. Superdense water. 223:5, 52–71.
A discussion of a new form of water that has a density almost 1.5 times that of normal water and exhibits some unusual properties.

MacIntyre, F. 1970. Why the sea is salt. 223:5, 104–15.
A summary of what is known of the processes that add and remove the elements dissolved in the ocean.

7

AIR-SEA INTERACTION

In this first chapter on physical oceanography we further consider the physical properties of ocean water, the Coriolis effect, the heat budget of the ocean, and the process by which solar energy is converted into kinetic and heat energy of moving masses of ocean water.

PHYSICAL PROPERTIES OF OCEAN WATER

Light

The color of ocean water ranges from a deep indigo blue in tropical and equatorial regions of little biological productivity to a yellow-green in the coastal waters of high-latitude areas where biological productivity occurs seasonally at a very high rate. The blue of the tropical waters where there is little particulate matter is due to the molecular scattering of solar radiation. This process is also responsible for the blue color of the sky. Higher concentrations of particulate matter, especially where phytoplankton is abundant, result in a greater amount of scattering and absorption and decreased transmission of solar radiation. This produces the greenish color characteristic of such waters.

The many forms of electromagnetic energy radiated by the sun may be seen in figure 7–1, the *electromagnetic spectrum.* The very narrow segment of this spectrum designated as visible light may be broken down on the basis of wavelength into violet, blue, green, yellow, orange, and red energy levels. Combined, these different wavelengths of light produce white light. The shorter wavelength forms of energy to the left of visible light represent highly dangerous forms of electromagnetic energy including X rays and gamma rays. To the right of the visible segment of the electromagnetic spectrum, we find longer wavelength forms of energy that are the basis for the technological fields dealing with heat transfer and communication. We will further consider the electromagnetic spectrum when we examine the role of solar energy in setting the ocean masses in motion. We now consider only that portion of the spectrum that is manifested as visible light.

Typhoons (from left) Gerald, Freda, and Holly in the west Pacific, September 8, 1987. Photograph courtesy of NOAA.

FIGURE 7–1

The Electromagnetic Spectrum and Transmission of Visible Light in Water. Wavelength of components of the spectrum increase from the left to right. The shorter wavelengths of the visible portion of the spectrum, especially blue, penetrate deeper into the ocean than the longer wavelengths.

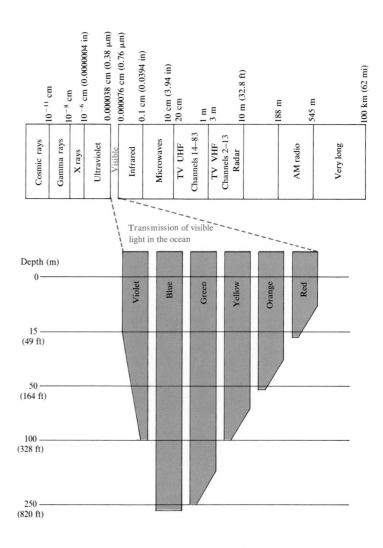

The reason we see things in color is because objects reflect wavelengths of light that represent the colors of the visible spectrum. If those wavelengths of light are not present in the light that falls upon an object, colors cannot be seen. In the ocean the absorption of visible light is greater for the longer wavelength colors, thus the shorter wavelength portion of the visible spectrum is transmitted to greater depths. As a result of this condition, the red wavelengths are absorbed within the upper 15–20 m (49–66 ft), the yellow disappears before a depth of 100 m (328 ft) has been reached, but green light can still be perceived down to 250 m (820 ft). Only the blue and some green wavelengths extend beyond these depths, and their intensity becomes low. It is because of this pattern of absorption that objects in the ocean usually appear blue-green. Only in the surface waters can the true colors of objects be observed in natural light, since only in the surface waters can all wavelengths of the visible spectrum be found.

To measure the transmission of visible light in the ocean, a simple device about 30 cm (12 in.) in diameter, the *Secchi disc,* is attached to a rope marked off in meters. After the disc is lowered into the ocean, the depth at which it can last be seen gives an indication of the *turbidity*—the amount of suspended material in the water. Increased turbidity increases the rate of absorption, thus decreasing the transmission of visible light.

Density

Density is expressed in grams per cubic centimeter and ranges from 1.02300 to 1.03000 g/cm^3 in the open ocean. Density of ocean water is increased by an increase in salinity or pressure or by a decrease in temperature. Oceanographers convert density to a term more conveniently used, σ_t (sigma t). Normally the density of ocean water is expressed to five decimal places. We will frequently use σ_t in place of density in our following discussions. To derive σ_t, ocean water density is first converted to specific gravity by dividing it by the density of pure water at 4°C (39.2°F):

$$\text{Specific gravity} = \frac{\text{Density of ocean water}}{\text{Density of pure water}}$$
$$= \frac{1.02567 \text{ g/cm}^3}{1.00000 \text{ g/cm}^3} = 1.02567$$

This eliminates the density units, g/cm³. The specific gravity of ocean water is then inserted into the following equation to convert it to σ_t:

$$\sigma_t = (\text{Specific gravity} - 1) \times 1000$$
$$25.67 = (1.02567 - 1) \times 1000$$

It can be determined by observing the temperature-salinity-density relationships in figures 7–2 and 7–3 that temperature change can be expected to have a much greater effect on density in the high-temperature, low-latitude areas than in polar regions. It can be noted that the lines of equal σ_t (density) are more nearly parallel to the temperature axis for low temperatures than for high temperatures, showing a greater density change per unit of temperature change in the high-temperature range.

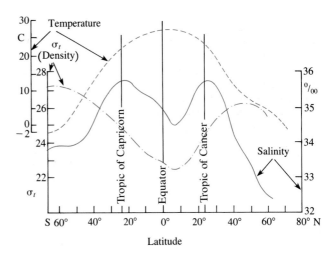

FIGURE 7–3

Average Temperature, Salinity, and σ_t (Density) of World Ocean by Latitude. σ_t increases from about 22 near the equator to maxima of 26–27 near 50°–60° latitude in the surface water of the ocean. At higher latitudes it decreases slightly. Salinity maxima in the tropics do not seem to affect density, pointing out the importance of temperature in controlling density in low latitudes. (After G. L. Pickard, *Descriptive physical oceanography.* © 1963. By permission of Pergamon Press Ltd., Oxford, England.)

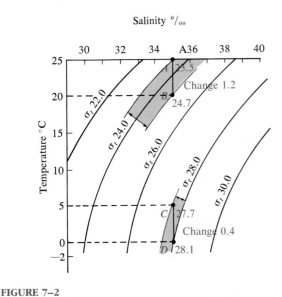

FIGURE 7–2

Temperature-Salinity-σ_t (Density) Relationship (*T-S* Diagram). In high-latitude areas characterized by low-temperature water, temperature has less effect on density (σ_t) than in the high temperature, low-latitude areas. Measured at a salinity of 35 ‰, σ_ts determined for points A, B, C, and D show that the density change is greater over a 5°C (9°F) temperature span at a higher temperature range, 20°–25°C (68°–77°F), than at a lower temperature range, 0°–5°C (32°–41°F). (After G. L. Pickard, *Descriptive physical oceanography.* © 1963. By permission of Pergamon Ltd., Oxford, England.)

Density and the variables that most affect it, temperature and salinity, are important water characteristics defined as *conservative properties* used in the identification of water masses. Conservative properties of water are those that are not significantly altered by any processes other than mixing and diffusion once the water sinks beneath the surface.

A *nonconservative property* of water is changed by some process other than mixing and diffusion after the water sinks beneath the ocean surface. An example is dissolved oxygen, which will be changed by biological processes.

Since the gain or loss of heat energy at the ocean surface is of primary importance in determining density characteristics of the ocean water, we find, as with most water properties, that the rate of change in density is greater in a vertical direction than in a horizontal direction.

The density of surface water is affected primarily by temperature changes in the open ocean. In the extreme high-latitude areas of the ocean, however, where temperatures remain relatively constant, salinity changes can have a significant effect on density. Figure 7–4 shows the vertical distribution of density in various regions of the ocean. In the equatorial region, there is a shallow zone of low-density water near the surface, and below this lies a zone where the increase in density is very rapid. This zone of rapidly increasing density is called the *pycnocline* (density slope) and represents a very stable barrier to the mixing of the low-density water above and the high-density water below. The pycnocline is considered to have a high gravitational stability because it would require a greater amount of energy to move a given mass of water from some point in the pycnocline either up or down than would be required to move an equal mass an equivalent distance in regions where density change occurred very slowly with increasing depth.

The pycnocline develops as a result of the combined effects of zones of rapid vertical changes in temperature, the *thermocline* (temperature slope), and salinity, the *halocline* (salinity slope). The interrelationship of these three zones, which determine the degree of separation between the *upper water* and *deep-water* masses, is shown in figure 7–4.

The pycnocline is lacking in the high latitudes. It can be seen that there is not a great amount of difference in the density of the surface waters as compared to the bottom waters in these high-latitude areas, thus these water columns would be considered to be less stable than the columns possessing a pycnocline that developed in the low latitudes. We will consider the importance of the low-stability columns of water in the high latitudes to mixing the waters of the world ocean in the next chapter.

Sound

Sound can be transmitted much more efficiently through water than through air. This has made it possible to develop systems for determining the position of objects in the ocean as well as determining the distance to the bottom of the ocean by *sonar* (sound navigation and ranging). The average velocity of sound in the ocean is 1450 m (4756 ft)/s, compared with 334 m (1095 ft)/s in a dry atmosphere at 20°C. Velocity of sound in-

FIGURE 7–4
Vertical Density Profiles for Latitudinal Regions. (After G. L. Pickard, *Descriptive physical oceanography* © 1963. By permission of Pergamon Press Ltd., Oxford, England.)

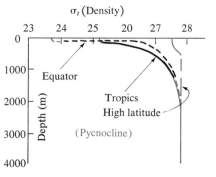

A. Because of natural tendencies toward stability, density can normally be expected to increase with depth. In low-latitude equatorial and tropical regions a thin layer of low-density surface water is separated from high-density deep water by a zone of rapid density change, the pycnocline. The pycnocline is poorly developed or missing in high-latitude waters. It is the absence of the stable pycnocline that facilitates vertical mixing in some high-latitude areas.

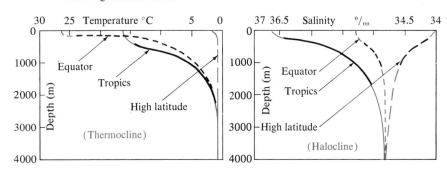

B. The gravitationally stable pycnocline described above is primarily the result of and coincides with a zone of rapid vertical decrease in temperature, the thermocline. The thermocline is also more highly developed in low latitudes. A zone of rapid vertical change in salinity, the halocline, may also affect density. The halocline usually represents a decrease in salinity with increasing depth in low latitudes but may represent the opposite condition in high-latitude areas (note that in B the temperature increases to the left. Although this is unconventional, it was done to make the curves easier to compare).

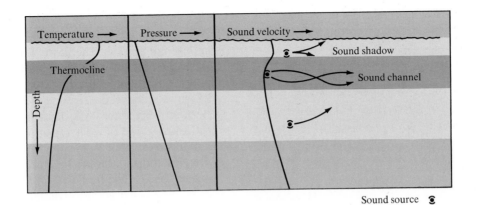

FIGURE 7–5

Transmission of Sound in the Ocean. Velocity of sound in the ocean increases with increases in the temperature, salinity, and pressure. Generally salinity changes are not as important in affecting the velocity of sound transmission as are the changes in temperature and pressure. Pressure increases steadily with depth, which increases the velocity of sound. However, the rapid decrease in temperature represented by the thermocline offsets the pressure increase and produces a low-velocity sound channel. Wave phenomena, such as sound, bend (refract) into low-velocity areas. This produces shadow zones when velocity maxima exist and channels that trap sound energy in low-velocity zones. The sound source beneath the sofar channel illustrates how refraction bends sound. If we take a band of the sound that is traveling to the right from the sound source, the sound at the bottom of the band will travel faster than the sound at the top of it. If the sound band is moving as a coherent unit, it will be bent upward into the sofar channel. As it passes through the sofar channel, the top of the band will be moving faster than the bottom and bend it back down into the channel—trapping the sound in the sofar channel.

creases with increases in temperature, salinity, and pressure. Therefore, the velocity with which sound travels through water in a particular area will vary on a seasonal basis. To determine the exact velocity of sound for any point in the ocean, these three variables—temperature, salinity, and pressure—must first be determined, or a *velocimeter* can measure it directly.

At a depth of around 1000 m (3280 ft) there exists a layer of ocean water where the combined effects of the values of these three variables result in relatively low velocity for the transmission of sound. Sound originating above and below this layer will be refracted, or bent, into the low-velocity layer and be trapped there. Thus, by following this channel, sound can be transmitted for unusually great distances (figure 7–5). This channel is called the *sofar* (sound fixing and ranging) channel because of its practical application in distance determination.

SOLAR ENERGY

Distribution of Solar Energy

Essentially all of the energy available to the earth comes from the sun. Solar energy strikes the earth at an average rate of 2 cal/cm²/min. Although energy radiated by the sun covers the full electromagnetic spectrum, most of the energy that reaches the earth's surface is con-

tained in wavelengths close to and including the visible portion of the spectrum.

Molecular oxygen (O_2), as it is broken into individual atoms of oxygen (O) that recombine with molecular oxygen to produce ozone (O_3), absorbs most of the solar radiation with wavelengths shorter than 0.29 μm (0.00001 in.). Water vapor absorbs a high percentage of solar radiation with wavelengths greater than 0.8 μm (0.00003 in.), leaving most of the solar radiation within the visible spectrum 0.38–0.76 μm (0.000015–0.00003 in.) to penetrate the atmosphere and reach the earth's surface. After scattering by atmospheric molecules and reflection by clouds in the atmosphere, about 47 percent of the solar radiation that is directed toward the earth is absorbed by the oceans and continents. Nineteen percent is absorbed by the atmosphere, and 34 percent is lost to space.

If it is true that the earth has maintained a constant average temperature over the years, the earth must be radiating energy back to space at the same rate it is absorbing solar energy. This energy that is radiated back into space from the earth's surface is characteristically of a longer wavelength than the energy received by the earth. The greatest amount of energy radiating from the earth falls within the infrared range (0.76 μm–0.1 cm, or 0.00003–0.04 in.). Figure 7–6 shows the intensity of energy radiated by the sun peaks at 0.48 μm (0.00002 in.) in the visible spectrum. It also shows that the radia-

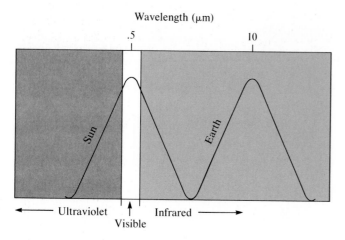

Wavelength (μm)

.5   10

Sun   Earth

Ultraviolet   Infrared
Visible

FIGURE 7–6

Comparison of Spectra of Energy Radiated by the Sun and the Earth. Most of the energy coming to the earth from the sun is within the visible spectrum and peaks at 0.48 μm (0.00002 in.). The atmosphere is transparent to most of this radiation, and it is absorbed at the earth's surface. When the earth reradiates this energy back toward space, it does so as infrared radiation with a peak at a wavelength of 10 μm (0.0004 in.). The CO_2 and H_2O in the atmosphere absorb this infrared radiation, producing the greenhouse effect.

tion from the earth peaks at 10 μm (0.0004 in.) in the infrared range. While the atmosphere absorbs only 19 percent of the short-wavelength solar radiation, the water vapor and carbon dioxide in the atmosphere absorb a large amount of the long-wavelength infrared radiation emitted by the earth. It is this change of wavelength that is the essence of the *greenhouse effect.* Some of the in-

frared energy absorbed in the atmosphere will be reabsorbed by the earth (the rest being lost to space) to continue the process. Therefore, the solar radiation received is retained for a time within our atmosphere to moderate the temperature fluctuations between night and day and between seasons. The amount of atmospheric heating varies with latitude because of the unequal distribution of solar radiation on the earth's surface and water vapor in the atmosphere.

Recent studies indicate the atmosphere is increasing its greenhouse effect because of increases in energy-absorbing gases such as carbon dioxide, methane, and others. Predictions have been made that the earth's atmospheric and oceanic temperatures will increase, which may cause a significant amount of glacial melting and rise in sea level within the next 50 years.

Considering the earth as a sphere, figure 7–7 shows that radiation reaching its surface will strike at an angle of 90° near the center of the sphere that is lit by this radiation, while it will be striking at 0° at the edge of this circle (solar rays will be tangent to the earth's surface). If we consider a 1-km² (0.4-mi²) cross-sectional area of solar radiation that falls near the center of the illuminated portion of the sphere, this ray of light will cover exactly 1 km² (0.4 mi²) of the earth's surface. By contrast, if we consider the same cross-sectional area falling on a high-latitude portion of the earth's surface where it does not strike the earth at a right angle, this square kilometer beam will be spread out over considerably more than 1 km² (0.4 mi²) of the earth's surface. The intensity of radiation available per unit of surface

FIGURE 7–7
Solar Radiation.

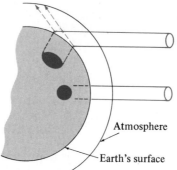

Atmosphere

Earth's surface

The amount of heat received at higher latitudes is less than that at lower latitudes for three reasons: (1) a ray of solar radiation that strikes the earth at a high latitude is spread over more area than an equal ray that is perpendicular to the earth's surface at a lower latitude, (2) the high latitude ray also passes through a greater thickness of atmosphere, and (3) more of its energy is reflected due to the low angle at which it strikes the earth's surface.

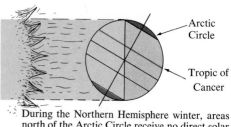

Arctic Circle

Tropic of Cancer

During the Northern Hemisphere winter, areas north of the Arctic Circle receive no direct solar radiation, while areas south of the Antarctic Circle receive continuous radiation for months.

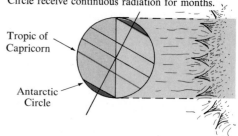

Tropic of Capricorn

Antarctic Circle

The reverse of the above situation is true during the Northern Hemisphere summer. Throughout the year, the sun is directly over some latitude between the tropics.

name implies, blow out of the southwest. Why do we observe this east-west component of motion in air masses that are moving north and south? To an observer watching these masses in motion from a stationary position in space above the earth's atmosphere, the motion of an individual air mass will be a straight north-south movement.

Figure 7–10 shows that as the earth rotates on its axis, points at different latitudes rotate at different velocities. The rotational velocity is proportional to the distance from the earth's surface to the axis of rotation and ranges from 0 km/h at the poles to over 1600 km/h (994 mi/h) at the equator. The next paragraphs will discuss why objects in motion that would be seen to move in a straight line by observers outside the earth's frame of reference appear to move along curved paths to an observer on the earth.

The trade winds appear to blow out of the northeast because, as the air mass moves south from the subtrop-

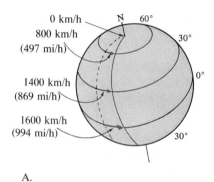

A.

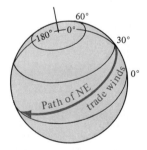

B.

FIGURE 7–10

Coriolis Effect. *A,* points at different latitudes on the earth's surface rotate at different velocities, ranging from 0 km/h at the poles to 1600 km/h (994 mi/h) at the equator. *B,* an air mass that leaves the 30° latitude where it is rotating with the earth at 1400 km/h (870 mi/h) and heads south toward the equator at 32 km/h will not reach the equator until 100 hours later. Since the equator is rotating east at 200 km/h (124 mi/h) faster than the air mass, it will meet the equator at a point 20,000 km (12,420 mi) west of the point on the equator that was directly south of the air mass when it left its position on 30° latitude. This air mass could represent the northeast trade winds, which follow a similar path.

ical region near 30°N latitude, it is also moving with the rotating earth in an easterly direction at about 1400 km/h (870 mi/h). The air mass is moving toward a point on the equator; this point is moving east at 1600 km/h (994 mi/h). The distance that the air mass must cover to reach the equator is about 3200 km (2000 mi). If it moves to the south at 32 km/h (20 mi/h), it will require 100 hours to arrive at the equator (figure 7–10).

The air mass is moving south at 32 km/h (20 mi/h) and east at 1400 km/h (870 mi/h). For every hour that the air mass moves in the southerly direction, the point on the equator toward which it started moves east 200 km (124 mi) farther than does the air mass (1600 km/h − 1400 km/h). In the period of 100 h it takes the air mass to make the trip, the point on the equator toward which it started will be 20,000 km (100 h × 200 km/h) east of the air mass when it reaches the equator. While the air mass was moving the 3200 km (2000 mi) in a southern direction, it appears as a result of the earth's rotation to have moved 20,000 km (12,420 mi) in a westerly direction. Stated another way, as the air mass covered 30° of latitude from north to south, it appeared to move across 180° of longitude in a westerly direction. Certainly, if we were to encounter this air mass aboard ship at some point between the equator and 30°N latitude, it would appear to be coming out of the east-northeast.

A westerly wind air mass moves from 30° latitude in a northerly direction to 60° latitude. This mass also is moving in an easterly direction at 1400 km/h (870 mi/h). The point at 60° latitude toward which the air mass is headed is moving in an easterly direction at only 800 km/h. Unlike the situation discussed in regard to the trade winds, the westerly air mass is moving in an easterly direction at a velocity greater than that of the point at 60°N latitude toward which it started. For every hour that this mass moves in a northerly direction, it moves 600 km (373 mi) farther east than the point at 60° latitude toward which it started. If we were to encounter this air mass at some point between 30° and 60°N latitude, it would appear to come out of the west-southwest.

If the air mass moves north at the same velocity as a trade wind mass moves south, the amount of deviation to the right in the case of the westerlies will be greater than for the trade winds. This results from the fact that the condition responsible for the intensity of the Coriolis effect is the different rates at which points at different degrees of latitude rotate about the earth's axis. The rotational velocity of these points ranges from 0 km/h at the poles to more than 1600 km/h (994 mi/h) at the equator, but the rate of change per degree latitude is not constant. As we approach the pole from the equator, the rate of change of rotational velocity per degree of latitude change increases. As a comparison, we can note that there is a difference of 200 km/h

(124 mi/h) in velocity across 30° of latitude from the equator to 30°N latitude, while there is a difference of 600 km/h (373 mi/h) in the velocity of rotation from 30°N to 60°N latitude. If we carry this consideration over another 30° of latitude from 60°N latitude to the pole, where the velocity is zero, the difference is over 800 km/h (497 mi/h). This explains the fact that the Coriolis effect increases with increased latitude.

The primary factor affecting the amount of deflection resulting from the Coriolis force is not, however, the magnitude of the force, since it is very small, even at high latitudes. It is the length of time a particle is in motion that will have the greatest influence on how much it is deflected. Thus, even at low latitudes, a large Coriolis deflection is possible if an object is in motion for a long time.

The air masses we have discussed essentially move north and south. Although we will not attempt to discuss it here, a complete mathematical treatment of the rotating earth shows that the effects described for these north-south moving masses are valid for movements in any direction. Objects veer to the right of their intended path in the Northern Hemisphere and to the left in the Southern Hemisphere.

HEAT BUDGET OF THE WORLD OCEAN

As a result of variations in the absorption of solar energy, ocean temperatures will vary from place to place and from time to time. We previously discussed the fact that the temperature difference between polar and equatorial regions remains constant as a result of a transfer of heat energy from the equator to the higher latitudes by moving air and ocean masses. Driven by the moving air masses, large surface current systems are set up in the world's oceans and play an important role in

transfer of heat energy. The temperatures at various places in the ocean will depend upon the rate at which heat flows in and out of these areas. Quantitative considerations of heat transfer in ocean water masses constitute the *heat budget* of these masses (figure 7–11.)

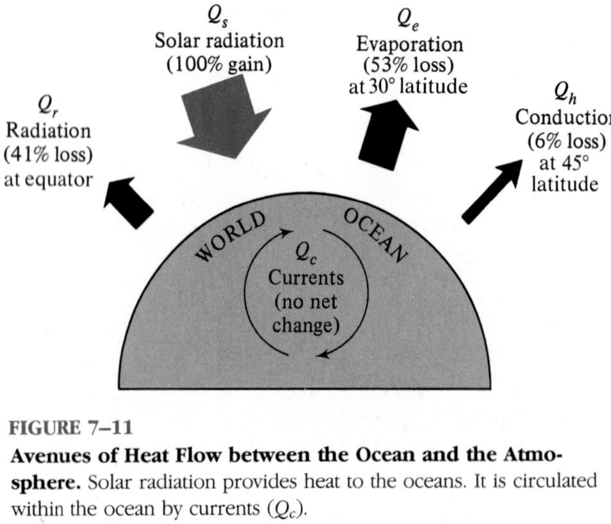

The heat budget for any *locality* in the ocean is expressed by the following equation:

Rate of heat gain − Rate of heat loss =
Net rate of heat loss or gain

$$(Q_s + Q_c) - (Q_r + Q_e + Q_b) = Q_t$$

In which

Q_s = the rate of heat gain from solar radiation
Q_c = the rate of heat gain (or loss) through ocean current
Q_r = rate of heat loss through radiation into space
Q_e = rate of heat loss through evaporation
Q_b = rate of heat loss through conduction into the atmosphere
Q_t = net rate of heat loss or gain in the locality

If Q_t is a positive value, it means the temperature of the ocean in that locality is rising. If it is a negative value, the temperature in that locality is falling, and if it is zero, the temperature of the mass of water is not changing.

Theoretically, Q_c and Q_t should be zero when we consider the world ocean over a long period of time. Current flow is internal, and seasonal increases and decreases in temperature should average out so that Q_t becomes zero. Therefore, the rate at which heat is gained in the world ocean through solar radiation should equal the rate at which heat is lost through radiation, evaporation, and conduction into the atmosphere (figure 7–11). Symbolically the heat budget of the *world ocean* is expressed using average values as:

$$Q_s = Q_r + Q_e + Q_b$$
$$100\% = 41\% + 53\% + 6\%$$

The amount of heat lost by radiation, primarily of infrared wavelengths, is greatest near the equator, where the greatest amount of solar energy is absorbed. The greatest heat loss by evaporation is in dry subtropics, while the heat lost through conduction reaches a maximum along the western margins of ocean basins where warm water currents carry water into regions where the atmosphere is much colder than the water.

FIGURE 7–11

Avenues of Heat Flow between the Ocean and the Atmosphere. Solar radiation provides heat to the oceans. It is circulated within the ocean by currents (Q_c).

THE OCEANS, WEATHER, AND CLIMATE

The idealized pressure belts and consequent wind systems previously discussed are significantly modified as a result of two factors. First, the tilt of the earth's axis of

mer. The fast ice, which is firmly attached to the shore, attains a winter thickness in excess of 2 m (6.5 ft).

In the Southern Hemisphere, where the polar region is covered by a continental mass, we might characterize all the sea ice that forms around the margin of the Antarctic continent as pack ice and fast ice that have a rather temporary existence. The pack ice rarely extends north of 55°S latitude and breaks up rather completely from October to January except in the very quiet bays (figure 7–17B). Winds that are frequently very strong help prevent the formation of a greater pack ice accumulation around the Antarctic continent.

Icebergs

Icebergs are derived from land ice in the form of glaciers that cover the continent of Antarctica and most of the island of Greenland and part of Ellesmere Island in the Arctic (figure 7–18). These vast sheets of ice that grow from the snow accumulation near the center of the landmasses spread outward toward the margins until they push their edges out into the marginal sea. There the ice sheets are buoyed up by the water and break up under the stress of current, wind, and wave action. This breakup that produces icebergs is called *calving*.

In the Arctic, the icebergs that are produced originate primarily from the tons of ice that follow valleys into the sea along the western coast of Greenland. Icebergs are also produced along the east coast of Greenland and the east coast of Ellesmere Island. The East Greenland Current and the West Greenland Current carry the icebergs at rates of up to 20 km (12.5 mi) per day into the North Atlantic, where the Labrador Current may move them into North Atlantic shipping channels. They are seldom carried south of 45°N latitude, but during some seasons large numbers of icebergs will move as far south as 40°N latitude and become shipping hazards.

Such an accumulation of icebergs developed when the *Titanic* sank in April 1912. After receiving repeated warning of the existence of such a hazard, the 46,000-tn ship containing 2224 passengers proceeded at an excessive speed of 41 km/h (25.4 mi/h) until it came to a grinding halt with buckled hull plates below the water line on the starboard side. The sinking of the *Titanic* at 41°46′N latitude, 50°14′W longitude near the Grand Banks cost the lives of 1517 people and was the event that brought about the formation of the ice patrol that has prevented further loss of life from such accidents. The U.S. Navy began this patrol immediately following the *Titanic* disaster, and it became an international patrol in 1914. The patrol concentrates its efforts between 40°30′ and 48°N latitude and 43° and 54°W longitude.

The shelf ice that represents the edges of glaciers pushing into the marginal seas of Antarctica produces glaciers that have received less attention than those of the North Atlantic, due to the fact that they interfered less with shipping. Vast tabular bergs that break from the edges of the shelf ice have lengths in excess of 100 km (62 mi). They may stand as much as 600 m (1970 ft) above the ocean surface, though most are probably less than 200 m (656 ft) above sea level. Most of the calving occurs, as it does in the Arctic region, during the summer months (figure 7–18C). When the sea ice breaks up and allows the swells driven by the strong westerly winds to reach the edge of the shelf ice, large bergs are calved and move to the north into the warmer and rougher water in which they disintegrate. Carried by the strong West Wind Drift, these bergs move in an easterly direction around Antarctica, rarely moving farther north than 40°S latitude.

RENEWABLE SOURCES OF ENERGY

The potential of energy extraction from the motions and heat distribution patterns of the atmosphere and ocean, which are maintained by the sun, is attractive for the following reasons:

1. Work can be achieved without significant pollution.
2. The amount of energy available at any time is far greater than that in fossil and nuclear fuels.
3. Such forms of energy are renewable and will not be depleted.

Considered in order of decreasing energy potential, the sources of renewable energy are (1) heat stored in the oceans, (2) kinetic energy of the winds, (3) potential and kinetic energy of waves, and (4) potential and kinetic energy of tides and currents. In later chapters we will consider the use or potential use of winds, waves, tides, and currents. Here we will consider only the renewable source with the greatest store of potential energy, that of the surface layers of the tropical oceans.

Ninety percent of the earth's surface between the Tropic of Cancer and Tropic of Capricorn is ocean. What makes this warm tropical surface water such an important source of energy is the presence of much colder water beneath the thermocline. With a temperature difference as small as 20°C (36°F), useful work can be done by Ocean Thermal Energy Conversion (OTEC) systems (figure 7–19).

The system works in the opposite way of a typical refrigeration system and on a much larger scale. The warm surface water heats a fluid such as propane or ammonia that is under pressure in evaporating tubes. The fluid is vaporized and passes through a turbine that drives an electrical generator. After passing through the turbine, the fluid is condensed by cold water that has been pumped up from the deep ocean. It is again ready for heating by warm surface water that will cause it to vaporize and pass through the turbine.

The only region with such potential along the conterminous United States coast is a strip about 30 km (18.6 mi) wide and 1000 km (621 mi) long. It extends north from southern Florida along the western margin of the Gulf Stream. The National Science Foundation once proposed an experimental OTEC system to be established 25 km (15.5 mi) east of Miami, Florida.

Mini-OTEC, a unit mounted on a U.S. Navy barge, was tested off Ke-Ahole Point, Hawaii. Designed primarily to test equipment under ocean conditions, it generated 50 kw of gross power and 15 kw of net power. A much larger unit, OTEC 1, designed to generate 1 mw proved the soundness of heat exchanger designs and biofouling systems under ocean conditions. This system was installed on the tanker *Chepachet* and also operated in Hawaiian waters. OTEC systems are much more expensive to build than con-

A.

B.

C.

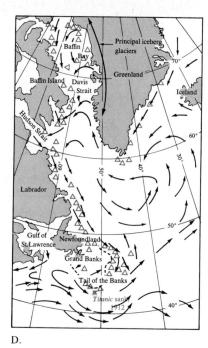
D.

FIGURE 7–18

Icebergs. *A,* North Atlantic icebergs off Cape York, Greenland. They are generally smaller and more irregular in shape than Antarctic icebergs. *B,* the USS *Wyandotte* moves past a tabular iceberg in the Weddell Sea, Antarctica. Many of the icebergs are kilometers in length and have been mistaken for land. (Photographs *A* and *B* from Oceanographer of the Navy, Public Affairs Office.) *C,* B-1 Iceberg. This infrared image shows the huge iceberg that broke from the Ross Ice Shelf in October 1987. The longitude and latitude lines intersecting near the east end of B-1 are 160°W and 78°S. This monster is 136 km (85 mi) long, but two larger icebergs formed the year before in the Weddell Sea. (Infrared image in *C* courtesy of Robert Whritner, Manager, Antarctic Research Center, Scripps Institution of Oceanography, University of California, San Diego.) *D,* a map showing North Atlantic currents and icebergs (△).

A.

B.

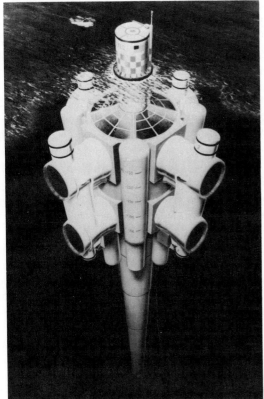

FIGURE 7–19

Ocean Thermal Power. *A,* Ocean Thermal Energy Conversion (OTEC) system with crew quarters and maintenance facilities. Attached around the outside are turbine-generators and pumps. It is over 75 m (246 ft) in diameter, 485 m (1590 ft) long, and weighs about 300,000 tn. This unit is designed to generate 160 million W of power, which is enough to meet the needs of a city with a population of 100,000. (Photo courtesy of Lockheed Missiles and Space Co., Inc.) *B,* OTEC plant for production of ammonia. Moving slowly through tropical waters, this plant could produce 1.4 percent of the ammonia requirements for the United States each year and save 22.6 billion ft^3 of natural gas. (Courtesy of U.S. Department of Energy.)

ventional steam-powered plants and will not likely be built until the economic climate changes in their favor.

Another proposed use of the OTEC concept involves the production of ammonia. A floating factory that could produce 586,000 tn of ammonia per year has been described in a feasibility report from the Johns Hopkins Applied Physics Laboratory to the U.S. Maritime Administration. The proposed demonstration ship would weigh 68,000 tn, have a width of almost 60 m (197 ft) and a length of over 144 m (472 ft). It would move at speeds less than 2 km/h (1.2 mi/h) through tropical waters using energy derived from the temperature difference between the warm surface waters and cold deep waters to produce ammonia. Each such plant could produce about 1.4 percent of our nation's ammonia requirements—75 percent of which is used for fertilizer—and save 22.6 billion ft^3 of natural gas each year (figure 7–19).

SUMMARY

In ocean water of low biological productivity, the molecular-sized particles scatter the short wavelength of visible light, producing a blue ocean. Larger sized particles in more productive ocean water scatter more green light, which, along with the chlorophyll pigmentation of the plants, produces a green ocean. With increasing depth in the ocean, the colors of the **visible spectrum** are absorbed by ocean water in a way that removes red, yellow, and violet at relatively shallow depths. The blue and green wavelengths of light are the last to be removed. High **turbidity,** the measure of the amount of suspended material in water, greatly reduces the depth to which light will penetrate.

Density of ocean water is increased by increased salinity and decreased temperature. Density change per unit of temperature change is greater in warmer water. Density, temperature, and salinity are **conservative properties** of water, properties that are affected only by mixing and diffusion once a surface water mass sinks. Zones of rapid change in the density, temperature, or salinity in the water column are called a **pycnocline, thermocline,** and **halocline,** respectively.

Velocity of sound transmission in the ocean increases with increases in temperature, salinity, and pressure. A low-velocity sound channel, **sofar,** is caused by a thermocline, in which temperature decreases with increased depth.

Radiant energy reaching earth from the sun is mostly in the **ultraviolet** and **visible light** range, while that radiated back to space from earth is primarily in the **infrared** spectrum. Some of this infrared radiation is recycled between the atmosphere and the earth's surface before it escapes to space. This phenomenon, called the **greenhouse effect,** helps warm the earth.

Because more energy is received and radiated back into space at low latitudes, and water vapor, which absorbs infrared radiation well, is unevenly distributed in the atmosphere, the atmosphere is unevenly heated and set in motion. **Belts of low pressure,** where air rises, are generally found at the equator and at about 60° latitude. **High-pressure regions,** where dense air descends, are located at the poles and at about 30° latitude. The air at the earth's surface that is moving away from the subtropical highs produces the **trade winds** moving toward the equator and the **westerlies** moving toward higher latitudes.

Because the earth's surface rotates at different velocities at different latitudes, increasing from 0 km/h at the poles to over 1600 km/h (994 mi/h) at the equator, objects in motion tend to veer to the right in the Northern Hemi-sphere and to the left in the Southern Hemisphere. This is called the **Coriolis effect.**

In the **heat budget** of the world ocean, the only source of heat gain is the sun, while heat may be lost by **radiation, evaporation,** and **conduction** into the atmosphere.

The tilt of the earth's axis of rotation and the distribution of continents modify the idealized pressure belts discussed above. High-pressure cells form over continents in winter and are replaced by low-pressure cells in summer. There is a **cyclonic** movement of air around the low-pressure cells and an **anticyclonic** movement around high-pressure cells. Cold air masses of high latitudes meet warm air masses of lower latitudes and create **cold and warm fronts** that move from west to east across the earth's surface at midlatitudes. Nonetheless, ocean climate patterns are closely related to the idealized pressure belts previously discussed.

Sea ice forms as sea water is frozen in high latitudes. The process usually involves the formation of a slush, which breaks into **pancakes** that ultimately grow into **floes.** Sea ice develops as **pack ice** that forms each winter and melts almost entirely each summer, **polar ice** that is a permanent accumulation in polar regions of the Arctic Ocean, and **fast ice** that forms frozen to shore during the winter. **Icebergs** form as large chunks of ice break away from the large continental **glaciers** that form on Ellesmere Island and Greenland in the Northern Hemisphere and Antarctica to the south.

The motions and patterns of heat distribution in the atmosphere and oceans offer nonpolluting **renewable energy sources.** These include heat stored in the ocean and potential and kinetic energy stored in the **winds, waves, tides,** and **currents. Ocean Thermal Energy Conversion (OTEC)** is a process developed to use the difference in temperature between warm tropical surface water and the cold water below the thermocline to produce electricity and ammonia.

QUESTIONS AND EXERCISES

1. How is the color of the ocean surface water related to biological productivity? Why does everything in the ocean at depths below the shallowest surface water take on a blue-green appearance?

2. Describe the relative effect of temperature change on water density at high versus low temperature ranges.

3. The position of the pycnocline in the water column is determined by the combined effects of the thermocline and halocline. Describe these relationships in tropical waters (see figures 7–3 and 7–4).

4. How does the development of a thermocline produce a sofar or sound channel below the ocean's surface?

5. Explain why the atmosphere is heated primarily by back-radiation from the earth rather than by direct radiation from the sun.

6. Describe the effect on the earth as a result of the earth's axis of rotation being tilted at an angle of 23.5° from perpendicular relative to the ecliptic.

7. Since there is a net annual heat loss at high latitudes and a net annual heat gain at low latitudes, why does the temperature difference between these regions not increase?

8. Why are there high-pressure caps at each pole and high pressure belts at 30° latitudes, compared with low-

pressure belts in the equatorial region and at 60° latitudes?

9. Describe the Coriolis effect in the Northern and Southern hemispheres and include a discussion of why the effect increases with increased latitude.

10. In considering the heat budget of the world ocean over a long period, why should Q_c (rate of heat gain or loss through ocean currents) and Q_t (net rate of heat gain or loss) be considered to be zero?

11. Describe the ocean regions where the maximum rates of heat loss by radiation, evaporation, and conduction occur and explain why.

12. Discuss why the idealized belts of high and low atmospheric pressure shown in figure 7–9 are modified by the continents.

13. How are the ocean's climatic belts related to the broad patterns of air circulation described in figure 7–9?

14. Describe the formation of sea ice from the initial freezing of the ocean surface water through the development of polar ice in the Arctic Ocean.

15. From the information given in the discussion of polar ice, determine the residence time of polar ice.

16. What is the difference in the average size and shape of icebergs in the Arctic and Antarctic? Why do these differences exist?

17. Construct your own diagram of how an Ocean Thermal Energy Conversion unit might generate electricity, or make a flow diagram representing the steps of the process.

REFERENCES

Changing climate and the oceans. 1987. *Oceanus* 29:4, 1–93.

Miller, A. 1971. *Meteorology.* Columbus, Ohio: Charles E. Merrill.

Ocean energy. 1979. *Oceanus.* 22:4, 1–68.

Oceans and climate. 1978. *Oceanus.* 21:4.

Pickard, G. L. 1975. *Descriptive physical oceanography,* 2nd ed. New York: Pergamon Press.

SUGGESTED READING

Sea Frontiers

Charlier, R. H. 1981. Ocean-fired power plants. 27:1, 36–43.
The potential for using the temperature difference between warm surface water and cold deep water to produce energy is considered.

Emiliani, C. 1971. Climate cycles. 17:2. 108–20.
Factors that affect climate and the recorded climatic changes over the last 5000 years are discussed. Sheds light on the problem of understanding the earth's climatic changes.

Land, T. 1976. Europe to harness the power of the sea. 22:6, 346–49.
An article emphasizing the clean, safe, permanent nature of ocean tides and waves as a source of power.

Mayor, A. l988. Marine mirages. 34:1, 8–15.
The nature of marine mirages and the research efforts that lead to our understanding them are considered.

Rush, B., and Lebelson, H. 1984. Hurricane! The enigma of a meteorological monster. 30:4, 233–39.
An overview of the nature of hurricanes and the problem of predicting where they will go.

Scheina, R. L. 1987. The *Titanic's* legacy to safety. 33:3, 200–209.

A brief summary of the *Titanic* sinking and an overview of the ice patrol and iceberg collision history subsequent to the sinking of the "unsinkable" luxury liner.

Smith, F. G. W. 1974. Planet's powerhouse. 20:4, 195–203.
A description of how the earth's "heat engine" works, with an emphasis on the nature of tropical cyclonic storms.

———. 1974. Power from the oceans. 20:2, 87–99.
A survey of the many tried and untried proposals for extracting energy from the oceans.

Sobey, E. 1979. Ocean ice. 25:2, 66–73.
The formation of sea ice and icebergs as well as their climatic and economic effects are included.

———. 1980. The ocean-climate connection. 26:1, 25–30.
The increasing amount of knowledge of the effect of the oceans on the earth's climate may be used to predict climatic trends of the future.

Scientific American

Gregg, M. 1973. Microstructure of the ocean. 228:2, 64–77.
A discussion of the methods of studying the detailed movements of ocean water by observing temperature and

salinity changes over distances of 1 cm (0.4 in.) and the motions they reveal.

MacIntyre, F. 1974. The top millimeter of the ocean. 230:5, 62–77.
Processes that are confined to a thin film at the surface of ocean water and their role in the overall nature of the oceans are discussed.

Penney, T. R., and Bharathan, D. 1987. Power from the sea. 256:1, 86–93.
A prediction that generating electricity by ocean thermal energy conversion will be competetive with fossil fuel plants as the price of oil rises.

Revelle, Roger. 1982. Carbon dioxide and world climate. 247:2, 35–43.
Continued increase in CO_2 content of the atmosphere could make parts of the world warmer and wetter.

Stanley, Steven M. 1984. Mass extinctions in the oceans. 250:6, 64–83.
Geologic evidence suggests most major periods of species extinction over the last 700 million years occurred during brief intervals of ocean cooling.

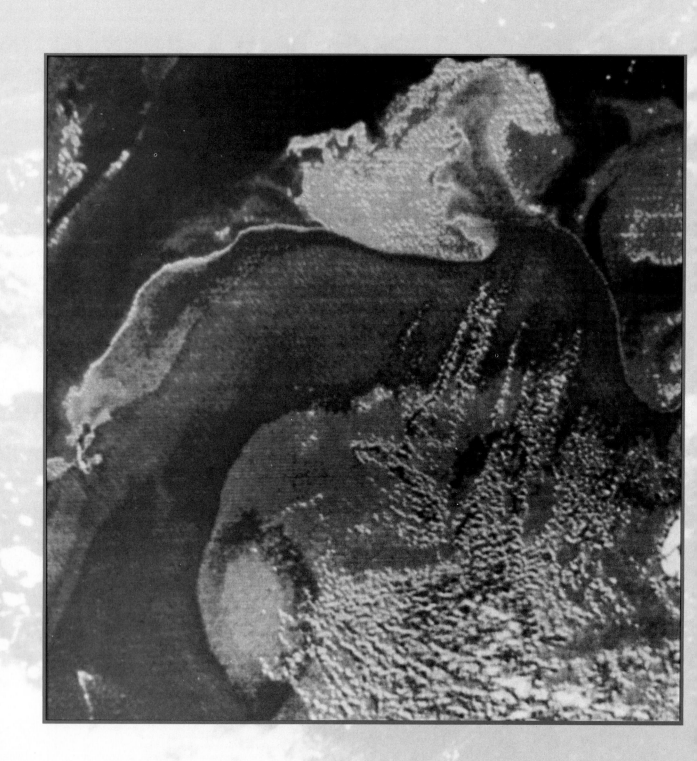

8
OCEAN CIRCULATION

Currents, or water masses in motion, are driven ultimately by energy derived from the sun. The circulation of these masses can be categorized as being either *wind-driven* or *thermohaline.* The wind-driven currents are set in motion by the moving air masses discussed in the previous chapter, and this motion is confined primarily to horizontal movement in the upper waters of the world ocean. The thermohaline circulation has a significant vertical component and accounts for the thorough mixing of the deep masses of ocean water. Thermohaline circulation is initiated at the ocean surface by temperature and salinity conditions that produce a high-density mass, which sinks and spreads slowly beneath the surface waters.

The *upper water* mass of the world ocean includes the well-mixed surface layers of the ocean and the pycnocline, a zone of rapidly changing density. The base of the pycnocline will usually be encountered above a depth of 1000 m (3280 ft). The *deep water* includes the rest of the water column below the pycnocline to the bottom of the ocean, throughout which the temperatures are low (figure 8–1).

HORIZONTAL CIRCULATION

Wind-driven horizontal circulation in the surface water develops as a result of the transfer of wind energy to the surface water. The trade winds blowing out of the southeast in the Southern Hemisphere and out of the northeast in the Northern Hemisphere may be thought of as providing the backbone of the system of ocean surface currents. Setting the water masses between the tropics into motion, the trade winds develop the equatorial currents that can be found in all the world's oceans. Due to the Coriolis effect, these currents move west parallel to the equator, and they are deflected by continents into clockwise gyres in the Northern Hemisphere and counterclockwise gyres in the Southern Hemisphere. A *gyre* is a horizontal circulation cell, and these major oceanic gyres are called *subtropical gyres* because they are centered in the subtropics (30° lati-

Satellite image of the Gulf Stream moving out into the Atlantic off Chesapeake Bay. Photo Courtesy of NOAA.

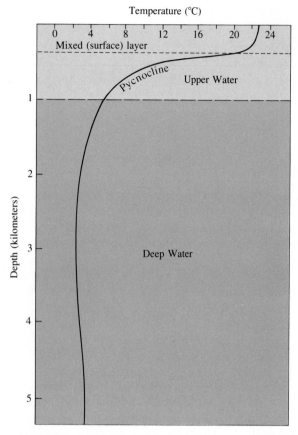

Temperature (°C)

FIGURE 8–1

Major Water Masses of the Open Ocean. Extending from the surface down through the mixed, or surface, layer and the pycnocline is the relatively light *upper water.* It is well developed throughout the low- and midlatitudes. It is underlain by the denser, cold *deep-water* mass that extends to the deep-ocean floor. The mixed layer represents water of uniform temperature resulting from mixing waves.

tude). As these deflected masses, *warm currents,* move along the western margin of their respective ocean basins toward higher latitudes, the westerly wind systems blowing out of the northwest in the Southern Hemisphere and the southwest in the Northern Hemisphere add more energy and push the circulating masses on in an easterly direction. Deflection of *cold currents* toward the equator along the eastern margin of their respective basins completes the path of circulation (figure 8–2).

Ekman Spiral

To explain the reasons for the water masses moving as they do relative to the wind direction, we will recall the observation made by Fridtjof Nansen during the voyage of the *Fram.* Nansen described his observations to V. Walfrid Ekman, a physicist who developed the mathematical relationships associated with the observation.

Ekman developed a circulation model that has been called the *Ekman spiral.* The model assumes that a homogeneous water column is being set in motion by wind blowing across its surface. Due to the Coriolis force the surface current moves in a direction 45° to the right of the wind in the Northern Hemisphere, and this surface mass of water moving as a thin *lamina,* or sheet, sets another layer beneath it in motion. The energy of the wind is passed through the water column from the surface down, with each successive layer of water being set in motion with a lower velocity than, and in a direction to the right of, the one that set it in motion. At some depth the momentum imparted by the wind to the moving water laminae will be lost and there will be no motion as a result of wind stress at the surface. Though it depends on wind speed and latitude, this normally occurs at a depth of possibly 100 m (330 ft), where the water may be moving in a direction opposite to the direction of the wind that set the water in motion. Figure 8–3 shows the spiral nature of movement with increasing depth from the ocean's surface. The length of each arrow in the figure is proportional to the velocity of the individual lamina, and the direction of each arrow is related to the direction of movement of the lamina. Under these theoretical conditions, the surface current should flow at an angle of 45° to the direction of the wind, and the net water movement, the *Ekman transport,* will be at right angles to the direction of the wind.

There are, however, no ideal conditions in existence in the ocean, and movements actually occurring as a result of wind stress on the ocean surface will deviate from this idealized picture. Generally, we can say that the surface current will move at an angle of less than 45° to the direction of the wind and that the Ekman transport will be at an angle less than 90° to the direction of the wind. This is particularly true in shallow coastal waters, where all the movement may be in a direction very nearly that of the wind and the turning with increased depth occurs at a very low rate.

Geostrophic Currents

If we consider a subtropical gyre in the Northern Hemisphere—in the North Atlantic Ocean, for example—and remember the effect of the Coriolis force (Ekman transport), it can be seen that a clockwise rotation in the Northern Hemisphere will tend to produce a convergence of water in the center of that gyre. Within all such ocean gyres, we find hills of water that rise as much as 2 m (6.5 ft) above the water level at the margins of the gyres. The water piles up on these hills until the downslope component of gravitational force acting on individual particles of water balances the Coriolis force (up-

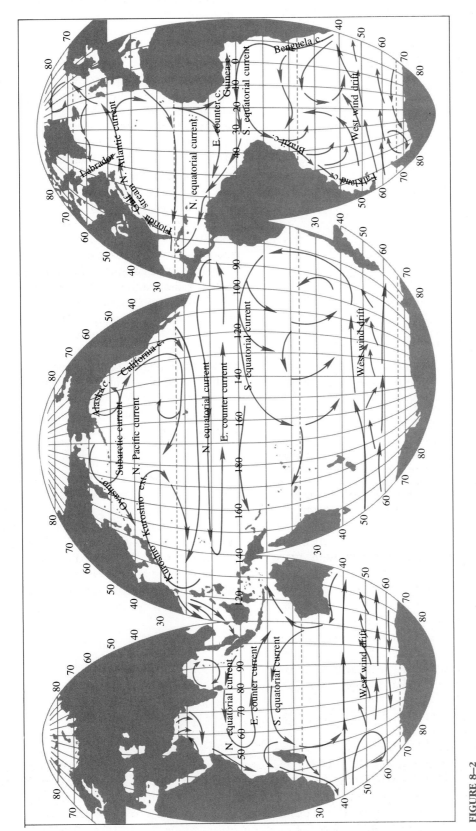

FIGURE 8–2
Wind-Driven Surface Currents in February and March. (After Sverdrup et al. 1942.)

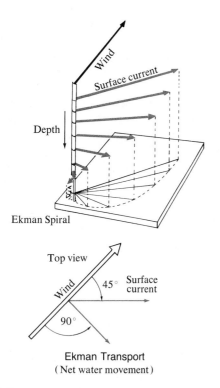

Depth

Ekman Spiral

Top view

Wind 45° Surface
current

90°

Ekman Transport
(Net water movement)

FIGURE 8–3

Ekman Spiral. Wind drives surface water in a direction 45° to the right of the wind in the Northern Hemisphere. Deeper water layers continue to deflect to the right and move at a slower speed with increased depth. The net water movement, or Ekman transport, is at right angles to the wind direction.

slope). As the water piles up on these hills, the gravitational force acts to move particles down the slope, but the Coriolis force deflects the water to the right in a curved path (figure 8–4). When these two forces finally balance, there is no further down-slope movement, and the *geostrophic current* moves around the hill.

Again, this idealized flow would exist if there were no friction. Due to the friction between water mole-

cules, the water follows a path that moves it gradually down the slope of the hill.

Westward Intensification

The hills formed within the rotating gyres do not have their apex in the center of the gyre. The highest part of each hill is located closer to the western boundary of the gyre. The causes for this are complex, but a good part of the cause can be explained in terms of the Coriolis effect. Remember that this effect increases in strength toward higher latitudes. Thus, the water that is driven eastward by westerly winds tends more strongly to veer toward the equator than the water that is driven westward along the equator tends to veer toward higher latitudes. This causes a broad equatorward flow of water across most of the subtropical gyre.

Assuming we have the same volume of water rotating around the apex of the hill, we can see by observing figure 8–4 that the velocity with which the water moves to higher latitudes confined to a narrow band along the western margin will be much greater than that of the broader equatorward flow around the eastern side of the hill. These relationships can be seen particularly well in the North Atlantic and North Pacific gyres, where the western boundary currents move with speeds in excess of 5 km/h (3 mi/h) in a northerly direction, while a flow along the eastern margin of each basin is better characterized as a drift moving at velocities well below 1 km/h (0.6 mi/h). Western boundary currents commonly flow 10 times faster and to greater depths than eastern boundary currents. The warm western boundary currents are usually less than one-twentieth the width of the broad, cool drifts flowing toward the equator on the east side of the subtropical gyres.

Directly related to this difference in speed is the steepness of the hill's slope. The slope of the hill on the side of the slow-moving eastern boundary current is

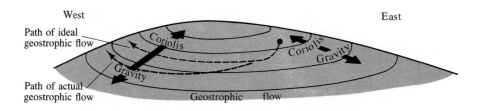

West   East

Path of ideal
geostrophic flow

Coriolis

Coriolis   Gravity

Gravity

Path of actual
geostrophic flow

Geostrophic   flow

FIGURE 8–4

Geostrophic Current. As the earth's wind systems set ocean water in motion, circular gyres are produced. Water piles up inside the gyre with the apex of the "hill" closer to its west margin due to the earth's rotation to the east. The theoretical geostrophic current flows parallel to the contour of the hill and represents an equilibrium between the Coriolis force pushing water toward the apex through Ekman transport and the down-slope component of gravity acting to move the water down the slope. Because of friction between water molecules, the path of the current is gradually down the slope of the hill.

gin of the South Atlantic. The *Falkland Current,* an important cold current, moves up the coast of Argentina as far north as 25°–30°S latitude, wedging its way between the continent and the Brazil Current. The Brazil Current has a much smaller volume than its Northern Hemisphere counterpart, the Gulf Stream. This results partly from the splitting of the South Equatorial Current by the configuration of South America (figure 8–7).

The *North Equatorial Current* moves parallel to the equator in the Northern Hemisphere, where it is joined by that portion of the South Equatorial Current that is shunted toward the north along the South American coast. This flow splits into two masses, the *Antilles Current* that flows along the Atlantic side of the West Indies and the *Caribbean Current* that passes through the Yucatan Channel into the Gulf of Mexico. These masses reconverge as the water that entered the Gulf of Mexico exits between Florida and Cuba as the *Florida Current.* The Florida Current flows close to shore over the continental shelf, carrying a volume that at times exceeds 35 sv. As it moves off Cape Hatteras and flows across the deep ocean in a northeasterly direction, it becomes the *Gulf Stream,* flowing at velocities up to 9 km/h (5.6 mi/h). The western margin of the Gulf Stream can frequently be defined as a rather abrupt boundary that moves periodically closer to and farther away from the shore. The eastern boundary becomes very difficult to identify, as it is usually masked by filamentous meandering masses that change their position continuously. Its character gradually merges with the water of the Sargasso Sea. A volume transport of over 90 sv off Chesapeake Bay indicates a large volume of Sargasso Sea water has been added to the flow provided by the Florida Current. However, by the time the Gulf Stream reaches the Tail of the Banks, south of Newfoundland, the volume carried by the Gulf Stream has been reduced to 40 sv. This would indicate that most of the Sargasso Sea water that joined the Florida Current and made up the Gulf Stream has returned to the diffuse flow of the Sargasso Sea.

Just how this loss of water occurs is yet to be determined, but much of it may be achieved by meanders (sinuous curves in the course of a current) pinching off to form large eddies. As is shown in figure 8–8, meanders north of the Gulf Stream pinch off and trap warm Sargasso Sea water in eddies rotating in a clockwise direction. These eddies move southwest at speeds of 3–7 km/day (1.9–4.3 mi/day) toward Cape Hatteras, where they rejoin the Gulf Stream. South of the Gulf Stream and west of 50°W longitude, eddies with a core of cold slope water rotating counterclockwise move south and west. These large eddies, ranging in diameter from 100 to 500 km (62–310 mi) and extending to depths of 1 km (0.62 mi) in the warm rings and as deep as 3.5 km (2.2 mi) in the cold rings, could remove large volumes of water from the Gulf Stream.

CURRENTS

A–Antillean	G–Guinea
Bg–Benguela	GS–Gulf Stream
Br–Brazil	I–Irminger
C–Canary	L–Labrador
CC–Caribbean	NA–North Atlantic
EG–East Greenland	NE–North Equatorial
EW–East Wind Drift	N–Norwegian
EC–Equatorial Counter	SE–South Equatorial
Fa–Falkland	WW–West Wind Drift
F–Florida	

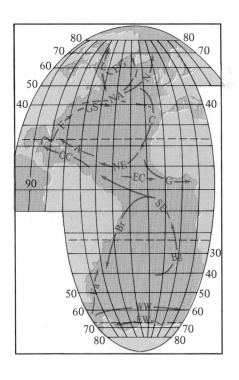

FIGURE 8–7
Atlantic Ocean Surface Currents. (Base map courtesy of National Ocean Survey.)

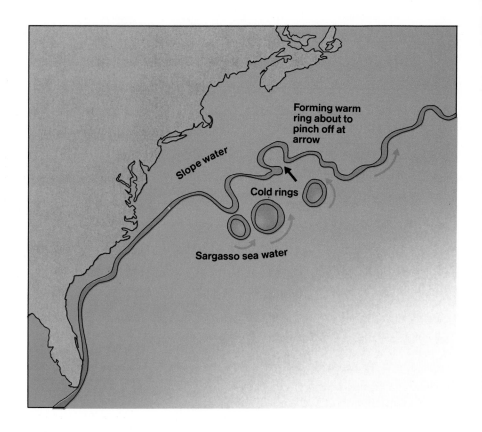

FIGURE 8–8

Gulf Stream Eddies. This diagram is an overlay of the satellite image in plate 5. It attempts to clarify the process of ring formation by showing the core of the Gulf Stream flow. It shows the warm ring and three cold rings as having formed by the pinching off of meanders. The warm ring and the middle cold ring have separated from the main Gulf Stream flow, but the other two cold rings have just formed as the meanders that flowed around them pinched off from the main flow.

At the Tail of the Banks, about 40°N latitude and 45°W longitude, the Gulf Stream becomes the *North Atlantic Current* and continues in a more easterly direction across the North Atlantic. This current is composed of numerous branches into which the Gulf Stream flow is ultimately broken. Two branches that are composed of water produced through the mixing of the *Labrador Current* and the Gulf Stream are the *Irminger Current,* flowing up along the west coast of Iceland, and the *Norwegian Current,* which moves north along the coast of Norway. The other major branch of the North Atlantic Current follows the 45°N latitude across the North Atlantic and turns south as the *Canary Current* passing between the Azores and Spain. This southward flow is very diffuse and spreads over a broad area as it moves southward and eventually joins the North Equatorial Current.

The effects of the westward intensification of current flow in the North Atlantic Ocean can be seen from the surface temperatures shown in figure 8–14. From 20° to 40°N latitude off the coast of North America, we can see a 20° temperature change from waters near the Dominican Republic to those near New York in February. By contrast, on the eastern side of the North Atlantic only a 5°–6° range in temperature can be observed between 20° and 40°N latitude from a point off the coast of Mauritania to waters off the coast of Spain.

PACIFIC OCEAN CIRCULATION

The surface circulation in the Pacific Ocean is generally similar to that which we described for the Atlantic Ocean, except that the *Equatorial Countercurrent* is much better developed in the Pacific Ocean than in the Atlantic (figure 8–9).

Embedded within the South Equatorial Current is a thin ribbonlike undercurrent, the *Cromwell Current.* This current flows east from the Samoa Islands in the western Pacific to the Galapagos. Only 0.2 km (0.12 mi) thick and approximately 300 km (186 mi) wide, the Cromwell Current at depths of 70–200 m (440–1256 ft) flows with velocities approaching 5 km/h (3 mi/h) in an easterly direction. Its volume transport is approximately 40 sv.

INDIAN OCEAN CIRCULATION

Surface circulation in the Indian Ocean, which extends to only about 20°N latitude, varies considerably from that which we have discussed for the Atlantic and Pacific oceans. From November to March the equatorial circulation is similar to that which exists in the other oceans, with two westward-flowing equatorial currents separated by an Equatorial Countercurrent. In contrast to the wind systems in the Atlantic and Pacific oceans, which are shifted to the north of the geographical equator, we find that in the Indian Ocean the meteorological equator is shifted to the south of the geographical equator. The *Equatorial Countercurrent* flows between 2° and 8°S latitude, bounded on the north by the North Equatorial Current, which extends as far as 10°N latitude, and on the south by the South Equatorial Current, which extends to 20°S latitude. During winter the typical northeast trade winds are developed and are referred to as the *northeast monsoon.* They are reinforced by the fact that rapid cooling of the air during the winter months over the Asian mainland creates a high pressure cell that forces atmospheric masses off the continent out over the ocean, where the air pressure is lower (figure 8–10).

The Asian mainland warms up faster than the oceanic water due to the relatively lower heat capacity of continental crustal material compared to water. As a result, a low-pressure cell develops over the continent during summer, which "sucks" in the air masses overlying the ocean. This gives rise to the *southwest monsoon,* which may be thought of as a continuation of the southeast trade winds across the equator. During this season the North Equatorial Current disappears and is replaced by the *Southwest Monsoon Current,* which flows from west to east across the North Indian Ocean. In September or October, the northeast trade winds are reestablished, and the North Equatorial Current reappears (figure 8–10).

The surface circulation in the southern Indian Ocean is similar to the counterclockwise circulation observed in other southern oceans. During the time the northeast trade winds blow, the South Equatorial Current provides water for the Equatorial Countercurrent and the *Agulhas Current,* which flows south along the eastern coast of Africa and joins the West Wind Drift. Turning north out of the West Wind Drift is the *West Australian Current,* which completes the gyre by merging with the South Equatorial Current.

During the southwest monsoon, a northward flow from the equator along the coast of Africa, the *Somali Current,* develops with velocities approaching 4 km/h (2.5 mi/h). The surface water characteristics in the southern Indian Ocean are typical.

VERTICAL CIRCULATION

Lateral surface movements of water may bring about vertical circulation within the upper water mass. We will refer to this shallow vertical circulation system as *wind-induced.* Of greater oceanwide importance in producing the thorough mixing within the deep ocean is the vertical circulation resulting from density changes at the ocean surface, which cause the sinking of water masses. Since the changes that increase the density are normally changes in temperature and salinity, this circulation is referred to as *thermohaline.*

Wind-Induced Circulation

There are regions throughout the world ocean where water moves vertically to the surface or away from the surface as a result of wind-driven surface currents car-

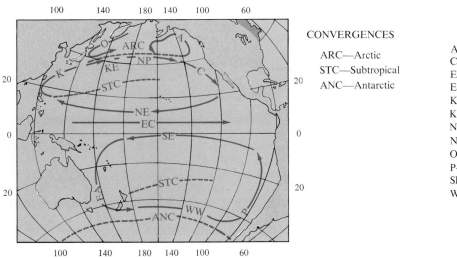

FIGURE 8–9
Pacific Ocean Surface Currents. (Base map courtesy of Open University Press.)

CONVERGENCES

ARC—Arctic
STC—Subtropical
ANC—Antarctic

CURRENTS

A—Alaskan
C—California
EA—East Australian
EC—Equatorial Counter
K—Kuroshio
KE—Kuroshio Extension
NE—North Equatorial
NP—North Pacific
O—Oyashio
P—Peru
SE—South Equatorial
WW—West Wind Drift

FIGURE 8–10

Indian Ocean Surface Currents.

(Base map courtesy of National Ocean Survey.)

A. WINTER: November–March, Northeast monsoon wind season

B. SUMMER: May–September, Southwest monsoon wind season

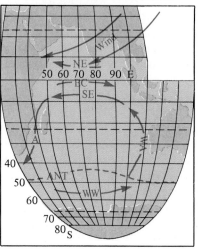

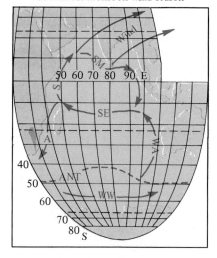

From November to March equatorial circulation in the Indian Ocean is similar to that of other oceans, except that the Equatorial Countercurrent is shifted south.

Low pressure over the mainland during summer draws the Southwest Monsoon winds over the North Indian Ocean. This wind produces the Southwest Monsoon Current which flows in an easterly direction and replaces the North Equatorial Current.

CURRENTS	A–Agulhas	EC–Equatorial Counter	NE–North Equatorial	S–Somali

SE–South Equatorial   SM–Southwest Monsoon   WA–West Australian   WW–West Wind Drift

ANT–Antarctic Convergence

rying water away from or toward these regions. *Upwelling* occurs in areas where the surface flow of water is away from the area. If volume is to be conserved and surface flows cannot bring sufficient replacement water into the area, water must come from beneath the surface to replace that which has been displaced. This condition may occur in the open ocean or along the margins of continents.

As the North Equatorial and South Equatorial currents flow in a westerly direction on either side of the

equator, the water on the north side of the equator tends to veer to the right, moving to a higher latitude, while that in the Southern Hemisphere veers to the left to a higher latitude. The net effect is a water deficiency at the surface between the two currents. Water from deeper within the upper water mass comes to the surface to fill the void (figure 8–11). This phenomenon is called *equatorial upwelling.*

Coastal upwelling is common along the margins of continents where the wind conditions are such that the

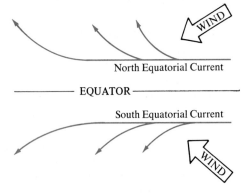

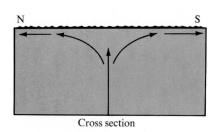

FIGURE 8–11

Equatorial Upwelling. The Coriolis effect acting on trade-wind-driven westward-flowing equatorial currents pulls surface water away from the equatorial region. This water is replaced by subsurface water.

surface waters adjacent to the continents are carried out to the open ocean as a result of Ekman transport. The replacement of this water comes from the lower portions of the upper water. As is the case with the Peru Current, such areas are characterized by low surface temperatures and high concentrations of nutrients that make them highly productive areas (figure 8–12). The

reversal of the direction of coastal winds that are causing upwelling tends to push water toward shore, where it piles up and causes *coastal downwelling*, a sinking of surface water.

Near the center of the rotating gyres of the major ocean current systems, the winds are relatively weak, and the water rotates at a very low rate. The winds that do blow in these regions may be steady in direction and are known to set up convection cells in the upper water mass. This phenomenon was first observed by Irving Langmuir while crossing the Sargasso Sea in 1938. He observed straight rows of seaweed parallel to the direction of the wind and concluded that the plants were trapped in zones of convergence between cells. Not only macroscopic plant material is concentrated in these regions but also microscopic plants and dissolved organic material. By contrast, in the regions of divergence, where water is surfacing as a result of convection, the concentrations of organic material are relatively low (figure 8–13). This phenomenon is called *Langmuir circulation*.

Thermohaline Circulation

The major vertical circulation in the ocean results primarily from surface density changes in oceanic water. Vertical mixing of ocean water is achieved primarily through the sinking and rising of water masses in high latitudes. It is only in the high latitude areas that the absence of the pycnocline allows vertical movements to the extent that surface masses may sink to the ocean bottom and deep water masses may rise to the surface. The relatively strong density stratification throughout the lower-latitude regions (the pycnocline) serves as a very effective barrier to such movements.

Since the intensity of solar radiation is greater in the equatorial region, we might expect that heating of surface water here may cause it to expand and move away from the equatorial region toward the poles, spreading over the colder, more dense high-latitude waters (figure 8–14). We may further expect that at some depth a return flow from the high latitudes toward the equator would be set in motion to replace the surface water moving away from the equatorial region. This does not seem to occur in the oceans, although it is theoretically proper to assume that such a pattern might develop. The energy imparted to the surface waters by winds is greatly in excess of the energy that is developed as a result of density changes due to temperature differences. Thus, the effect of the wind overcomes any tendency that may exist for development of such a pattern of circulation.

Another variable that affects the density of surface water is salinity. It appears that salinity has a very minimal

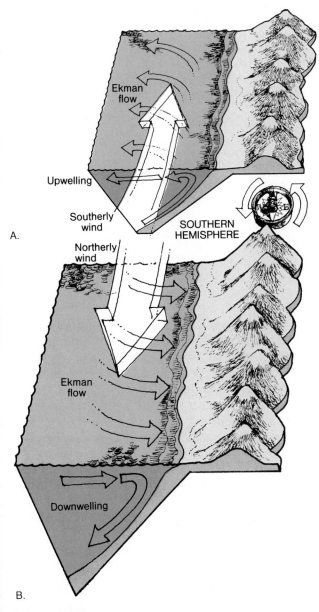

FIGURE 8–12
Coastal Upwelling and Downwelling. *A*, in areas where wind-driven coastal currents flow along the western margin of continents and toward the equator, the Ekman transport carries surface water away from the continent. An upwelling of deeper water replaces the surface water that has moved away from the coast. *B*, a reversal of the direction of the winds that cause upwelling will cause water to pile up against the shore and force downwelling.

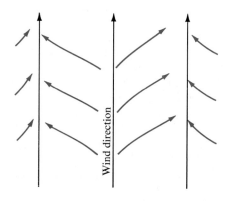 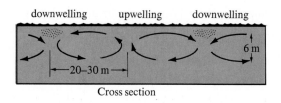

downwelling   upwelling   downwelling

←20–30 m→   6 m

Cross section

FIGURE 8–13

Langmuir Circulation. Steady winds blowing across the ocean surface create convection cells with alternate right- and left-hand circulation. The axes of these cells run parallel to wind direction. Organic debris accumulates in downwelling zones of convergence and produces windrows that run parallel to the wind direction for great distances.

effect on the movement of water masses in the lower latitudes, and density changes resulting from salinity changes are of importance only in the very high latitudes, where water temperature remains constantly relatively low. For example, the highest salinity water in the open ocean is found in the subtropical regions, but there is little sinking of water in these areas due to the fact that water temperatures are high enough to maintain a low density for the surface water mass and prevent it from sinking. In such areas, a strong halocline, or salinity gradient, may develop with a relatively thin surface layer of water having salinities in excess of 37‰. Salinity decreases rapidly with increasing depth to typical ocean water salinities below 35‰ (figure 8–15).

An important sinking of cold surface waters that become deep-water masses occurs in the subpolar regions of the Atlantic Ocean. In the North Atlantic, the major sinking of surface water is thought to occur in the Norwegian Sea. From there it flows as a subsurface current into the North Atlantic, becoming the *North Atlantic Deep Water.* Additional surface water may sink at the margins of the Irminger and the Labrador seas. In the southern subpolar latitudes, the most significant area of deep-water-mass formation is the Weddell Sea, where rapid winter freezing produces high-density water that sinks down the continental slope of Antarctica and becomes *Antarctic Bottom Water,* the densest water in the open ocean (figure 8–16).

On a broad scale, there are latitudinal regions where surface water masses converge and may cause sinking (figure 8–16). These regions are in the Arctic and Antarctic. A major water mass formed from sinking at the Antarctic convergence is the *Antarctic Intermediate Water.*

Figure 8–17 shows what is thought to be the general pattern of deep-water circulation in the world ocean. The most intense flow occurs as western boundary currents because of the rotation of the earth. In figure 8–18, the temperature, salinity, and dissolved oxygen characteristics of the Atlantic water masses are shown.

Sandwiched between the Antarctic Intermediate and the Antarctic Bottom water masses at southern latitudes is the *North Atlantic Deep Water,* which moves southward with its core at about 2200 m (7200 ft) depth. It can be identified on the basis of its temperature and salinity, which are slightly above those of the Intermediate Water and the Antarctic Bottom Water. This deep water of North Atlantic origin has been away from the surface for hundreds of years and is sandwiched between two water masses that have more recently left the surface, so it is possible to identify it on the basis of a dissolved oxygen minimum. Although low in dissolved oxygen content, it still has a very positive effect on the biological productivity of the Antarctic Zone due to its high concentration of nutrients that accumulated during the hundreds of years it was beneath the photosynthetic zone. The residence time of deep water in the Atlantic Ocean is approximately 275 years.

The deep water of the Pacific and Indian oceans has widespread average properties because it is a mixture of water masses from the North Atlantic and South Atlantic oceans that has been modified somewhat while being transported to the Indian and Pacific oceans by the Circumpolar Current. This deep water, much of which originated as North Atlantic Deep Water or Antarctic Bottom Water, has an average temperature of 1.5°C (34.7°F) and an average salinity of 34.7‰. Because of the uniform character of this large mass of water, it is called the *Oceanic Common Water.*

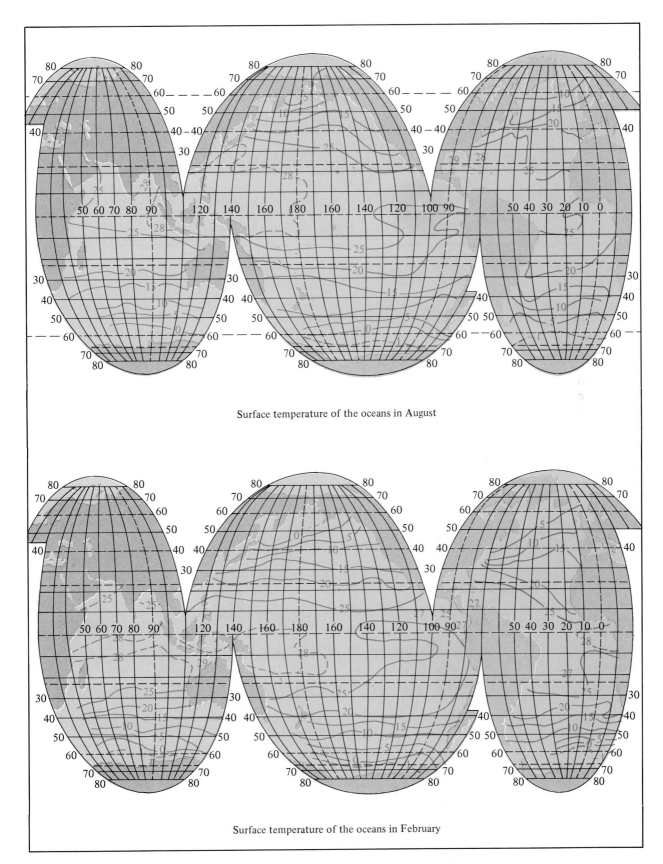

Surface temperature of the oceans in August

Surface temperature of the oceans in February

FIGURE 8–14
Surface Temperature of the World Ocean (°C). (After Sverdrup et al., 1942. Base map courtesy of National Ocean Survey.)

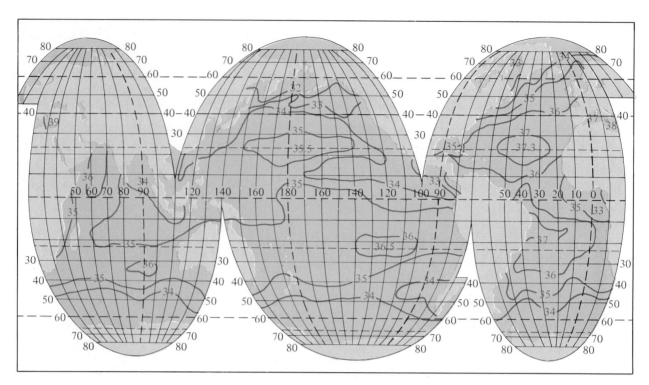

FIGURE 8–15
Surface Salinity of the Oceans in August (‰). (After Sverdrup et al.,1942. Base map courtesy of National Ocean Survey.)

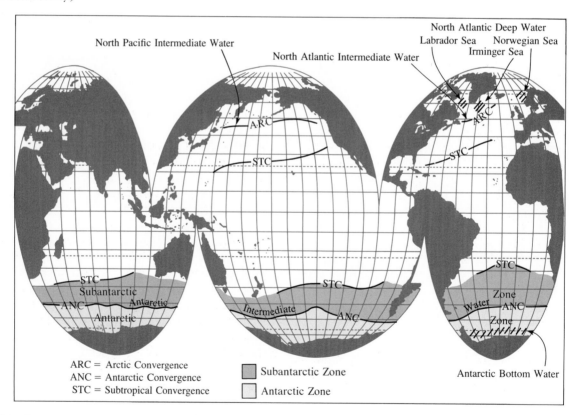

FIGURE 8–16
Regions where Intermediate and Deep Water Masses of the World Ocean Sink. (Base map courtesy of National Ocean Survey.)

FIGURE 8–19

Ocean Acoustic Tomography. *A,* during the 1981 demonstration of ocean acoustic tomography, transmitters (T1 to T4) emitted low-frequency sound signals that were received by the receivers (R1 to R5). Temperature and currents were measured by conventional means at moorings E1 and E2. *B,* results of the 1981 ocean acoustic tomography demonstration. Source of data: 1 and 3, ship surveys; 2, tomography. During the demonstration, data from a depth of 700 m showed a large cold eddy in the center of the survey area at the start of the survey. It was separated from warm water to the northeast by a well-defined temperature front. The cold water eddies observed moved off to the northwest, while the front progressed to the southwest during the demonstration. Cold eddies were indicated by the slowing of sound signals that passed through them. (Courtesy of Woods Hole Oceanographic Institution.)

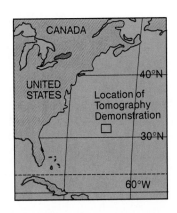

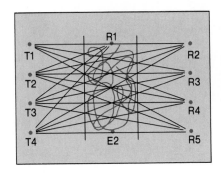

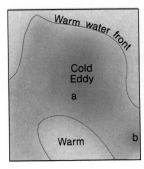

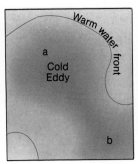

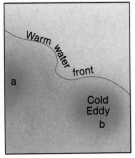

A.

B.

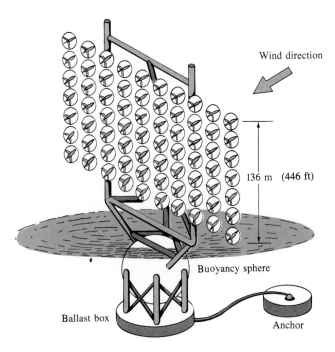

FIGURE 8–20

Offshore Windpower System. (Courtesy of Woods Hole Oceanographic Institution.)

sider this possibility and concluded that some 2000 MW of electricity could be recovered along the east coast of southern Florida. Devices proposed for extraction ranged from underwater "windmills" to a Water Low-Velocity Energy Converter (WLVEC) operated by parachutes attached to a continuous belt. Again, calculations indicate such systems can be economically competitive. The Coriolis Program was proposed in 1973 by William J. Moulton of Tulane University. Hydroturbines with diameters of 170 m (560 ft) were envisioned that could generate 43 MW of electricity from the movement of ocean currents (figure 8–21). An array of 242 units covering an area 30 km (18.6 mi) wide and 60 km (37 mi) long could generate 10,000 MW—the equivalent of 130 million barrels of oil. Tests are now being conducted to establish an optimum size, identify suitable sites, and confirm engineering, economic, and environmental estimates. What remains to be seen is if the designs for ocean wind and current generation of electricity actually function as expected in the often hostile marine environment.

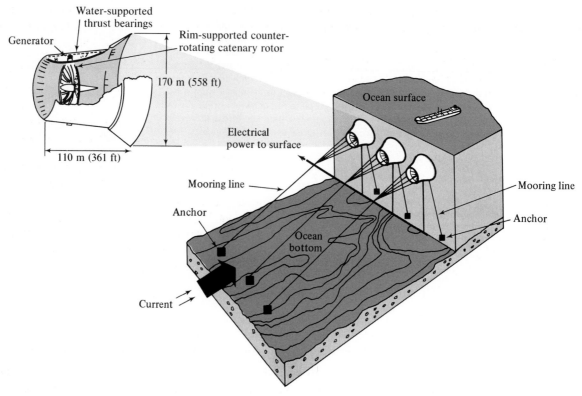

FIGURE 8–21

Coriolis Program. Buoyant Coriolis hydroturbines anchored in ocean current. It is estimated that an array of 242 such units placed in the Gulf Stream could provide 10 percent of the present electricity needs of Florida. (Courtesy of U.S. Department of Energy.)

SUMMARY

Horizontal currents set in motion by wind systems are characteristic of the surface waters of the world ocean. According to the **Ekman spiral** concept, winds set the surface waters in motion in a direction 45° to the right of the wind in the Northern Hemisphere and 45° to the left of the wind in the Southern Hemisphere. The net water movement, called **Ekman transport**, is at right angles to the wind direction. As a result of the Ekman spiral phenomenon, water is pushed in toward the center of clockwise gyres in the Northern Hemisphere, forming "hills." As water in the Northern Hemisphere runs down-slope on the resulting hills of water, the Coriolis effect causes it to turn right and into the clockwise flow pattern. As a result, a **geostrophic current** flowing parallel to the contours of the hill is maintained. The apex of the hill is always located to the west of the geographical center of the gyre due to the earth's rotation from west to east. Using **dynamic topography**, oceanographers can map the surface currents in ocean basins by determining the average density of water columns at lo-

cations throughout the basin and mapping this distribution. The speed of the current increases with increased slope of the hills, which are steepest on the west side.

Vertical circulation produced by wind is confined to the surface water mass. An example is **equatorial** or **coastal upwelling**, which is created by the Ekman transport pulling wind-driven water masses away from the equator or coastal areas. Water surfaces from a depth of no more than a few hundred meters to replace the surface water. **Coastal downwelling** can be produced by winds blowing in the opposite direction of those that cause coastal upwelling. An even shallower vertical movement of water is represented by the **Langmuir circulation** of long, shallow convection cells running parallel to the gentle, steady winds that create them. These cells may extend no deeper than 6 m (20 ft). The mixing of the water in the deepest ocean basins is caused by **thermohaline circulation**, which is initiated by the sinking of dense surface water in high latitudes. The **North Atlantic Deep Water** and the **Antarctic Bottom Water** carry most of the oxygen-bearing water to the deep basins of all the oceans to maintain their life-sustaining capacity.

The **Circumpolar Current** flows in a clockwise direction around the continent of Antarctica. It is the largest current in the world ocean, transporting over 190 sv of water. The surface portion of this flow is often referred to as the **West Wind Drift** because it is driven by strong westerly winds. Two major deep-water masses form in Antarctic waters. The Antarctic Bottom Water, the densest water mass in the oceans, forms primarily in the **Weddell Sea** and sinks along the continental shelf into the Atlantic Ocean. Farther north, at the **Antarctic Convergence**, the low-salinity **Antarctic Intermediate Water** sinks to a depth of about 900 m (3000 ft). Sandwiched between these two masses is the North Atlantic Deep Water, rich in plant nutrients after hundreds of years in the deep ocean. The highest-velocity current in the oceans is the **Gulf Stream** as it flows along the southeast Atlantic coast. The best development of an equatorial countercurrent system is found in the Pacific, where large eastern movement of water is achieved by the **Pacific Countercurrent** and the **Cromwell Current**, a ribbonlike subsurface flow. The Indian Ocean circulation is dominated by the **monsoon** wind systems that blow out of the northeast in the winter and the southwest in the summer.

Research activities such as MODE, GEOSECS, TTO, and Ocean Acoustic Tomography rapidly increase our understanding of ocean circulation and mixing processes, while continued investigation of the energy potential of ocean winds and currents may lead to development of a significant addition of renewable energy that can be exploited by society.

QUESTIONS AND EXERCISES

1. Compare the forces directly responsible for creating horizontal and deep vertical circulation in the oceans. What is the ultimate source of energy for driving both circulation systems?

2. On a base map of the world, plot the major surface circulation gyres of the oceans, the meteorological equators of each ocean, and the Subtropical, Arctic, and Antarctic Convergences. Superimpose the major wind belts of the world on the gyres. Label the currents using the symbols used in figures 8–7, 8–9, and 8–10A.

3. Diagram and discuss how the Ekman transport produces the "hill" of water within major ocean gyres that causes geostrophic current flow. As a starting place on the diagram, use the prevailing wind belts, the trade winds, and the westerlies.

4. What causes the apex of these geostrophic "hills" to be offset to the west of the center of the ocean gyre systems?

5. Explain how oceanographers determine the dynamic topography of the ocean waters.

6. Describe the relationships among the wind, the surface current it creates, and the development of equatorial and coastal upwellings and coastal downwelling.

7. Discuss why thermohaline vertical circulation is driven by sinking of surface water that occurs only in high latitudes.

8. Name the two major deep-water masses and give the locations of their formation at the ocean's surface.

9. The largest current in the world ocean in terms of volume transport is the Antarctic Circumpolar Current. Explain why its surface portion is referred to as the West Wind Drift. What is its average volume transport as compared to the maximum volume transport measured for the Gulf Stream?

10. Why does the North Atlantic Deep Water enhance biological productivity in the Antarctic Zone as it surfaces between the Antarctic Intermediate Water and the Antarctic Bottom Water?

11. Explain why the Gulf Stream eddies that develop northeast of the Gulf Stream rotate clockwise and have warm water cores, while those that develop to the southwest rotate counterclockwise and have a core of cold water.

12. The Antarctic Intermediate Water is identifiable throughout much of the South Atlantic based on a temperature minimum, salinity minimum, and dissolved oxygen maximum. Why should it be colder, less salty, and contain more oxygen than the surface water mass above it and the North Atlantic Deep Water below it?

13. What evidence is there that the bottom water in the Pacific Ocean has been away from the surface much longer than that in the Atlantic Ocean?

14. Discuss the changes in current flow in the North Indian Ocean and their relationship to the monsoon winds.

15. Where is the potential for use of ocean windpower systems and ocean current power systems greatest, and why?

REFERENCES

Defant, A. 1961. *Physical oceanography,* 2 vols. New York: Macmillan.

Gill, A. 1982. *Atmosphere-ocean dynamics.* New York: Academic Press.

Hammond, A. L. 1971. Oceanography: geochemical tracers offer new insight. *Science* 195:164–66.

MacLeish, W., ed. 1974. Energy and the sea. *Oceanus* 17:52.

Montgomery, R. 1940. The present evidence of the importance of lateral mixing processes in the ocean. *American Meteorological Society Bulletin* 21:87–94.

Munk, W. 1955. The circulation of the oceans. *Scientific American* 191:96–108.

Pickard, G. L. 1975. *Descriptive physical oceanography,* 2nd ed. New York: Pergamon Press.

Stommel, H. 1987. *A view of the sea.* Princeton, N.J.: Princeton University Press.

Stommel, H. 1964. The Gulf Stream, 2nd ed. Berkeley, Calif.: University of California Press.

Stuiver, M., Quay, P. D., and Ostlund, H. G. 1983. Abyssal water carbon-14 distribution and the age of the world oceans. *Science* 219:4586, 849–51.

Sverdrup, H. U., Johnson, M. W., and Fleming, R. H. 1970. *The oceans.* Englewood Cliffs, N.J.: Prentice Hall.

SUGGESTED READING

Sea Frontiers

Frye, J. 1982. The ring story. 28:5, 258–67.
An interesting discussion of the Gulf Stream eddies that break away from the main current.

Gruber, M. 1971. The great ocean sweepstakes. 17:3, 146–59.
The problems related to oil pollution of the ocean are discussed. Devices and methods designed to aid in confining and cleaning up oil spills are also covered.

Mason, P. 1975. The changeable ocean river. 21:3, 171–77. A description of the Gulf Stream Current.

Miller, J. 1975. Barbados and the island-mass effect. 21:5, 268–72.
A discussion of the phenomenon by which waters around tropical islands are much more productive than the surface waters of the open ocean.

Smith, F. G. W. 1972. Measuring ocean movements. 18:3, 166–74.
Discussed are some practical problems related to current flow and some methods used to determine current direction, speed, and volume.

Smith, F. G. W., and Charlier, R. H. 1981. Turbines in the ocean. 27:5, 300–305.

A discussion of the use of such currents as the Gulf Stream as a source of energy.

Scientific American

Baker, D. J., Jr. 1970. Models of ocean circulation. 222:1, 114–21.
Discusses observations of a model depicting a segment of the surface of the earth over which fluids move. Helps explain geostrophic flow within ocean gyres. Some knowledge of basic physics is necessary for full comprehension of material presented.

Hollister, Charles D., and Nowell, Arthur. 1984. The dynamic abyss. 250:3, 42–53.
Submarine "storms" are associated with deep, cold currents flowing away from polar regions toward the equator.

Stewart, R. W. 1969. The atmosphere and the ocean. 221:3, 76–105.
The exchange of energy between the atmosphere and ocean and the resulting phenomena, currents, and waves are covered in this readable, comprehensive article.

9
WAVES

Ocean waves are one of the most obvious phenomena of the ocean, yet it was not until well into the 19th century that some understanding of what caused waves and how they behaved developed. We will first take a look at the character of waves in general before proceeding to discuss oceanic waves. Wave phenomena involve the transmission of energy and momentum by means of oscillatory motions through the various states of matter. Theoretically, the medium itself does not move as the energy passes through. The particles that make up the medium simply move in a back-and-forth or orbital pattern, transmitting energy from one to another.

Simple progressive waves shown in figure 9–1A may be described as *longitudinal* and *transverse*. In longitudinal waves such as sound waves, the particles that are in vibratory motion move back and forth in a direction parallel to the direction of energy travel. Energy may be transmitted through all states of matter—gaseous, liquid, or solid—by longitudinal movement of particles.

Transverse wave phenomena involve the transmission of energy at right angles to the direction of particle vibration. An example of such a wave would be that created when one end of a rope is tied to a doorknob while the other end is moved up and down with the hand. A waveform is set up and progresses along the rope, and energy is transmitted from the motion of the hand to the doorknob. If one was to pick out a particular segment of the rope and watch it, the segment would be seen to move up and down at right angles to a line drawn from the doorknob to the hand that is putting energy into the rope. We generally consider that this type of wave can transmit energy only through solids, since it is only in a solid that particles are strongly attached to one another.

Waves on the ocean surface, as all waves that transmit energy along an interface between two fluids of different density, have particle movements that are neither longitudinal nor transverse. We may say more correctly that the movement of particles along such an interface involves components of both, since the parti-

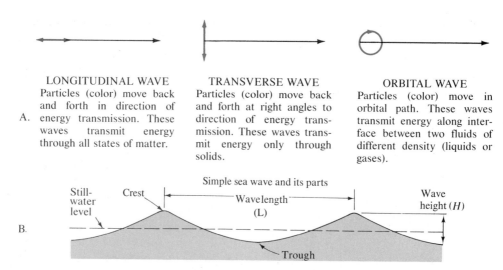

LONGITUDINAL WAVE
Particles (color) move back and forth in direction of energy transmission. These waves transmit energy through all states of matter.

TRANSVERSE WAVE
Particles (color) move back and forth at right angles to direction of energy transmission. These waves transmit energy only through solids.

ORBITAL WAVE
Particles (color) move in orbital path. These waves transmit energy along interface between two fluids of different density (liquids or gases).

A.

B.

FIGURE 9–1
Types of Progressive Waves.

cles move in circular orbits at the interface between the atmosphere and ocean.

WAVE CHARACTERISTICS

When we observe the ocean surface, we see that there are waves of various sizes moving in various directions, resulting in a complex wave pattern that is constantly changing. We will, for the sake of introducing some characteristics that must be used to discuss waves, look at a much simpler form of an idealized wave representing the transmission of energy created by a single source and traveling along the ocean-atmosphere interface. Such a series of waves will have very uniform characteristics. We use such an idealized form in figure 9–1B.

As an idealized *progressive wave* (a phenomenon in which the waveform can be observed to move) passes a

permanent marker, such as a pier piling, we will notice a succession of high parts of the waves, *crests,* separated by low parts, *troughs.* If we were to mark the water level on the piling when the troughs pass and then do the same for the crests, the vertical distance between the marks would be the *wave height, H.* The horizontal distance between corresponding points on successive waveforms, such as from crest to crest, is the *wavelength, L.* The time that elapses during the passing of one wavelength is the *period, T.* Since the period is the time required for the passing of one wavelength (figure 9–2),

$$\text{Speed} = \frac{L}{T}$$

For example,

$$\text{Speed} = \frac{L}{T} = \frac{156 \text{ m}}{10 \text{ s}} = 15.6 \text{ m/s}$$

FIGURE 9–2

Speed—Deep-Water Waves. The theoretical relationships between wave speed, wavelength, and period for deep-water waves. Speed is equal to wavelength divided by period. Because of these relationships, if one can determine one of the characteristics of a wave, the other two can be calculated. The easiest to determine directly is the period.

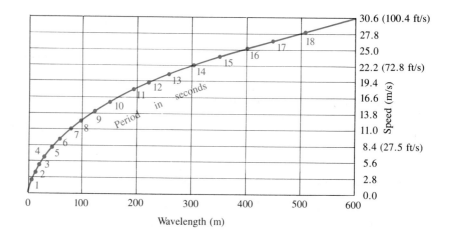

Another characteristic related to wavelength and speed is *frequency, f.* Frequency is the number of wavelengths that pass a fixed point per unit of time. If six wavelengths pass a point in 1 min, and we have the same wave system as in our previous example, then

$$\text{Speed} = Lf = 156 \text{ m} \times \frac{6}{\text{min}} = 936 \text{ m/min}$$

$$\frac{936 \text{ m}}{1 \text{ min}} \times \frac{1 \text{ min}}{60 \text{ s}} = 15.6 \text{ m/s}$$

Since the speed and wavelength of ocean waves are such that less than one wavelength passes a point per second, the preferred unit of time for scientific measurements, period (rather than frequency) is the more practical measurement to use when calculating speed. Wave height is not related to period, wavelength, or speed, so it must always be determined by measurement. The ratio H/L is the *wave steepness.*

Returning to the particle motion associated with ocean waves, we see that the circular orbits followed by the water particles at the surface have a diameter equal to the wave height. While a particle is in the crest of a passing wave, it is moving in the direction the waveform is traveling. While it is in the trough, it is moving in the opposite direction. The half of the orbit that is accomplished in the trough is at a lower speed than in the crest half of the orbit. Therefore, there is a small net transport of water in the direction the waveform is moving. This condition results from the fact that particle speed decreases with increasing depth below the still water line. Also, the diameters of particle orbits decrease with increased depth until particle motion associated with our idealized wave ceases at a depth of one-half wavelength, $L/2$ (figure 9–3A).

Deep-Water Waves

The ocean waves with the characteristics just discussed belong to a category called *deep-water waves.* Such waves travel across the ocean where the water depth (d) is greater than one-half the wavelength (figure 9–3A). Included in deep-water waves are all wind-generated waves as they move across the open ocean. As can be seen by the speed equations previously discussed, speed of deep-water waves is related to wavelength (L) and

A. DEEP-WATER WAVE

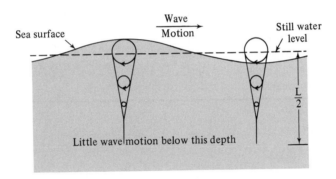

B. SHALLOW-WATER WAVE

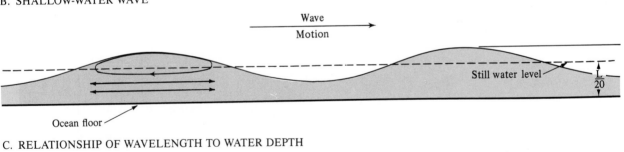

C. RELATIONSHIP OF WAVELENGTH TO WATER DEPTH

Shallow-water wave (d < L/20)	Transitional wave (d < L/2 but > L/20)	Deep-water wave (d > L/2)	> = greater than < = less than

FIGURE 9–3
Deep-Water and Shallow-Water Waves. *A,* wave profile and water-particle motions of a deep-water wave. Note the diminishing size of the orbits with increasing depth below the surface. Wave motion dies out at a depth of $L/2$. *B,* motions of water particles in shallow-water waves. Water motion extends to ocean floor because depth is less than $L/20$.

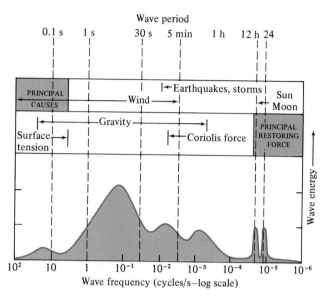

FIGURE 9–4

Wave Classification by Wave Period. This figure shows that most of the energy possessed by ocean waves is in wind-generated waves with a period of about 10 seconds. The long-period peak above the frequency, 10^{-3}, represents tsunamis, while the two sharp peaks to the right represent the tides with semidaily and daily periods. (After Blair Kinsman, *Wind waves: their generation and propagation on the ocean surface.* © 1965. Reprinted by permission of Prentice Hall, Englewood Cliffs, N.J.)

period (T). The easier of these characteristics to measure is period. Thus the following equation is most commonly used for computing wave speed.

$$\text{Speed (m/s)} = 1.56T$$

Shallow-Water Waves

Waves in which $d < L/20$ are classified as *shallow-water waves* (long waves). Included in this category are wind-generated waves that have moved into shallow nearshore areas, *tsunami* (seismic sea waves) generated by

disturbances in the ocean floor, and *tide waves* generated by gravitational attraction of the sun and moon (figure 9–3B). In these waves, the wavelength is very great relative to water depth, and speed is determined by water depth, d.

$$\text{Speed (m/s)} = 3.1 \sqrt{d\text{(m)}}$$

Particle motion in shallow-water waves is in a very flat elliptical orbit approaching horizontal oscillation, and the vertical component of particle motion decreases with increasing depth. The presence of shallow-water waves can be detected to the ocean bottom.

Transitional Waves

Transitional waves have wavelengths greater than twice, but less than 20 times, the water depth. The speed of transitional waves is determined partially by wavelength and partially by water depth. For deep-water waves generated by winds at the ocean surface, the transition to a shallow-water wave usually begins when their periods reach 10–12 seconds.

Since it is easier to measure the period of a wave than any of its other characteristics, Kinsman has used the system presented in figure 9–4 as a means of classifying waves. The illustration also shows the principal causes of waves of different periods.

WIND-GENERATED WAVES

Sea

As the wind blows over the ocean surface, the pressure and stress deform the ocean surface into small rounded waves with wavelengths less than 1.74 cm (0.7 in.). These waves are called *capillary waves* (ripples) because the dominant force, *restoring force,* that works to destroy them and smooth the ocean surface is *surface tension.* Capillary waves characteristically have rounded crests and V-shaped troughs (figure 9–5).

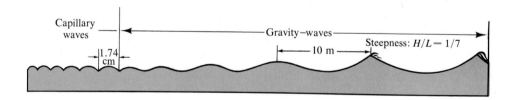

FIGURE 9–5

Capillary and Gravity Waves. As energy is put into the ocean surface by wind, small rounded waves with V-shaped troughs develop (capillary waves). As the water gains energy, the waves increase in height and length. When they exceed 1.74 cm (0.7 in.) in length, they take on the shape of the sine curve and become gravity waves. Increased energy increases the steepness of the waves. The crests become pointed and the troughs rounded. As the steepness reaches 1/7, the waves become unstable, and whitecaps form as they "break." (Not drawn to scale.)

special features of wave motion—*interference patterns*. An interference pattern produced by the superposition of two or more wave systems in a medium will be the algebraic sum of the disturbance each wave would have produced individually. The result may be a larger or smaller trough or crest, depending on whether the individual disturbances meet in or out of phase (figure 9–7).

For instance, if swells are moving away from two storm areas and come together as shown in figure 9–7, the interference pattern may be one of constructive interference, destructive interference, or, more likely, mixed interference. *Constructive interference* results if the wave trains with the same wavelength come together in phase, crest to crest and trough to trough. Adding the disturbances that would result from the wave individually, we see that the interference pattern produced will be a wave with the same length as the two wave systems that are converging but with a wave height that is equal to the sum of the wave heights of the original wave systems.

Destructive interference results if the crest that was produced by the waves from one generating area coincides with the troughs produced by the waves from the second generating area. If the waves have identical characteristics, the algebraic sum of the crest plus the trough is zero, and the ocean develops a calm surface as the wave systems cancel each other out.

It is more likely that the two systems possess waves of different heights and lengths and come together with both destructive and constructive interference. Thus, a more complex *mixed interference* pattern develops. It is such interference that explains the occurrence of a sequence of high waves followed by a sequence of lower waves and other irregular wave distribution patterns observed as the swell approaches the margins of the continents.

Free and Forced Waves   In the swell we see what may be referred to as a *free wave,* which is moving with the momentum and energy imparted to it in the sea area but is not experiencing a maintaining force that keeps it in motion. In the wave-generating area, free and forced waves are present. A *forced wave* is one that is maintained by a force that has a periodicity coinciding with the period of the wave. This force is the wind, and due to the variability of the wind, many wave systems in the sea area alternate between being forced waves and free waves.

Surf

Most waves that are generated in the sea area by the force of storm-speed winds move across the ocean as swell. They release their energy at the margins of the continents in the *surf zone,* where the swell forms breakers. As the deep-water waves making up the swell move toward the margin of the continent over gradually shoaling water, they eventually encounter water depths which are less than one-half wavelength.

These shoaling depths interfere with the particle movement at the base of the wave, and the wave slows down. As one wave is slowed down, the following waveform, which is still moving at an unaffected speed, tends to "catch up" with the wave that is being slowed, thus reducing the wavelength. Wave height increases as the water mass represented by the wave has to be confined to a wavelength that is decreasing. The crests become narrow and pointed, while the troughs become wide curves, a form that was previously described for high-energy waves in the sea. The increase in wave height accompanied by a decrease in wavelength increases the steepness (H/L) of the waves. As the wave steepness reaches $1/7$, the waves break as *surf.*

If the surf is composed of swell that has traveled from distant storms, breakers will develop relatively near shore in shallow water, the shoaling of which is primarily responsible for their formation. The surf will be characterized by parallel lines of relatively uniform breakers. However, if the surf is composed of waves that have been generated by local storms, the waves may not have been sorted out into swell. The surf may be more nearly characterized by unstable, deep-water, high-energy waves with steepness already near $1/7$. They will break shortly after feeling bottom some distance from shore, and the surf will be rough and choppy with an irregular nature.

Ideally, when the water depth is about 1.3 times the breaker height, the crest of the wave topples over in the surf. When the water depth becomes less than $1/20$ the wavelength, the waves in the surf zone begin to behave as shallow-water waves. Particle motion is greatly interfered with by the bottom, and a very significant transport of water toward the shoreline occurs (figure 9–8).

The breaking experienced in the surf results from the fact that the particle motion near the bottom of the wave has been interfered with greatly by contact with the ocean bottom, and this tends to slow down the waveform. However, the individual particles that are orbiting near the ocean surface have not been slowed down as much as the entire waveform because they have experienced no contact with the bottom. This causes the top of the waveform to lean toward the shore as the bottom of the waveform is held back by friction. The slowing down of the entire waveform compared to the increasing speed of motion in the circular orbits near the ocean surface because of increasing wave height results in the water particles at the interface between the ocean

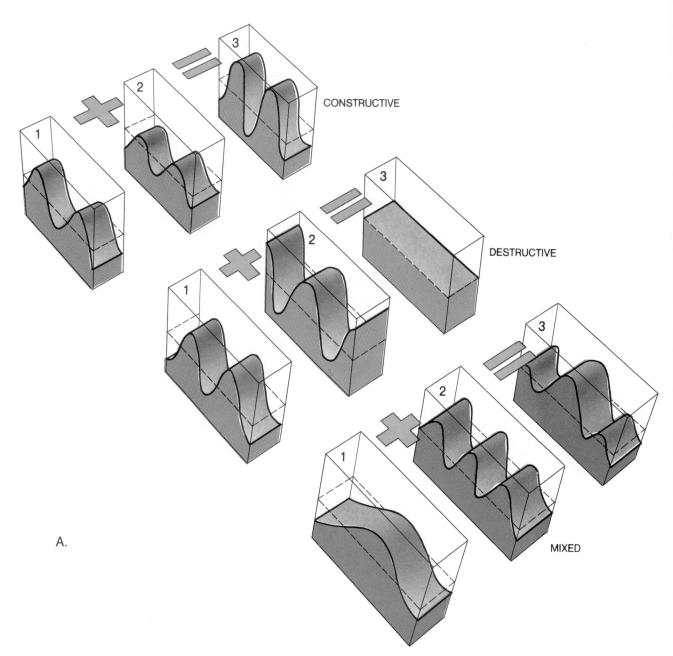

CONSTRUCTIVE

DESTRUCTIVE

MIXED

A.

FIGURE 9–7

Wave Interference. *A,* as wave trains come together from different sea areas (1 and 2), three possible interference patterns may result (3). Very rarely, if the waves have the same length and come together in phase (crest to crest and trough to trough), totally constructive interference may occur. The algebraic sum of amplitudes produces a wave of the same length but of greater height. If two sets of waves have identical characteristics and come together 180° out of phase, the result will be no wave at all (destructive). More commonly, waves of different lengths and heights encounter one another and produce a mixed interference pattern. *B,* as swell from seas A and B cross, wave interference occurs. The waves from Sea A are larger than those from Sea B, and a varied pattern of interference occurs. If the waves

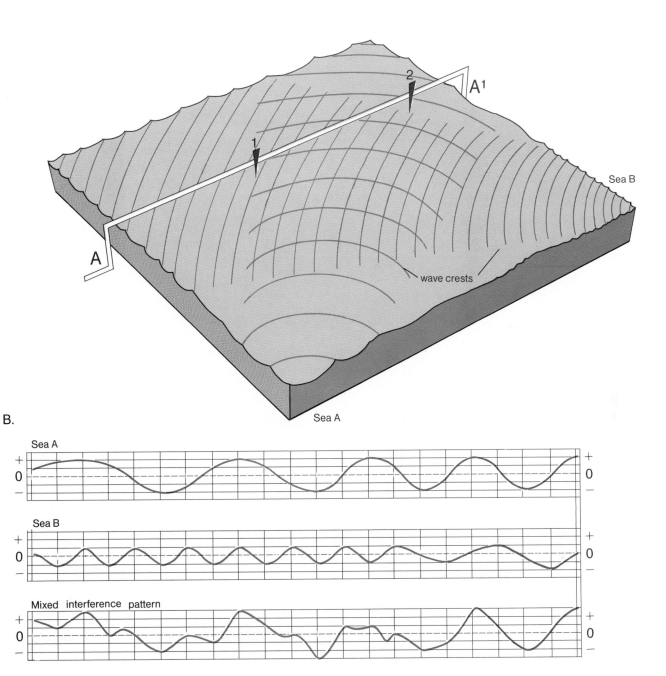

B.

Sea A

Sea B

Mixed interference pattern

C.

come together crest to crest or trough to trough, *constructive interference* occurs and produces a higher crest or deeper trough. If the waves come together crest to trough, *destructive interference* occurs. Whichever represents the smaller displacement from the still-water line is subtracted from the other, and the resulting crest or trough is reduced in size. The overall pattern of interference includes many degrees of constructive and destructive interference and is called *mixed interference. C,* the individual crests and troughs of waves from Sea A and Sea B along line *A–A'* of part *B* are shown separately and in a pattern of mixed interference. Points along the line where constructive and destructive interference occurred are indicated by the lines that run vertically through all three wave patterns.

FIGURE 9–8

Surf Zone. *A,* as waves approach the shore and encounter water depths less than one-half a wavelength (*L*/2), friction removes energy and the waves slow down, wavelength decreases, and wave height increases. When the water depth is 1.3*H,* the wave reaches a steepness of ⅐ and breaks on the shore. *B,* spilling breakers (above) and plunging breakers (below). (Photos courtesy of Charles H.V. Ebert.)

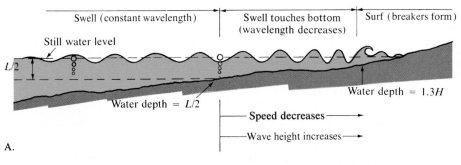

A.

B.

and the atmosphere moving faster toward shore than the waveform itself.

This motion can be seen particularly well in *plunging breakers* (figure 9–8), which have a curling crest that seems to move over an air pocket, resulting from the fact that the curling particles have outrun the wave and there is nothing beneath them to support their motion. Plunging breakers form on moderately steep beach slopes. The more commonly observed breaker is the *spilling breaker* that results from a relatively gentle slope of the ocean bottom, which more gradually extracts the energy from the wave, producing a turbulent

mass of air and water that runs down the front slope of the wave instead of a spectacular cresting curl. Because of the gradual extraction of energy associated with spilling breakers, they have a longer life span and give surfers a longer, if less exciting, ride than do the plunging breakers.

Recalling the particle motion of ocean waves in figure 9–3, we can see that in front of the crest the water particles are moving up into the crest. It is this force along with the buoyancy of the surfboard that helps maintain a surfer in front of a cresting wave. As the individual water particles move past the surfer into the crest of the wave, more water is carried from the shoreward side of the crest to maintain the hill along the front of the wave. When this upward motion of water particles is interrupted by the wave passing over water that is too shallow to allow this movement to continue, the ride is over. A skillful surfer, by positioning the board properly on the wave front, can regulate the degree to which the propelling gravitational forces exceed the buoyancy forces and can obtain high speed while moving along the face of the breaking wave.

Wave Refraction

We previously discussed that waves begin to bunch up, and wavelengths become shorter as swell begins to "feel bottom" upon approaching the shore. It is seldom that the swell will approach the shore at right angles. Some segment of the wave can be expected to feel bottom first and therefore be slowed down before the rest of the wave. This produces *refraction,* or bending, of the waves as they approach the shore. The slowing down of a portion of a wave may result from its angular approach to a straight shoreline. As shown in figure 9–9, an irregular shoreline may result from an irregular bottom topography that can slow down portions of a wave approaching the shore at right angles. Any combination of these factors, of course, is possible.

The effect of portions of a wave moving toward the shore at greater speeds than other portions of the same wave is to unevenly distribute energy along the shoreline (figure 9–9). *Orthogonal lines* are constructed perpendicular to the wave fronts and spaced in such a way that the amount of energy between lines is equal at all times. They are of great assistance in illustrating how energy is distributed along the shoreline by breaking waves. Orthogonals can be seen to converge on headlands jutting out into the ocean and diverge in bays, indicating there is a concentration of energy released against the headlands, while energy released in the bays is spread more thinly. This condition produces erosion on the headlands, while deposition may occur in the bays.

Wave Reflection

It is possible for swell to be reflected back into the ocean with little loss of energy by a vertical barrier such as a sea wall. For this ideal reflection to occur without energy loss, the wave would have to strike the barrier at a right angle. Such a condition is rare in nature. Less ideal reflections will nonetheless produce *standing waves,* which are the product of two waves of the same length moving in opposite directions.

The standing wave is a special interference pattern where no net momentum is carried because the waves are moving in opposite directions. The particles continue to move vertically and horizontally, but there is no more of the circular motion that we saw in the progressive wave. Standing waves are characterized by lines along which there is no vertical movement. There may be one or more such *nodes. Antinodes,* crests that alternately become troughs, represent the points of the greatest vertical movement within a standing wave (figure 9–10). There is no particle motion when an antinode is at its greatest vertical displacement, and the maximum particle movement occurs when the water surface is level. At this time, the maximum movement of the water is in a horizontal direction directly beneath the nodes. The movement of water particles beneath the antinodes is entirely vertical.

All bodies of water have a natural period at which standing waves develop in them. It is related to the length and depth of the basin containing the water. If this natural period is similar to the period of waves that are produced by some natural process such as tides in the area, *resonance* (constructive interference) of the two waves can cause unusually large standing waves in the basin.

For the most part, reflection of wind-generated waves from coastal barriers will be at an angle equal to the angle at which the wave approached the barrier, as shown in figure 9–11. This type of reflection may produce a small-scale interference pattern similar to those we previously discussed.

An outstanding example of reflection phenomena is the Wedge that develops west of the 400-m (1312-ft) jetty at Newport Harbor, California. As the incoming waves strike the jetty and are reflected, a constructive interference pattern develops. The crests of the incoming waves merge with the crests of the reflected waves, causing plunging wedge-shaped breakers that may reach heights in excess of 8 m (26 ft). These waves present a fierce challenge to the most experienced body surfers. Attracting some who were not up to the challenge, the Wedge has killed and crippled many who have come to try it.

Storm Surge

Large cyclonic storms that develop over the ocean will, due to their low pressure, produce a hill of water beneath them that moves with the storm across the open ocean. Additionally, as the storm approaches shallow water near shore, the portion of the hill over which the wind is blowing shoreward frequently will produce a *storm surge,* a mass of wind-driven water that produces an abnormal increase in sea level that may be extremely

FIGURE 9–9
Refraction. *A,* as the waves "feel bottom" first in the shallow areas off the headlands, they are slowed. The segments of the waves that move through the deeper water leading into the bay are not slowed until they are well into the bay. As a result, the waves are refracted (bent), so the release of wave energy is concentrated on the stacks and headlands. Erosion is active on the headlands, while deposition occurs in the bay, where the energy level is low. Orthogonal lines, spaced so that equal amounts of energy are present between any two adjacent lines, help to show the distribution of energy along the shore. *B,* wave refraction along the California coast. Note the changes in wavelength that occur in the bay on the right. (Photo from U.S. Department of Agriculture.)

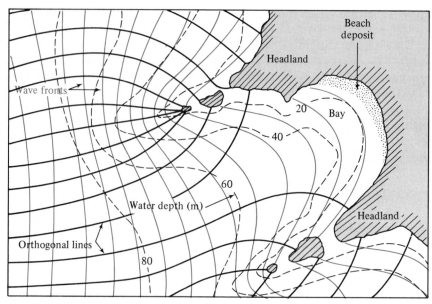

A.

B.

FIGURE 9–10

Reflection—Standing Waves. An example of water motion at quarter period intervals. Water is motionless when antinodes reach maximum displacement. Water movement is maximum when the water is level. Movement is totally vertical beneath the antinodes, and maximum horizontal movement occurs beneath the node. The circular motion of particles in progressive waves does not exist.

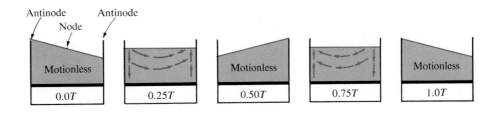

destructive to low-lying coastal areas (figure 9–12 and plates 16 and 17). Storm surges can be particularly destructive when accompanied by high tide. The coincidence of a storm surge with high tide in areas that have particularly high tides, such as the North Sea coastal region, frequently produces major catastrophies with great loss of life and property damage.

One of the most outstanding examples of storm surge is that which causes the windows of a lighthouse atop a 90 m (295 ft) cliff at Dunnet Head, Scotland, to be broken by stones tossed by waves breaking on the cliff. The lighthouse marks the western entrance to Pentland Firth, which connects the Atlantic Ocean with the North Sea between the Orkney Islands to the north and Scotland to the south.

TSUNAMI

The large waves referred to by the Japanese word *tsunami* (seismic sea waves) originate as a result of disturbances within the earth's crust. We commonly hear them referred to as *tidal waves,* which implies that they are related to the tides. They have no relationship to the tides, however. Tsunami are usually caused by fault movement, a displacement in the earth's crust along a fracture, that causes a sudden change in water level at

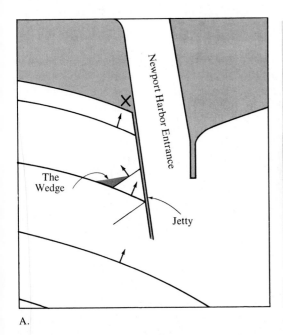

A.

B.

FIGURE 9–11

Reflection—The Wedge. *A,* the Wedge, a wedge-shaped crest that may reach heights in excess of 8 m (26 ft) develops as a result of interference between incoming waves and reflected waves near the jetty protecting the entrance to Newport Harbor, California. *B,* a view of a wedge crest taken from the landward end of the jetty (marked by the "x" on the diagram). The three dots in the water in front of the wave are the heads of body surfers waiting to catch the wave.

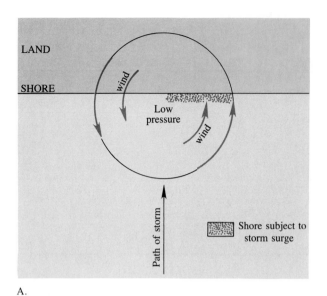

A.

B.

FIGURE 9–12

Storm Surge. *A,* as a cyclonic storm in Northern Hemisphere moves ashore, the low-pressure cell around which the storm winds blow and the onshore winds associated with the storm will produce a high-water phenomenon called a *storm surge. B,* low pressure and high onshore winds of Hurricane Camille produced a storm surge along Mississippi coast in 1969. (Photo Courtesy of NOAA.)

the ocean surface. Secondary events, such as underwater avalanches, produced by the faulting may also produce tsunami (figure 9–13). The ocean that is most plagued by the tsunami is the Pacific. It is ringed by a series of trenches that represent the unstable margins of crustal plates along which fault activity is common.

One of the most destructive tsunami ever generated was associated with the greatest release of energy from the earth's interior observed during historical times. On August 27, 1883, the island of Krakatoa in the Sundra Strait, between Sumatra and Java, exploded and essentially disappeared. The sound of the explosion was heard 4800 km (2980 mi) away at Rodriguez Island in the western Indian Ocean. The dust that ascended into the atmosphere circled the earth and produced unusual and beautiful sunsets for nearly 1 year. A tsunami that rose to heights of more than 30 m (98.4 ft) devastated the coastal region of the Sundra Strait, taking more than 36,000 lives. The energy carried by this wave was still detectable after crossing the Indian Ocean, passing the southern tip of Africa, and moving north in the Atlantic Ocean into the English Channel.

Since the wavelength of a typical tsunami is in excess of 200 km (124 mi), it is obviously a shallow-water wave, the speed of which is determined by water depth. Moving at speeds well in excess of 700 km/h (435 m/h) and with wave heights of approximately 0.5 m (1.6 ft) in the

open ocean, tsunami are not readily observable until they reach shore, are slowed down, and the water begins to pile up to form crests that may exceed heights of 30 m (98.4 ft).

Unlike the hurricane that represents a great hazard to ships at sea and may send them seeking the protection of a coastal harbor, a tsunami sends them from their coastal moorings into the open ocean. Until 1948, there was no possible warning of the coming of a tsunami that would be adequate to allow shippers to take the appropriate precautions to avoid the destructive waves. Since that time and as a result of a destructive wave that struck Hawaii in 1946, a tsunami warning system has been established throughout the Pacific Ocean. With every seismic disturbance that occurs beneath the ocean surface with the potential of producing a tsunami, observations are made at the closest tide-measuring stations to see if there is any indication that such a wave has been created. Should one be detected, warnings are sent to all the coastal regions that might conceivably encounter a destructive wave, along with the time at which the wave is to be expected. This warning allows for the removal of ships from harbors before the waves arrive if the disturbance has occurred at a great enough distance.

Prior to the establishment of such a warning system, the first notice that most observers had of an impending tsunami would be the rapid seaward recession of the

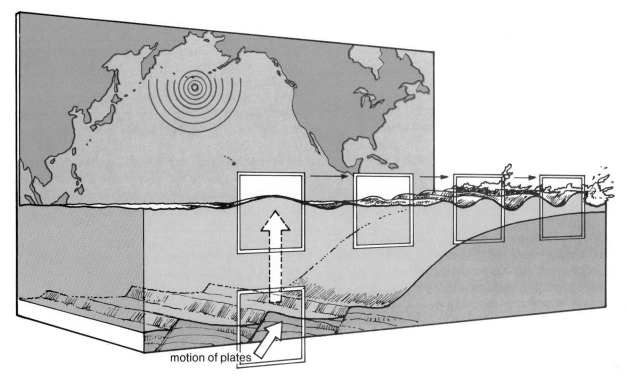

A.

B.

C.

FIGURE 9–13

Tsunami. *A,* movement along a fault in the earth's crust sends energy to the ocean surface. This energy then is distributed laterally along the atmosphere-ocean interface in the form of a tsunami. The energy is transmitted across the open ocean by undetectable waves over 200 km (124 mi) long and about 0.5 m (1.6 ft) high. They release their energy after reaching the shore and developing heights that may be in excess of 30 m (98 ft). Vertical uplift of ocean floor pushes up the ocean water column above the fault. The energy released into the water column by the earthquake travels away from the affected area as a tsunami. *B* shows Sandy Beach on the Island of Oahu moments before a tsunami generated by an earthquake in the Aleutian Trench area struck the beach. Note that the beach was exposed as the shoreline moved out to sea. The arrow indicates the location of *C,* which shows the beach a few minutes later, after the tsunami arrived. A man escaping the wave is circled at left; several automobiles swept off the highway are circled at right. (Photos by Y. Ishii. Reprinted by permission of *Honolulu Advertiser.*)

shoreline. The recession would then be followed in a few minutes by the destructive wave (figure 9–13C). Such a recession was observed in the port of Hilo, Hawaii, on April 1, 1946, as a result of an earthquake that occurred in the Aleutian Trench off the island of Unimak over 3000 km (1863 mi) away. The recession created by the formation of a trough preceding the first tsunami wave crest was followed by a wave that carried the waters to heights nearly 8 m (26 ft) above the normal high tide level. The tsunami also struck Scotch Cap, Alaska, on Unimak Island. The base of the lighthouse on the island stood 14 m (46 ft) above sea level and was totally destroyed by a wave that must have reached a height of 36 m (118 ft).

It was the wave produced by this disturbance in the Aleutian Trench that was recorded throughout the coastal regions of the Pacific Ocean on seismographs and in tide-recording stations and led to the development of what is now the International Tsunami Warning System (ITWS). The warning system was originally designed to warn only the Hawaiian Islands, but many countries bordering the Pacific Ocean later joined. Since its development, the system has worked well in warning areas some distance from the generating disturbances. In the areas adjacent to the disturbances, however, there is not enough time to provide official warning, because of the high speed with which the waves move. In 1957, warning could not be provided Adak Island of a wave that had been produced by a seismic disturbance in the nearby Aleutian Trench, nor could many of the coastal communities in Prince William Sound, Alaska, be warned in time to evacuate after the Alaskan earthquake in 1964.

INTERNAL WAVES

We have to this point discussed waves that occur at an obvious density discontinuity existing between the atmosphere and the ocean. From our previous discussions we should be aware that there are also such discontinuities within the ocean water column itself, for instance, in the pycnocline that is typically associated with the thermocline. Just as energy can be transmitted along the interface between the atmosphere and the ocean, it can also be transmitted along density interfaces beneath the ocean surface in what are called *internal waves*. Internal waves may have heights ranging above 100 m (328 ft) (figure 9–14).

There is still much to be learned about internal waves, but their existence is well documented and it is thought that many causes for them may be identified. Internal waves are known to have periods related to tidal forces, indicating that these forces may be a significant cause. Underwater avalanches in the form of turbidity currents and energy put into the water by moving vessels may also be causes of internal waves. It is thought that the parallel slicks seen on the surface waters may be related to underlying internal waves. Surface water above the crests may flow toward the troughs where the surface water column is thicker. This may transport surface debris, including an oil film to the troughs. This surface layer of debris can prevent the surface water above the troughs from being as disturbed by wind as is the water above the crests. Thus, the slicks may represent the position of internal wave troughs.

Internal waves reach greater heights from a small energy input than do the waves resulting from very large

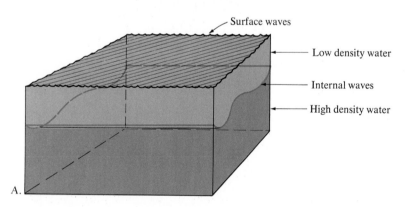

FIGURE 9–14
Internal Wave. *A,* a simple internal wave moving along the density interface below the ocean surface. *B, Seasat* Synthetic Aperture Radar image of internal waves in the Gulf of California. They are thought to be associated with tidal currents in the shallow water around the island of Angel de La Guarda located in the lower right. The image is about 100 km on each side.

across the open ocean with undetectable heights of about 0.5 m (1.6 ft) at speeds in excess of 700 km/h (435 mi/h). On approaching shore, they may increase in height to over 30 m (98 ft). Single-wave tsunami have been known to cause millions of dollars worth of damage and take tens of thousands of lives. **Internal waves** are not well understood but are thought to form at density interfaces beneath the ocean surface, especially in connection with the pycnocline. They may be expected to have heights of up to 100 m (328 ft) with periods from 5 to 8 min.

Ocean waves can potentially be focused to produce hydroelectric power, but there are significant environmental considerations that will have to be incorporated into any such plan.

QUESTIONS AND EXERCISES

1. Discuss longitudinal, transverse, and orbital wave phenomena including the states of matter in which they transmit energy.

2. Calculate the speed for deep-water waves with the following characteristics:
   A. $L = 351$ m, $T = 15$ s
   B. $T = 11$ s
   Express speed in meters per second (m/s).

3. Calculate the speed with which a shallow-water wave will travel across an ocean basin 4 km (2.5 mi) deep.

4. Describe the change in the shape of waves that occurs as they progress from capillary waves to increasingly larger gravity waves until they reach a steepness ratio of 1/7. A change in which variable will make gravity the dominant restoring force, H, L, S, or T?

5. What is the minimum fetch and duration of wind required to produce a fully developed sea with a wind speed of 45 km/h (28 mi/h)? What would be the average height, length, and period of waves in this fully developed sea?

6. Waves from separate sea areas move away as swell and produce an interference pattern when they come together. If the waves from Sea A and Sea B have wave heights of 1.5 m (4.9 ft) and 3.5 m (11.5 ft), respectively, what would be the height of waves resulting from constructive interference and destructive interference? Illustrate your answer with a diagram (refer to figure 9–7).

7. Describe the changes, if any, in wave speed (S), length (L), height (H), and period (T) that occur as waves move across shoaling water to break on the shore.

8. Using orthogonal lines, illustrate how wave energy can be distributed along the shore. Identify areas of high and low energy release.

9. Based on the fundamental characteristics of standing waves shown in figure 9–10, construct a similar diagram of a standing wave in which two nodes and three antinodes exist.

10. List three factors that may affect the height of storm surge. Make a diagram of a hurricane coming ashore from the south along an east-west shore and indicate the segment of shore along which you think the storm surge will reach maximum height. Explain why.

11. Why is it more likely that a tsunami will be generated by faults beneath the ocean along which vertical rather than horizontal movement has occurred?

12. How long will it take for a tsunami to travel across 2000 km (1242 mi) of ocean if the average depth of the ocean is 4500 m (14,760 ft)?

13. What ocean depth would be required for a tsunami with a wavelength of 220 km (137 mi) to travel as a deep-water wave? Is it possible that such a wave could become a deep-water wave any place in the world ocean?

14. Why is the development of internal waves likely within the thermocline?

15. Discuss some environmental problems that might result from the development of facilities for the conversion of wave energy to electrical energy.

REFERENCES

Bascom, W. 1959. Ocean waves. *Scientific American* 201:89–97.

Kinsman, B. 1965. *Wind waves: Their generation and propagation on the ocean surface.* Englewood Cliffs, N.J.: Prentice Hall.

Pickard, G. L. 1975. *Descriptive physical oceanography: An introduction,* 2nd ed. Oxford: Pergamon Press.

Pond, S., and Pickard, G. L. 1978. *Introductory dynamic oceanography.* Oxford: Pergamon Press.

Sverdrup, H. U., Johnson, M. W., and Fleming, R. H. 1970. Reprinted 1970. *The oceans: their physics, chemistry, and general biology.* Englewood Cliffs, N.J.: Prentice Hall.

von Arx, W. S. 1962. *An introduction to physical oceanography.* Reading, Mass.: Addison-Wesley.

SUGGESTED READING

Sea Frontiers

Changery, M. J. 1987. Coastal wave energy. 33:4, 259–62.
The wave energy resources of the world's shores are considered.

Ferrell N. l987. The tombstone twins: Lights at the top of the world. 33:5, 344–51.
A short history of the two lighthouses on Unimak Island, Alaska.

Land, T. 1975. Freak killer waves. 21:3, 139–41.
The British design a buoy that will gather data in areas where 30-m waves, which may be responsible for the loss of many ships, occur.

Mooney, M. J. 1975. Tragedy at Scotch Cap. 21:2, 84–90.
A recounting of the events resulting from an earthquake off the Aleutians April 1, 1946. The resulting tsunami destroyed the lighthouse at Scotch Cap, Alaska.

Pararas-Carayannis, G. 1977. The International Tsunami Warning System. 23:1, 20–27.
A discussion of the history and operations of the International Tsunami Warning System.

Robinson, J. P., Jr. 1976. Newfoundlander's disaster of '29. 22:1, 44–51.
A description of the destruction caused by a tsunami that struck Newfoundland on November 18, 1929.

———. 1976. Superwaves of southeast Africa. 22:2, 106–16.

A discussion of the formation and destruction caused by large waves that strike ships off the southeast coast of South Africa.

Smail, J. R. 1982. Internal waves: The wake of sea monsters. 28:1, 16–22.
An overview of the causes, effects, and nature of internal waves.

Smith, F. G. W. 1985. Bermuda mystery waves. 31:3, 160–63.
A discussion of the possible source of large waves that struck Bermuda on November 12, 1984.

———. 1971. The real sea. 17:5, 298–311.
A comprehensive and readable discussion of wind-generated waves.

———. 1970. The simple wave. 16:4, 234–45.
A very readable explanation of the nature of ocean waves. It deals primarily with the characteristics of deep-water waves.

Truby, J. D. 1971. Krakatoa—the killer wave. 17:3, 130–39.
The events leading up to the 1883 eruption of Krakatoa and the tsunami that followed are described.

Scientific American

Bascom, W. 1959. Ocean waves. 201:2, 89–97
An informative discussion of the nature of wind-generated waves, tsunami, and tides.

10

TIDES

Humans have undoubtedly observed the rise and fall of the tides since first inhabiting the coastal regions of the continents. But until Herodotus (450 B.C.) observed tides on the Mediterranean, there was no known record of them. The first Greek observers concluded that the tides were related to the motion of the moon since both followed a similar cyclic pattern, but it was not until Sir Isaac Newton (1642–1727) developed his universal law of gravitation that the tides could be explained adequately.

The tides are the ultimate manifestation of shallow-water wave phenomena, possessing lengths measured in thousands of kilometers and heights ranging from zero to more than 15 m (50 ft). The ocean tides generated by the gravitational attraction of the sun and moon on the mass of the ocean affect every particle of water from the surface to the deepest part of the ocean basin. Thus, the tide undoubtedly has a much farther reaching effect on ocean phenomena than we are able to observe from our position at its surface.

TIDE-GENERATING FORCES

Law of Gravitation

Isaac Newton published his *Philosophiae Naturalis Principia Mathematica* in 1686 and stated in his preface, ". . . I derive from the celestial phenomena the forces of gravity with which bodies tend to the sun and several planets. Then from these forces, by other propositions which are also mathematical, I deduce the motions of the planets, the comets, the moon, and the sea." What followed represented our first understanding of why tides behave as they do.

Newton's law of gravitation states that every particle of mass in the universe attracts every other particle of mass with a force that is proportional to the product of their masses and inversely proportional to the square of the distance between the masses. The greater the mass of the objects and the closer they are together, the greater will be the gravitational attraction. Mathematically this can be expressed as:

The Bay of Fundy at high tide (top) and low tide (bottom). Photographs by Breck P. Kent.

FIGURE 10-1

Relative sizes of the Earth, Moon and Sun The earth has a diameter of 12,682 km (7876 mi). The diameter of the moon is 3478 km (2160 mi), which is 0.27 that of the earth. The diameter of the sun is 1,392,000 km (864,432 mi), which is 109 times the diameter of the earth. The relative sizes are shown to scale. See table 10-1 for a comparison of the masses of the sun and moon and their distances from the earth. These factors are important in determining their tide-generating effect on the earth.

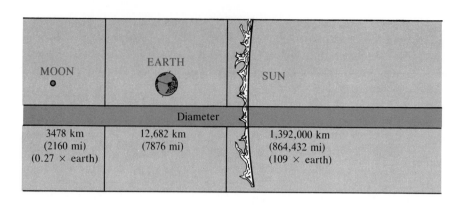

$$\text{Gravitational force} = G \frac{m_1 m_2}{r^2}$$

Here, G is the universal gravitational constant, m_1 and m_2 are the masses, and r is the distance between the masses. For spherical bodies, all the mass can be considered to exist at the center of the sphere. Thus, r will always be the distance between the centers of bodies being considered.

Actually, tide-generating forces vary inversely as the *cube* of the distance from the center of the earth to the center of the tide-generating object instead of varying inversely as the square of the distance, as does the gravitational attraction. This tells us that the tide-generating force, although it is derived from the force of gravitational attraction, is not proportional to it. Distance becomes a more highly weighted variable when we

consider the tide-generating forces than when we determine gravitational attraction force.

$$\text{Tide-generating force} \propto \frac{m_1 m_2}{r^3}$$

Although the gravitational attraction between the earth and sun is over 177 times that between the earth and moon, the moon dominates the tides, as will be seen from the following comparison of the tide-generating force of the sun with that of the moon. Since the sun is 27 million times more massive than the moon, it should have a tide-generating force 27 million times greater than the moon based solely on comparative masses. However, we must also consider the distance that exists between the earth and the moon as compared with the distance between the earth and the sun. Since

TABLE 10-1

Relative Masses of Sun and Moon and Distances from Earth—Relative Tide-Generating Effect.

Tide-Generating Body	Distance from Earth (avg)	Mass (Metric tons)	Relative Tide-Generating Effect
Moon	384,835 (km) (238,483 mi)	7.3×10^{19}	Based on relative masses, the sun is 27 million times more massive than the moon and has 27 million times the tide-generating effect. However, since the sun is 390 times farther than the moon from the earth, its tide-generating effect is reduced by 390^3, or 59 million times.
Sun	149,758,000 km (93,016,845 mi)	2×10^{27}	

Moon

Earth   Distance of moon and sun from earth shown approximately to scale.   Sun

DETERMINATION OF TIDE-GENERATING FORCE OF SUN RELATIVE TO MOON

$$\text{Tide-generating force} \propto \frac{\text{Mass}}{(\text{Distance})^3} \propto \frac{\text{Sun—27 million times more mass}}{(\text{Sun—390 times farther away})^3}$$

$$(390)^3 = 59,000,000 \quad \text{Thus,} \quad \frac{27 \text{ million}}{59 \text{ million}} = 0.46 \text{ or } 46\%$$

The sun has 46% the tide-generating force of the moon.

the sun is 390 times farther from the earth than the moon, its tide-generating force is reduced by 390^3, or about 59 million times compared with that of the moon. These conditions result in the sun's tide-generating force being $^{27}/_{59}$, or about 46 percent, that of the moon (figure 10–1 and table 10–1).

Although the mathematics of the explanation is complex, we will attempt to explain how the tide-generating force decreases in proportion to the cube of the increase in distance between the particle of the earth that is affected and the center of the tide-generating bodies (moon and sun). We confine our consideration to the moon, but a similar explanation applies to the sun.

The tidal pattern we see on the earth is primarily the result of the rotation of the earth-moon system around its center of mass. Figure 10–2A shows the path followed by the center of the earth as it rotates around the center of gravity for the earth-moon system, which is a point about 4700 km (2918 mi) from the center of the earth. It also shows that the *zenith* (point on the earth's surface closest to the moon) and *nadir* (point on the

earth's surface farthest from the moon) follow identical circular paths. Actually, all points on the earth must follow circular paths identical to that of the center of the earth. If we divide the earth into millions of tiny particles of equal mass, each must have an identical *centripetal force* act on it to keep it in this circular path.

As an example of centripetal force, if you tie a string to a rock and swing the tethered rock around your head, the string pulls the rock toward your hand—a point in the center of the circular orbit the rock is following. The string is providing a centripetal (center-seeking) force to the rock. If the string breaks, the rock can no longer maintain its circular orbit and will fly off in a straight line (figure 10–2B).

Figure 10–3A shows that that each particle is located at a different distance and/or direction from the center of the moon. This causes the force of gravitational attraction between each particle and the moon, which must provide the required centripetal force to each particle, to be different. As a result of the differences between the required centripetal force and the gravita-

FIGURE 10–2

Earth-Moon-System Rotation. *A,* the dashed line through the center of the earth is the path of the earth's center as it moves around the common center of gravity of the earth-moon system. The circular paths followed by the nadir (point *a*) and zenith (point *b*) have the same radius as that followed by the earth's center. *B,* if you tie a string to a rock and swing it in a circle around your head, it stays in the circular orbit because the string exerts a centripetal (center-seeking) force on the rock. This force pulls the rock toward the center of the circle. If the string breaks with the rock at the position shown, the rock will fly off along a straight path.

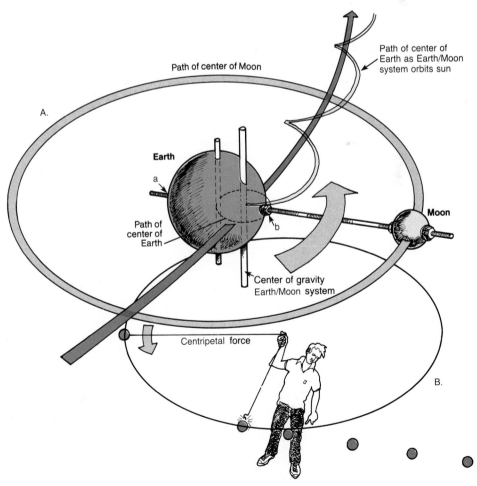

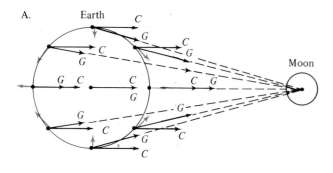

A.

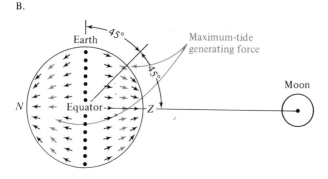

B.

FIGURE 10–3

Tide-Generating Force. *A,* the length and direction of the arrows in this figure represent the magnitude and direction of the forces related to each of the eight identical masses located at each of the points shown on the figure. *C arrows*—The centripetal force required to keep each particle in the identical circular orbit it follows as a result of the earth-moon rotation system. *G arrows*—The gravitational attraction between particles and the moon. This force provides the required centripetal force, but it is identical to it only for the particle at the center of the earth. It is in the required direction only for particles on a line connecting the centers of the earth and moon. For all particles on the half of the earth facing the moon, the force is larger than required, and for the particles on the half of the earth facing away from the moon, it is smaller. *Blue arrows*—For all particles except at the center of the earth, there is a resulting residual force because the gravitational force varies from the required centripetal force. This force is small—averaging about 10^{-7} of the magnitude of the earth's gravity. Therefore, where forces act perpendicular to the earth's surface, as does gravity, they do not have any tide-generating effect. However, where they have a significant horizontal component—tangent to the earth's surface—they aid in producing tidal bulges on the earth. Since there are no other large horizontal forces on the earth with which they must compete, these small, ever-present residual forces can push water across the earth's surface. *B,* along the intersection of the earth's surface and a plane running through the earth's center perpendicular to a line connecting the centers of the earth and moon, the residual forces are directed toward the center of the earth and have no horizontal component—tide-generating forces are zero (dots). There is also no horizontal component at the zenith (*Z*) and nadir (*N*) because the residual forces point away from the center of the earth. Elsewhere on the earth's surface, there are horizontal tide-generating forces. The magnitude of the tide-generating force varies. It reaches maximum values along two circles that lie 45° on either side of the previously described circle of zero tide-generating force.

tional force of attraction between each particle and the moon, small horizontal or tide-generating forces are produced to push the ocean's water into bulges at the zenith and the nadir.

A plane passed through the center of the earth and perpendicular to a line through the centers of the earth and moon intersects the earth's surface along a circle where tide-generating forces are zero. There is also no tide-generating force at the nadir and zenith. At all other points on the earth's surface there is a horizontal component to the gravitational force between the particle it represents and the moon. It is these tiny horizontal forces, the tide-generating forces, that push the water into the bulges at the zenith and nadir. The magnitude of the tide-generating forces increases from zero along the circle previously discussed to a maximum at an angle of 45° on either side of the circle and decreases again to zero at the zenith and nadir. These two bulges provide the framework within which the *equilibrium tide* operates (figure 10–3*B*).

EQUILIBRIUM THEORY OF TIDES

We will now consider the kinds of tide observations that would be predictable on an earth with two tidal bulges, one toward the moon and one away from the moon as the earth rotates on its axis. In the following discussion, we assume that an ideal ocean of uniform depth covers the earth and that there is no friction between the water in the ocean and the ocean floor. We will call this theoretical tide the *equilibrium tide.* Although this oversimplification will not give us an accurate means of predicting the types of tides that will occur throughout the earth's surface, it will give us the gross characteristics of tides in the world ocean. Only after we have established an understanding of the equilibrium theory will we attempt to consider the *dynamical theory* of tides that deals with changes in ocean depth, the existence of continents, and friction between the ocean water and ocean floor.

The Rotating Earth

This ideal ocean is modified only by the tide-producing forces that cause bulges on opposite sides of the earth as shown in figure 10–4. Let us assume that a stationary moon is directly above the equator so that the maximum bulge will occur on the equator on opposite sides of the earth. Since the earth requires 24 h for one complete rotation, an observer on the equator would experience two high tides per day. The time that would elapse between high tides, the tidal period, would be 12 h. An observer at any latitude north or south of the equator would experience a similar period, but the high tides would not be so high, since the observer would be at the edge of the bulge rather than the apex.

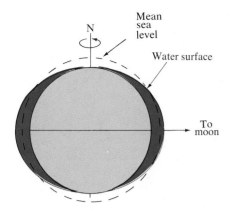

FIGURE 10–4
Equilibrium Tide. Assuming an ocean of uniform depth covering the earth and the moon located above the equator, the tide-generating forces will produce two bulges on the ocean surface. One will extend in the direction of the moon and the other away from the moon. As the earth rotates, all points on the earth except the poles will experience two high tides each day as the earth rotates beneath the two bulges.

But high tides do not occur every 12 h on the earth's surface, because the earth-moon system is rotating about its center of mass while the earth is rotating on its axis.

The *lunar day,* the time that elapses between successive passages of the moon across the meridian (longitude line of an observer), must be somewhat longer than the solar day of 24 h. It is actually 24 h 50 min. If one observes the time at which the moon rises on successive nights, it can be seen to rise about 50 min later each night. Figure 10–5 shows that this results from the fact that as the earth has rotated on its axis in 24 h, the moon has moved 12.2° to the east. The earth must rotate another 50 min to have the moon again on the meridian of the observer. Therefore, if we wish to divide the lunar day of 24 h 50 min into 24 lunar hours, each lunar hour will be about 1 h 2 min of solar time in length.

We have so far considered the effects of the earth's rotation and the revolution of the earth-moon system about its center of mass on the prediction of tides. We have assumed that the moon remained above the equator at all times, and we have ignored the effect of the sun. In the following discussion we will consider the combined effect of the moon and sun on the earth's tides.

Combined Effects of Sun and Moon

Figure 10–6 shows the path of the earth and moon as the earth-moon system revolves around the sun. It can be seen that approximately every 29½ days, the moon is in the same phase. When the moon is between the earth and the sun it is said to be in *conjunction,* producing a new moon. The moon is in *opposition* when it is on the opposite side of the earth from the sun, causing a full moon. A quarter moon results when the moon is at right angles to the sun relative to the earth.

When the sun and moon are in opposition or conjunction, the tide-generating forces of the sun and moon are additive, and we experience maximum *tidal range,* the vertical difference between high and low tide. During quarter-moon phases, the tide-generating force of the sun is working at right angles to the tide-generating force of the moon, and we have minimum tidal range. The maximum tidal range condition that exists during the new- and full-moon phases is referred to as the *spring tide,* and during the 1st and 3rd quarter phases we have *neap tide* (figure 10–6). The time that elapses between successive spring tides (full moon and new moon) or neap tides (first and third quarters) is about 2 weeks.

Effects of Declination

Up to this point we have considered that the moon and the sun remain at all times above the equator, or at least we did not describe specifically any deviation from such a position. We now must consider the fact that this is not the case. As it revolves around the sun, the earth's axis of rotation is tilted 23½° from being perpendicular relative to the *ecliptic,* the plane of the earth's orbit about the sun. It is due to this tilt that we experience the seasons—spring, summer, fall, and winter. To further complicate our consideration, the plane of the

FIGURE 10–5
The Lunar Day. A lunar day is the time that elapses between successive appearances of the moon on the meridian directly over a stationary observer. As the earth rotates on its axis, the earth-moon-system rotation moves the moon in the same direction (to the east). During one complete rotation of the earth on its axis (the 24-h solar day), the moon moves east 12.2°, and the earth must rotate another 50 min to put the observer in line with the moon.

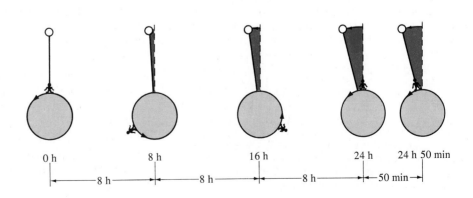

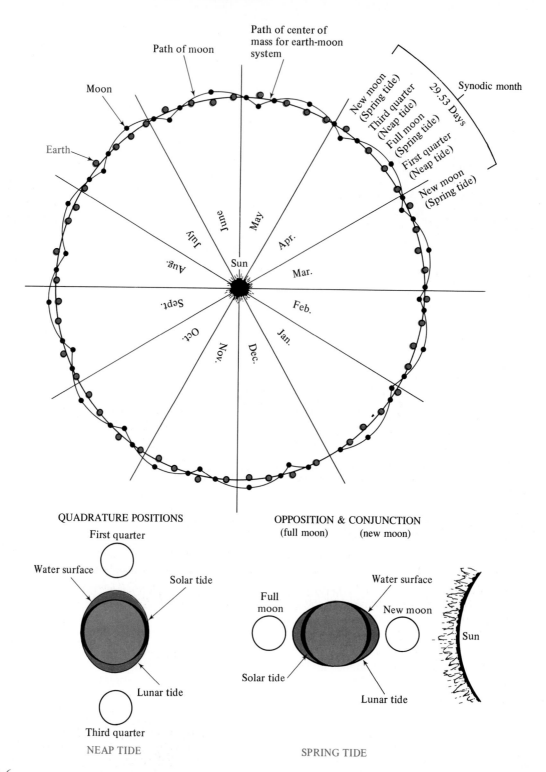

Path of center of
mass for earth-moon
system

Path of moon

Moon

Earth

New moon
(Spring tide)
Third quarter
(Neap tide)
Full moon
(Spring tide)
First quarter
(Neap tide)

New moon
(Spring tide)

Synodic month

29.53 Days

Sun

June
May
July
Aug.
Sept.
Oct.
Nov.
Dec.
Jan.
Feb.
Mar.
Apr.

QUADRATURE POSITIONS

First quarter

Water surface

Solar tide

Lunar tide

Third quarter

NEAP TIDE

OPPOSITION & CONJUNCTION
(full moon)   (new moon)

Water surface

Full
moon

New moon

Sun

Solar tide

Lunar tide

SPRING TIDE

FIGURE 10–6
Sun-Moon-Earth Motion and Tides. As the earth-moon system
moves around the sun, the centers of each body follow a wavy path
because they are also rotating about the center of mass of the earth-
moon system (see figure 10–2A). We can think of this system as a
sledgehammer, with the earth being the heavy business end and the
moon being a knob at the end of the handle. The "handle" that connects the earth and moon is gravity. If the hammer were thrown
through space, the center of the knob (moon) at the end of the handle would follow an orbit around the center of mass with a much
greater radius than that of the orbit of the center of the head of the
hammer (earth). Therefore, the wobble of the moon is more pronounced and is shown in exaggerated form. During a synodic month,
the moon moves from a position between the sun and the earth

(new moon) to a position that puts the earth between the moon and
the sun (full moon) and back to its original position. When the moon
is in the new or full position, the tidal bulges created by the sun and
moon are aligned, producing constructive interference and a larger
bulge. When the moon is in the positions halfway between the new
and full phases, the first and third quarters, the tidal bulge produced
by the moon is at right angles to the bulge created by the sun. The
bulges tend to "cancel" each other (destructive interference), and the
resulting bulge is smaller. New moon and full moon phases produce
spring tides with maximum tidal ranges, while the first and third
quarter phases of the moon produce neap tides with minimal tidal
ranges.

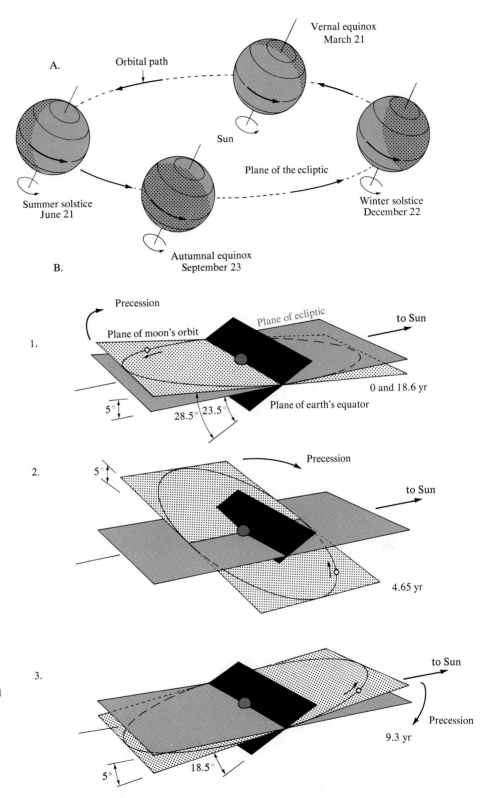

FIGURE 10–7

Declination—Orbital Planes of Earth and Moon. *A*, as the earth orbits the sun during one year, the declination of the sun varies from a minimum of 23½° north of the equator during the summer solstice (June 21) to 0° during the autumnal equinox (September 23) to a maximum declination south of the equator of 23½° during the winter solstice (December 22) and back to 0° during the vernal equinox (March 21). This occurs because the axis of the earth's rotation is tilted 23½° from vertical relative to the ecliptic and assumes a constant direction in space throughout the yearly cycle. *B*, if you spin a top on the floor, the axis of the top's rotation slowly precesses (wobbles) through a complete cycle while the top is spinning. A similar phenomenon occurs with all objects that rotate around an axis. As the moon moves through its orbit around the axis of the earth-moon system, the plane of this orbit precesses. However, in this case the wobble requires a long time, 18.6 years. While the plane of the moon's orbit precesses, it maintains a tilt of 5° relative to the ecliptic. Note that during this period of time, the plane of the earth's equator maintains the same orientation relative to the ecliptic. Part 1 of this figure shows the orientation of the moon's plane of orbit relative to the earth's equatorial plane at the beginning and end of one of these 18.6-year precession cycles. The plane of the earth's equator and the plane of the moon's orbit are on opposite sides of the ecliptic, so the maximum declination of the moon relative to the earth's equator is 23½° + 5° or 28½° north or south depending on where the moon is in its orbit. In part 2, 4.65 years later (one-fourth of the precession cycle later), the maximum declination of the moon is 23½°. In part 3, the precession is 9.3 years (one-half cycle) from part 1, and the plane of the earth's equator and the plane of the moon's orbit are on the same side of the plane of the ecliptic. Thus, the maximum declination of the moon is 23½° − 5° or 18½°. (*B*, from C. Hauge. Tides, currents, and waves. *California Geology,* July 1972. Reprinted by permission of the author.)

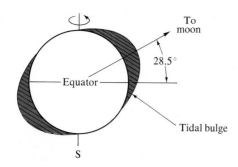

FIGURE 10–8

Maximum Declination of Tidal Bulges from Equator. The center of the tidal bulges may lie at any latitude from the equator to a maximum of 28.5° on either side of the equator.

moon's orbit is at an angle of 5° to the ecliptic. This angular distance of the sun or moon above or below the equatorial plane of the earth is called *declination.*

The tilt of the earth's axis relative to its plane of revolution around the sun is shown in figure 10–7A. It can be seen that the tilted axis assumes a constant direction in space throughout the yearly cycle that includes the equinoxes and solstices. At the *vernal equinox,* which occurs about March 21, the sun is directly above the equator and is moving from the Southern Hemisphere to the Northern Hemisphere. On about June 21 the *summer solstice* occurs. At this time the sun reaches its northernmost point in the sky, directly above the Tropic of Cancer, which is at 23½°N latitude. Following this occurrence, the sun moves farther south in the sky each day, and on about September 23 it is directly above the equator again and produces the *autumnal equinox.* During the next 3 months the sun appears to be more southerly in the sky until the *winter solstice* on about December 22, when it is directly over the Tropic of Capricorn at 23½°S latitude. Thus the sun may be found at declinations between 23½° north and 23½° south of the equator on a yearly cycle.

Since the plane of the moon's orbit intercepts the plane of the ecliptic at an angle of 5°, and since the plane of the moon's orbit *precesses,* or rotates, while maintaining its 5° angle with the ecliptic with a precessional cycle of 18.6 years, we have a relatively complex consideration regarding the declination of the moon relative to the plane of the earth's equator. In part 1 of figure 10–7B, the declination of the moon's orbit relative to the earth's equator is 28½°. It will move from 28½° south to 28½° north and back to 28½° south of the equator in a period of 1 month. Part 2 of the illustration shows the relationship of the ecliptic, the plane of the moon's orbit, and the plane of the earth's equator after one-fourth of the precession, or 4.65 years later. The maximum declination of the moon's orbit relative to the earth's equator still approaches 28½°. However,

in part 3, when one-half precession is completed, which is 9.3 years after the condition observed in part 1, it can be observed that the maximum declination of the moon relative to the earth's equator is 18½°.

Based on these considerations we must alter our previous concept of the predicted equilibrium tide. We must now expect that tidal bulges will rarely be aligned with the equator and will occur for the most part either north or south of the equator. Since the moon is the dominant force that creates tides in the earth's oceans, we would suspect that the bulges would follow the moon as it moves on its monthly journey across the equator, being found at a maximum of 28½° north and south of the equator (figure 10–8).

Effects of Distance

Additional considerations that affect the tide-generating force of the sun and moon on the earth are their changing distances from the earth. The earth ranges from *perihelion* with the sun, when the distance between the bodies is 148.5 million km (92.2 million mi) during the winter months in the Northern Hemisphere, to *aphelion,* when the distance between the bodies is 152.2 million km (94.5 million mi) during our summer months. The moon moves from *perigee,* its closest approach to the earth, of 375,200 km (233,000 mi) to *apogee* of 405,800 km (252,000 mi) and back to perigee during a period 27½ days—the *anomalistic month* (figure 10–9).

Because of these movements, tides have greater ranges during the Northern Hemisphere winter than in the summer. Also, as the result of changing distance between the earth and moon, tidal ranges become greater at perigee each anomalistic month. Keeping in mind that tide-generating forces vary inversely with the cube of the distance from the center of the earth to the center

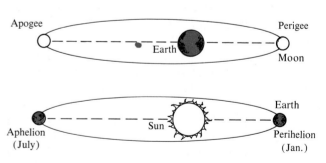

FIGURE 10–9

The Effects of Elliptical Orbits. The moon moves from perigee of 375,200 km (233,000 mi) to apogee of 405,800 km (252,000 mi) (top). Greater tidal ranges are experienced during perigean tides. Perihelion brings the earth within 148,500,000 km (92,200,000 mi) of the sun. Aphelion distance is 152,200,000 km (94,500,000 mi) (bottom). Greater tidal ranges are experienced during perihelion tides.

of the tide-generating body, it can be appreciated that these changes resulting from the elliptical nature of the earth's orbit around the sun and the moon's orbit around the earth can readily be observed in the tides.

Equilibrium Tide Prediction

To consider the tidal patterns we would predict for an idealized water-covered earth, let us return our attention to the effect of declination. Let us assume that the declination of the moon, which will determine the alignment of the tidal bulges, is 28° north of the equator. If we position an observer at this latitude on a permanent point, the observation of the tides that occur here will be different from observations at the equator.

If the observations begin when the moon is directly over the observer's head, high tide will be recorded. Six lunar hours (6 h 12½ min solar time) later a low tide will be recorded, followed by another high tide that will be much lower than the initial high tide (figure 10–10A–D). At the end of a 24-lunar-hour period, the observer will have passed through a complete lunar-day cycle of two low tides and two high tides. A representative curve of the type of tide observed can be seen in figure 10–10E. Tide curves showing the heights of the same tides during one lunar day at the equator and at 28°S latitude are also provided. The vertical difference in successive high tides or successive low tides that occurs as a result of the declination of the moon and the sun relative to the earth's equator is called the *diurnal inequality*.

To summarize the tides that we might predict as a result of the tide-generating forces of the moon and sun on an earth covered with a uniform depth of water for any point not on the equator, we would predict that the equilibrium tide would generally be observed at any location to have (1) two high tides and two low tides per lunar day. Except for the rare occasions when the sun and moon are simultaneously above the equator, we would expect that (2) neither the two high tides nor the two low tides would be of the same height because of the changing declination of the moon and the sun. We would expect (3) yearly and monthly cycles of tidal range related to the changing distances of the earth from the sun and moon. Lastly, we would expect that (4) each fortnight, half a lunar month, we would experience spring tides that would be separated by neap tides (figure 10–10F).

Considering the great number of variables that are involved in the prediction of tides, it is interesting to consider when the conditions might be right to produce the maximum tide-generating force. This will occur when the sun is at perihelion and in conjunction (new moon) or opposition (full moon) with the moon at perigee and when both the sun and moon have a zero declination. This con-

dition occurs only once each 1600 years, and the next occurrence is predicted for A.D. 3300.

An indication of the significance of even a near coincidence of the earth's perihelion with the perigee of the moon during a spring tide was made woefully clear during the January 1983 storms in the North Pacific. The introduction of slow-moving low-pressure cells that developed over the Aleutians caused strong northwest winds to blow across the ocean from Kamchatka to the U.S. coast. Averaging about 50 km/h (30 mi/h) the winds produced a near fully developed 3-m (10-ft) swell along the coast from Oregon to Baja California (see tables 9–1 and 9–2). The storm surge from this condition would have been trouble enough under average conditions, but the situation was made worse by the 2.25-m (7.4-ft) high spring tides.

The unusually high tides occurred because the earth was still near perihelion (January 2) when the moon reached perigee on January 28, 1983. Some of the largest waves came ashore January 26–28 when over $100 million in damage was done (figure 10–11). Twenty-five homes were destroyed, over 3500 homes were seriously damaged, and many commercial and municipal piers collapsed. At least a dozen lives were lost. With each such occurrence, we learn more of the cost of developing the shore.

DYNAMICAL THEORY OF TIDES

In our previous discussion of the equilibrium tide, we considered that the tidal bulges were directed at the moon and away from the moon on opposite sides of the earth. Maintaining this relationship with the moon as the earth rotates beneath the moon, the bulges (or wave crests), which are separated by a distance of half the earth's circumference (about 20,000 km or 12,420 mi), would be moving across the earth at a velocity of more than 1600 km/h (994 mi/h).

We previously stated that the tides were an extreme example of shallow-water waves whose velocity is proportional to the square root of the water depth. In order for the tidal wave to travel at 1600 km/h (994 mi/h), the depth of the idealized ocean would have to be 22 km (13.7 mi). Since the mean depth of the ocean is 3.9 km (2.4 mi), the tidal bulges move as forced waves whose velocity is determined by the depth of the ocean. Based on the mean depth of the ocean, the mean speed with which tidal waves can travel across the open oceans is about 700 km/h (435 mi/h). With this limitation on the speed of tidal waves, the theoretical equilibrium theory bulges that point toward and away from the tide-generating body cannot exist and break up into a number of cells.

In the open ocean, the crests and troughs of the tide wave actually rotate around an *amphidromic point* near the center of each cell. There is theoretically no tidal range at this point. Radiating from this point are *cotidal lines* that connect points along which high tide will simultaneously occur. Figure 10–12 shows the cotidal lines at 2-hour intervals for the world ocean. The cotidal lines are labeled with numbers indicating the time of high tide in hours after the moon crosses the Greenwich Meridian. They indicate that the rotation of the tide wave is counterclockwise in the Northern Hemisphere and clockwise in the Southern Hemisphere. The wave

FIGURE 10–10
Predicted Equilibrium Tides. (*F*, from Anikouchine & Sternberg. *The world ocean: An introduction to oceanography.* © 1973. Reprinted by permission of Prentice Hall, Englewood Cliffs, N.J.)

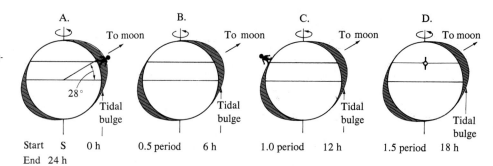

A. At start of lunar day the observer is at zenith in the center of the tidal bulge. He experiences a high tide. *B.* Six lunar hours later (0.5 period) the observer experiences low tide while located on the back side of diagram. *C.* After twelve lunar hours (1.0 period or 0.5 lunar day) the observer again experiences high tide. It is much lower than the first high tide, however, because the observer is now passing through the edge of the tidal bulge. *D.* Observer experiences low tide again eighteen lunar hours (1.5 period) after start. At the end of one lunar day (24 hours) the observer returns to *A* and experiences a high high tide.

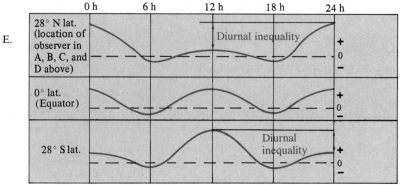

Tide curves for 28°N, 0°, and 28°S latitudes when the declination of the moon is 28° N (all curves for same longitude). Note that tide curves for 28°N and 28°S have identical highs and lows, but are out of phase by 12 hours. This results from the fact that the bulges in the two hemispheres occur on opposite sides of the earth.

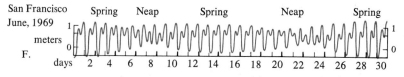

Along with the unequal heights for the two high and low tidal extremes occurring each lunar day, we expect to observe spring and neap tides. Thus the tide curve for the month of June 1969 demonstrates the general character of the predicted equilibrium tide.

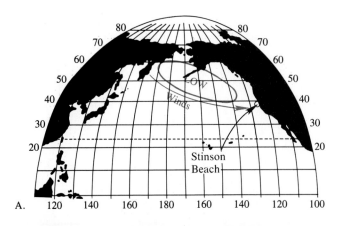

FIGURE 10–11
High Tides of January 1983. *A*, January 1983 storm winds blow uninterrupted across the North Pacific Ocean from Kamchatka to the U.S. coast. *B*, homes threatened by storm waves and unusually high tides on January 27, 1983, at Stinson Beach north of San Francisco. (Wide World Photos.)

makes one complete rotation during the tidal period. The size of the cells is limited by the fact that the tide wave must make one complete rotation during the period of the tide (usually 12 lunar hours). Within an amphidromic cell, low tide is occurring 6 hours behind high tide; for instance, if high tide is occurring along the cotidal line labeled 10, low tide is simultaneously occurring along the cotidal line labeled 4.

We must also consider the effect of the continents, which interrupt the free movement of the tidal bulges across the ideal unobstructed ocean surface we considered to exist in our discussion of the equilibrium tide. The ocean basins between continents have set up within them free-standing waves whose character modifies the forced astronomical tidal waves that develop within the basin. It is impossible for us to explain here the various tidal phenomena that occur throughout the world. For instance, high tide rarely occurs at the time the moon is

at zenith, and the amount of time that elapses between the passing of the moon and the occurrence of high tides varies from place to place as a result of the many factors that determine the characteristic of the tide at any given location.

Types of Tides

Equilibrium tidal theory tells us that we should expect two high tides and two low tides of unequal heights during a lunar day. Due to modifications resulting from varying depths, sizes, and shapes of ocean basins, the tide predicted by the equilibrium theory is replaced in many parts of the world by either a *diurnal* (daily), *semidiurnal* (twice daily), or *mixed* tide (figure 10–13).

The diurnal tide is characterized by a single high and low water each lunar day. These tides are common in the Gulf of Mexico and along the coast of Southeast Asia. Such tides have a tidal period of 24 h 50 min.

The semidiurnal tide has two high and two low waters each lunar day, and the heights of successive high waters and successive low waters are approximately the same. Since tides are always getting higher or lower at any location, due to the spring-neap tide sequence, successive high tides and successive low tides can never be exactly the same at that location. Semidiurnal tides are common along the Atlantic Coast of the United States. The tidal period is 12 h 25 min.

The mixed tide may have characteristics of both diurnal and semidiurnal tides. The diurnal inequality discussed earlier is a characteristic of this tide, as successive high tides and/or low tides will have significantly different heights. Mixed tides commonly have a tidal period of 12 h 25 min, which is a semidiurnal characteristic, but may also possess diurnal periods. This is the tide that is most common throughout the world and the type that is found along the Pacific Coast of the United States. Diurnal inequalities are greatest when the moon is at its maximum declination, and such tides are called *tropical tides* because the moon is over one of the tropic regions. When the moon is over the equator, the diurnal inequality is minimal, and tides with this characteristic are called *equatorial tides* (figure 10–14).

Tides in Narrow Bays

When tide waves enter coastal waters they are subject to reflection. In some cases the standing waves set up by reflections may have periods near that of the forced tidal wave. Under such conditions, *resonance* (constructive interference) can produce significant increases in the tidal range.

The Bay of Fundy is such a place. With a length of 258 km (160 mi), it has a wide opening into the Atlantic Ocean.

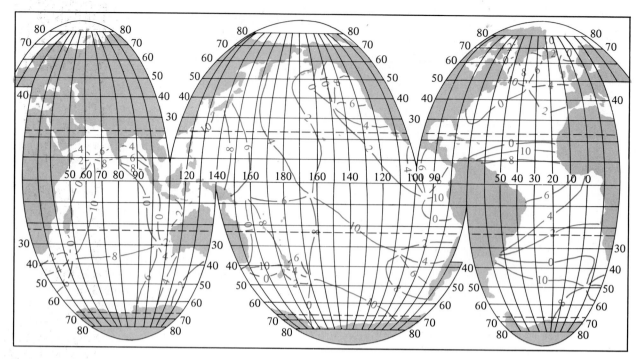

FIGURE 10–12

Cotidal Map of the World. Numbers indicate times of high tide in lunar hours after the moon has crossed the Greenwich Meridian. Tidal ranges generally increase with increasing distance along cotidal lines away from the amphidromic points. Where cotidal lines terminate at both ends in amphidromic points, maximum tidal range will be near the midpoints of the lines. (Base map courtesy of National Ocean Survey.)

The Bay of Fundy splits into two narrow basins at its northern end, Chignecto Bay and Minas Basin (figure 10–15). The period of free oscillation in the Bay of Fundy is very nearly that of the tidal period. The resulting resonance, which along with the narrowing of the bay toward the north end and the shoaling in that direction, produces maximum tidal ranges in the extreme northern end of Minas Basin. The maximum perigean tidal range of about 17 m (56 ft) occurs at the north end of the Minas Basis, and the minimum tidal range of about 2 m (6.6 ft) is found at the opening into the bay. The tidal range progressively increases from the mouth of the bay northward.

Coastal Tide Currents. The current that accompanies the slowly turning tide crest in a Northern Hemisphere basin will turn in a counterclockwise direction, producing a *rotary current* in the open portion of the basin. Because of increased effects of friction in shoaling, nearshore waters, the rotary current is changed to an alternating or *reversing current* that moves in and out rather than along the coast as would a rotary current. These reversing currents are of the greatest concern to navigators, since they are known to reach velocities of 44 km/h (27.6 mi/h) in restricted channels between islands of coastal British Colum-

bia. Velocities of the rotating currents in the open ocean are usually well below 1 km/h (0.6 mi/h).

The reversing tidal current that develops throughout a lunar day for a mixed tide is depicted in figure 10–16. Beginning at high tide, the current velocity is zero as the water has just reached its highest stage and is momentarily to begin its outward flow. Following this *high slack water,* the lowering of the tide begins, and the *ebb current* velocity increases and reaches a maximum about 3 lunar hours after high slack water. The velocity decreases and eventually reaches zero again at the first low *slack water.* Following the change in current velocity associated with the tidal phase throughout the day, it can be seen that the maximum current velocity is reached midway through the ebb current that occurs between the higher high water and lower low water.

The illustration shows that the lower low water (LLW) is below the datum of the chart, the zero mark. How can the tide be less than zero? The datum that is commonly used for mixed tides is the mean lower low water, the average height of the lower low tides at the locality. Since the lower low tide that we are recording is below the average lower low tide for this locality, this tide will have a negative value. In areas where mixed tides are not observed, the tide datum is very commonly the mean low tide recorded at that place—the average low tide. Thus, most tide extremes recorded, even low tides, will have positive values. Only during spring tides are negative low tides observed (figure 10–14).

Tides in Rivers

The Amazon River probably possesses the longest estuarine stretch that is affected by oceanic tides. Tides can be measured as far as 800 km (497 mi) from the mouth of the Amazon, although the effects are quite small at this distance. Tidal waves that move up river mouths lose their energy due to the decreasing depth of water and the flow of the river water against the tide during the flood interval. As the wave moves up the river, it becomes more and more asymmetrical, developing a steep front (figure 10–17). This front produces a rapidly rising tide that falls slowly. An extreme development of

this type produces a *tidal bore* in which a very steep wave front surges up the river. In the Amazon it is called *pororoca* and appears as a waterfall up to 5 m (16.4 ft) in height moving upstream at speeds up to 22 km/h (13.7 mi/h). Other rivers that experience bores are the Chientang in China, where they may reach 8 m (26.2 ft), the Petitcodiac in New Brunswick, Canada, the Seine in France, and the Trent in England.

TIDES AS A SOURCE OF POWER

The history of human efforts to harness the energy of the tides dates back at least to the middle ages. Interest in pursuing this renewable energy source waned during the time of cheap fossil fuel availability. But today, as we realize that the end of cheap fuel could arrive at any time, there is increased interest in assessing the prospects of generating electricity using tidal energy.

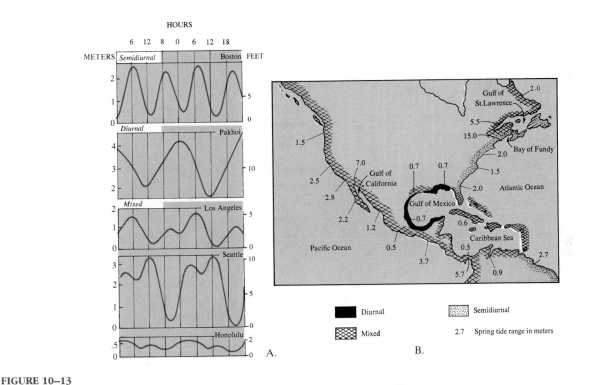

FIGURE 10–13
Types of Tides. *A*, types of tides. In a semidiurnal (twice daily) type of tide, there are two highs and lows during each lunar day, and the heights of each successive high and low are about the same. In the diurnal (daily) type of tide, there is only one high and one low each lunar day. In the mixed type of tide, both diurnal and semidiurnal effects are detectable, and the tide is characterized by a large difference in the high water heights, the low water heights, or both, during one lunar day. Even though a tide at a place can be identified as one of these types, it still may pass through stages of one or both of the other types. *B*, map showing the types that have been observed along portions of the coasts of North and South America. The numbers give the spring tide range in meters and are therefore near the maximum tidal range that can be expected. Storm waves, lower barometric pressure, ocean currents, and the concurrence of perigean and perihelion conditions could increase the range. (After C. Hauge. Tides, currents, and waves. *California Geology*, July 1972. Reprinted by permission of the author.)

FIGURE 10–14

Tidal Curves for Three Locations. The declinations of the moon and the sun, phase of the moon, and position of the moon in its orbit as noted across the top of the chart all contribute to the tidal variations that are depicted. Additional contributing factors include the position of the earth in its orbit around the sun and the configuration of the sea bottom and basin boundaries. Some names of water levels are listed along the right margin. Those that are used as the chart datum for a place are marked with an asterisk. MHWS, mean high water springs—the average height of the high water of the spring tides; MHW, mean high water—the average height of all the high tides at a place; MLW, mean low water—the average height of all the low tides at a place; MLWS, mean low water springs—the average height of all low waters of the spring tides; MHHW, mean higher high water—the average height of the higher high tides at a place where the tide is the mixed type and displays one inequality during a tidal day; MLLW, mean lower low water—the average height of the lower low tides at a place where the tide is of the mixed type. (From C. Hauge. Tides, currents, and waves, *California Geology,* July 1972. Reprinted by permission of the author.)

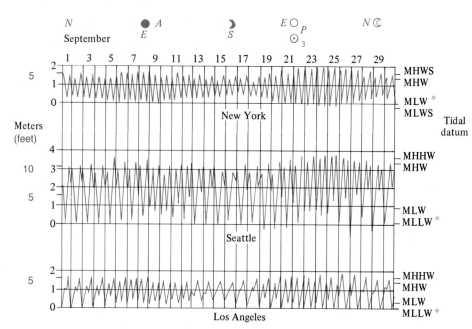

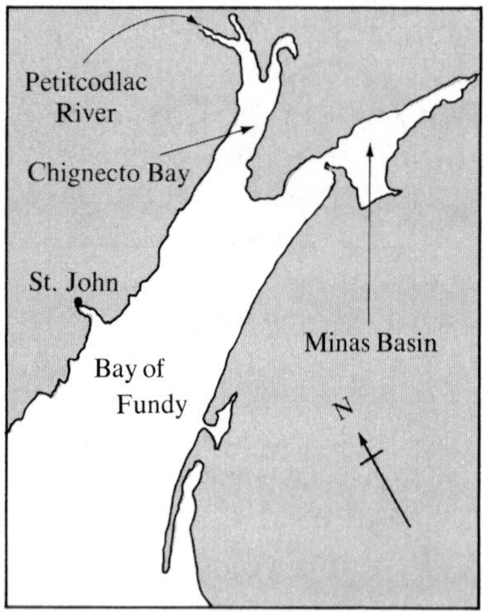

●, new moon; ☽, first quarter; ○, full moon; ☾, last quarter; *E*, moon on the equator; *N, S,* moon farthest north or south of the equator; *A, P,* moon in apogee or perigee; ₁⊙₃, sun at autumnal equinox; *, chart datum.

FIGURE 10–15

Bay of Fundy. The largest tidal range known in the world occurs at the north end of the Minas Basin in the Bay of Fundy. Because of its dimensions, this bay has a natural free-standing-wave period about equal to that of the forced tide wave. This combined with the fact that the bay narrows and becomes shallower toward its head causes a maximum tidal range at the northern end of 17 m (56 ft).

Some Basic Considerations The most obvious benefit of using the tides to generate electrical power would be in reduced operating costs as compared to the conventional thermal power plants that require radioactive isotopes or fossil fuels. Even though the initial cost of the tidal power-generating plant may be higher, there would be no ongoing fuel bill.

A negative consideration involves the periodicity of the tides. As generation of electricity would be our primary concern, power would be generated only throughout a portion of a 24-hour day unless special design features were included in the construction of the facility. Since the tides operate on a lunar period and our energy demand operates on a solar period, the energy available through generating power from the tides would only accidentally coincide with need.

Generators must run at constant speed, and we must utilize the flow of the tidal current in two directions. It is therefore obvious that we would benefit by using turbine blades whose angle could be varied and reversed. Adjusting the pitch of the turbine blades would allow them to be set flatter when the head of water was greatest, and the angle of the blades would be increased as the head of water decreased, so that the generators could be maintained at a constant speed. Since the maximum head would usually not be great, the flow channels constructed through dams would have to be of small diameter. This would require a series of small

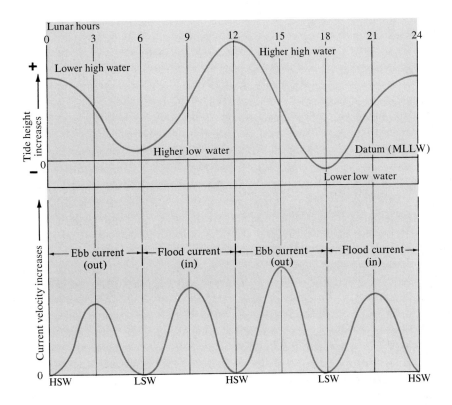

FIGURE 10–16
Reversing Current. Note that tidal current velocity is zero at high and low tidal extremes. Maximum tidal current velocity occurs midway between tidal extremes.

HSW—High slack water (velocity zero)

LSW—Low slack water (velocity zero)

channels and small turbines, which would be less efficient than one large machine.

Let us examine the tidal power plant constructed in the estuary of La Rance River off the English Channel in France. The estuary has a surface area of approximately 23 km² (8.9 mi²), and the tidal range at La Rance reaches a maximum of 13.4 m (44 ft). Useable tidal energy is proportional to the area of the basin and to the square of the amplitude of the tide. A barrier was built across the estuary a little over 3 km (1.9 mi) upstream (figure 10–18), where it is 760 m (2493 ft) wide, to protect it from storm waves. The deepest water ranges from just over 12 m (39.4 ft) at low tide to more than 25 m (82 ft) at high tide. To allow water to flow through the bar-

rier when the generating units are shut down, sluices (artificial channels) with adjustable vanes 10 m (33 ft) high and 15 m (49 ft) wide have been built into the barrier.

The power plant extending across the estuary has an arched roof over a tunnel that runs its entire length to help it withstand pressure from either side. Cranes travel back and forth within the tunnel to service 24 generating units located in conduits below the power plant. The bottom of the conduits containing the generating units are 10 m (33 ft) below the surface of the water at the lowest tide. The conduits containing the units are 53 m (174 ft) long, and the cross-sectional area at each end is about 93 m² (995 ft²).

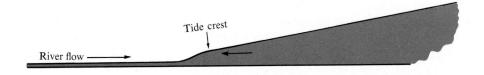

FIGURE 10–17
River Bores. As the tidal crest moves up-river, it develops a steep forward slope through resistance to its advance by the river flowing to the ocean. Such crests (bores) may reach heights of 5 m (16.4 ft) and move at speeds up to 22 km/h (13.7 m/h).

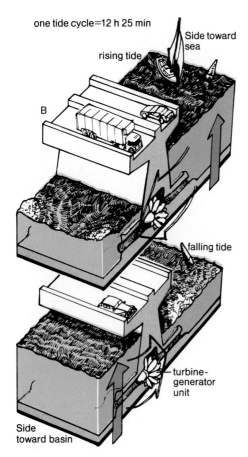

one tide cycle=12 h 25 min

Side toward
sea

rising tide

B

falling tide

turbine-
generator
unit

Side
toward basin

FIGURE 10–18

La Rance Tidal Power Plant. Block diagrams show the barrier between the open ocean to the right and the La Rance estuary to the left. The relative water levels during rising and falling oceanic tides are also shown.

The turbine-generator units resemble large fat torpedoes that are surrounded by water and held in place by radial struts. One of the struts is hollow to allow entry from the control room above by persons servicing the unit. Adjustable propellers with a diameter of 5.3 m (17.4 ft) fit snugly into the restricted throat of the flow channel conduit. Each unit has a generating capacity of 10,000 kW at 3.5 kV.

The plant generates electricity only during about one-half of the tidal period when sufficient head exists between the pool and the ocean. Annual power production of about 540 million kWh without pumping can be increased to 670 million kWh by using the turbine-generators as pumps at the proper times to increase the head of water.

Although some engineers think a tide-generating plant proposed for Passamaquoddy Bay near the U.S.-Canadian border at the south end of the Bay of Fundy could be made to generate electricity constantly, others who have studied the project are less optimistic. Potentially, the useable tidal energy seems great compared with La Rance because the volume of flow is about 117 times greater than that at La Rance. Whether or not tide-generating units are ever constructed on a large scale, this potential source of energy will receive increased attention as the cost of generating electricity by conventional means increases. Figure 10–19 shows the locations of some sites around the world that have potential for generation of tidal electrical power.

SUMMARY

The tides of the earth are derived from the gravitational attractions of the sun and moon. The moon has about twice the tide-generating effect of the sun. Small horizontal forces tend to push water into two bulges, one at the earth's **zenith** and one at the **nadir** relative to the tide-generating body—the sun or moon. Since the moon bulges are dominant, the tides we observe on earth have periods dominated by lunar motions and modified by the changing position of the solar bulges.

If the earth were a uniform sphere covered with an ocean of uniform depth, and if we could ignore some of the considerations of physical motion, the tides on earth would be those predicted by the **equilibrium the-**ory of tides. Such a tide would have a **period** of 12 h 25 min, or half a **lunar day**. A tide with maximum **tidal range** would occur each **new moon** and **full moon**, and tides with minimum range would occur with the **first and third quarter phases** of the moon. These would be the **spring** and **neap tides**, respectively. Since the moon may have a **declination** as much as 28.5° north or south of the equator, and the sun is directly over the equator only two times per year, the tidal bulges would usually be located so as to create two high tides of unequal height per lunar day. The same could be said for the low tides. Tidal ranges are greater when the earth is at **perihelion** in its orbit around the sun and when the moon is at **perigee** in its orbit around the earth

11

COASTAL GEOLOGY

Moving from the oceans onto the continent, one encounters the *shore,* which is the zone that lies between low tide and the highest elevations on the continent that are affected by storm waves. The *coast* extends from the landward limit of the shore inland as far as features that seem to be related to marine processes can be found. The width of the coast may vary from less than 1 km (0.6 mi) to many tens of kilometers. As the waves beat against the shore they cause erosion, which produces sediment that will be transported along the shore and deposited in the low-energy areas.

GENERAL DESCRIPTION OF THE COASTAL REGION

Erosional Shore Features

The landward limit of a shore that is dominated by erosion is commonly marked by a cliff. The *coastline,* which marks the boundary between the shore and the coast, is a line along the cliff, connecting points at which the highest effective wave action takes place.

The shore is divided into the *foreshore,* that portion exposed at low tide and submerged at high tide, and the *backshore,* which extends from the normal high tide to the coastline. The *shoreline* migrates back and forth with the tide and represents the water's edge. The *nearshore* zone is that region between the low tide shoreline and breakers. Beyond the low tide breakers is the *offshore* zone (figure 11–1).

The *beach* is a deposit that consists of the wave-worked sediment that moves along the wave-cut bench of the shore area. It may continue from the coastline across the nearshore region to the line of breakers.

Wave Erosion Due to refraction (bending of waves discussed in chapter 9), the wave energy is concentrated on *headlands* that jut out from the continent, while the amount of energy reaching the shore in bays is reduced. As the waves concentrate their energy on the headlands, erosion occurs and the shoreline re-

Photograph by Michael DiSpezio.

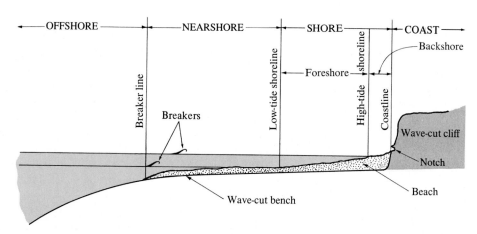

Landforms and Terminology of Coastal Regions. The coastline marks the most landward evidence of direct erosion by ocean waves. It separates the shore from the coast. The shore extends from the coastline to the low tide shoreline (water's edge). It is divided into the backshore, above the high tide shoreline, that is only covered with water during storms and the foreshore (intertidal or littoral zone). Never exposed but affected by waves which touch bottom is the nearshore that extends seaward to the low tide breaker line. The greatest amount of sediment transport as beach deposit occurs within the shore and nearshore zones. Beyond the nearshore lies the offshore region where depths are such that waves rarely affect the bottom.

treats. The greatest concentration of wave energy, on a day-to-day basis, is in the foreshore region. However, during the rare periods when storm waves batter the shore, more erosion may occur across the entire shore in one day than may be achieved by average wave conditions over a period of years.

The cliff shown in figure 11–1 is referred to as a *wave-cut cliff,* which has been produced by wave action cutting away its base. The cliff develops as the upper portions collapse after being undermined by wave action. The undermining may be evident in the form of a notch at the base of the cliff that may be characterized by *sea caves.* Further wave action eroding the softer portions of the rock outcrops may develop the caves into openings running through the headlands, *sea arches.* Continued erosion and crumbling of arches will produce *stacks.* Such remnants rise from the relatively smooth wave-cut bench cut into the bedrock by wave erosion (figure 11–2).

The rate of wave erosion is determined by a number of variables:

1. One of the most important variables is the *degree of exposure* of the coastal region to the open ocean.

Coasts that are fully exposed receive higher energy wave action and are likely to have rugged cliffs.

2. The *tidal range* is also an important variable affecting the amount of wave erosion. Given the same amount of wave energy, a region with a small tidal range will erode much more rapidly than one with a large tidal range that allows the wave energy to be spread over a much broader shore. Although high-velocity tidal currents may develop in areas where a large tidal range exists, these currents are of limited importance in eroding the coastline.

3. The *composition of coastal bedrock* is very significant. Crystalline igneous rocks, such as granite, are relatively resistant and generally produce rugged shoreline topography. Sedimentary rocks, such as sandstone and shale, are more easily eroded, and a gentler topography associated with more extensive beach deposits are characteristic of coasts with such bedrock. Extensive deposits of relatively young glacial till are the medium presently being eroded along some coasts north of New York.

Regardless of the rate of erosion, all coastal regions follow the same developmental paths. As long as there is no change in the elevation of the landmass relative to the ocean surface, the cliffs will continue to retreat, the benches will widen, and the eroded material will be carried from the high-energy areas and deposited in the low energy areas.

FIGURE 11–2

Coastal Erosion Features. Erosion of resistant bedrock that juts out into the ocean as headlands produces features such as sea caves, sea arches and stacks.

Depositional Shore Features

Longshore Current   The coastal erosion we have just discussed, as well as erosion caused by running water inland, produces large amounts of sediment that must be distributed along the continental margin. As waves strike the shore at an angle, they set up a longshore movement of water—the *longshore current*. The velocity of the longshore current increases with increasing beach slope, angle of breakers with the beach, wave height, and decreasing wave period.

This current moves parallel to the shore between the shoreline and the breaker line, carrying with it the materials that make up the beach. At the landward margin of the surf zone, the *swash,* a thin sheet of water, moves sediment onto the exposed beach at an angle, but the force of gravity causes the backwash to carry the sediment straight down the beach face. As a result, the pebbles and sand grains that are transported by the swash move in a zigzag pattern along the shore in the same direction as the longshore current within the surf zone (figure 11–3).

Longshore drift is a term that has been applied to the movement of beach sediment by the processes just described. The amount of longshore drift in any coastal region is determined by an equilibrium between erosional and depositional forces. Any interference with the movement of sediment along the shore will destroy this equilibrium and result in a new erosional and depositional pattern determined by the nature of the interference.

As water moves along in the longshore current, it must eventually return to the ocean. This backwash of water finds its way into the open ocean as a thin sheet flow across the ocean bottom called *undertow* or in local *rip currents* that occur perpendicular or at an angle to the coast where topographic lows or other conditions allow their formation. Rip currents may be less than 25 m (82 ft) wide and may attain velocities of 7 to 8 km/h (4.5–5 mi/h) but do not travel far from shore before they break up. If a light moderate swell is breaking, numerous rip currents moderate in size and velocity may develop. A heavy swell will usually produce fewer, more concentrated rips (figure 11–3).

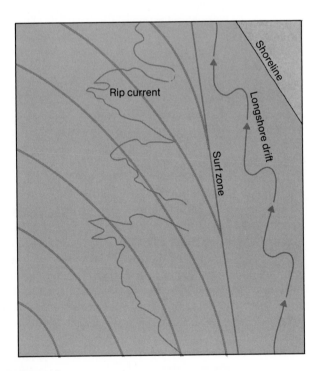

FIGURE 11–3
Longshore Currents and Rip Currents. As waves approach shore from a southerly direction, they produce a longshore current that flows north. Water in the current follows a zigzag path as waves push it up the beach slope from the direction of approach. The water runs back down the slope under the influence of gravity. Water of the longshore current finds its way offshore by passing through topographic lows as strong seaward flows. These rip currents are visible in the photograph as jets of water that appear light in color because of the turbidity resulting from sediment they have resuspended from the ocean floor. (Photo courtesy of Scripps Institute of Oceanography, University of California/San Diego.)

Beach Composition The material found in a beach deposit will depend on the source of the sediment that is transported as the longshore drift. In areas where the sediment is provided by granitic mountains, the beaches will be composed of quartz and feldspar fragments with smaller amounts of other common rock-forming minerals and may be of relatively coarse texture. If the sediment is provided primarily by rivers that drain lowland areas, sediment that reaches the coastal regions will normally be finer in texture, and in many cases mud flats develop along the shore because only clay- and silt-sized particles are emptied into the ocean. In low-relief areas such as Florida, most of the material of the beaches is derived from the remains of the organisms that live in the coastal waters. Beaches in these areas will be predominantly composed of shell fragments and the remains of microscopic animals, particularly foraminifera. Many volcanic island beaches in the open ocean will be composed of dark-colored fragments of the basaltic lava that make up the islands or of coarse fragments of coral debris from the reefs that develop around the margins of those islands that lie within the proper latitudes.

The slope of beaches is closely related to the size of the particles of which they are composed. Waves washing onto the beach carry particles that are deposited, thereby increasing the slope of the beach. If the backwash returns as much sediment as the waves carried in, the beach has reached equilibrium and will not steepen. A beach composed of fine-grained sand that is relatively angular will have a gently sloping, firm surface. Due to the fact these small grains interlock closely, little of the swash sinks down among the grains. Most of it runs back down the slope to the ocean, possessing enough energy to maintain equilibrium on a gentle slope. The backshores of such beaches are usually nearly horizontal.

Beaches composed of coarse sands or pebbles will usually contain more rounded particles that are more loosely packed. The swash can quickly percolate into such deposits when it moves up the beach slope. Deposition of particles by the swash will continue until the beach slope becomes steep enough that the backwash running to the ocean has sufficient energy to maintain equilibrium. Such beaches are usually much less firm than beaches composed of finer material, and the backshores will slope significantly toward the coastline (table 11–1).

Sedimentary Features Numerous depositional features that are partially or wholly separated from the shore of the continent are deposited by the longshore drift and other processes that are not well understood. A *spit* is a linear ridge of sediment attached at one end to land. The other end of the deposit points in the direction of longshore drift and ends in the open water. Spits are

TABLE 11–1
The Relationship of Particle Size to Beach Slope.

Wentworth Particle Size	(mm)	Mean Slope of Beach
	256	
Cobble	64	24°
Pebble	4	17°
Granule	2	11°
Very coarse sand	1	9°
Coarse sand	0.5	7°
Medium sand	0.25	5°
Fine sand	0.125	3°
Very fine sand	0.063	1°

Source: After Table 9, "Average beach face slopes compared to sediment diameters" from *Submarine geology,* 2nd ed. by Francis P. Shepard (Harper & Row, 1963), p. 171.

simply extensions of beaches into the deeper water near the mouth of a bay. The open-water end of the spit will normally curve into the bay as a result of current action.

If tidal currents or the currents initiated by river runoff are too weak to keep the mouth of the bay open, the spit may eventually extend across the bay and tie to the mainland to completely separate the bay from the open ocean. The spit then has become a *bay barrier*. A *tombolo* is a sand ridge that connects an island with another island or the mainland. Such a feature probably results from the growth of a spit (figure 11–4).

Barrier islands are long offshore deposits of sand lying parallel to the coast. The origin of a barrier island is complex, and it may be that there are several explanations for their existence. It appears, however, that many such structures developed during the rise in sea level that was initiated with the last melting of the glaciers some 18,000 years ago.

Barrier islands are rather continuous along the Atlantic coast of the United States. They extend around Florida and along the Gulf of Mexico coast, where they may be found well past the Mexican border. Barrier islands may attain lengths in excess of 100 km (62 mi) and have widths of several kilometers. Examples of such features are Coney Island off the New York coast and Padre Island off the coast of Texas. Barrier islands are an integral part of some estuarine environments.

A typical barrier island has the following physiographic features from the ocean to the lagoon behind it: (1) ocean beach, (2) dunes, (3) barrier flat, and (4) salt marsh (figure 11–4B). The *ocean beach* is typical of the beach environment discussed earlier in this chapter. During the summer, as gentle waves carry sand to the beach, it widens and becomes steeper. Higher-energy winter waves carry sand offshore and produce a narrow, gently sloping beach.

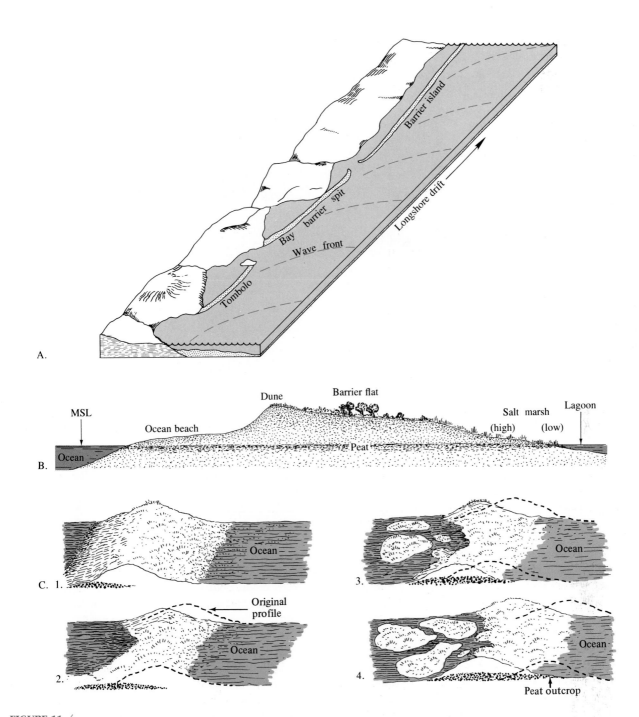

FIGURE 11–4

Coastal Depositional Features. *A,* a *tombolo* is a deposit that connects an island to the mainland or another island. A *spit* is a deposit that extends from land into open water. A spit that is extended across a bay and closes the bay off from open water is a *bay barrier.* A *barrier island* is a long deposit separated from the mainland by a lagoon. *B, cross section through a barrier island.* The major physiographic zones of a barrier island are the *ocean beach, dunes, barrier flat,* and *high* and *low salt marsh.* MSL is mean sea level. The peat bed represents ancient marsh environments that have been covered by the island as it migrates toward the mainland with rising sea level. *C,* (1) barrier island before overwash with salt marsh development on mainland side; (2) overwash erodes seaward margin of barrier island, cuts inlets, and carries sand into the lagoon, covering much of the existing marsh; (3) new marsh forms in the lagoon that has retreated toward the mainland; (4) with continued rise in sea level and migration of the barrier island toward the mainland, peat formed from previous marsh deposits may be exposed by erosion of the foreshore of the ocean beach.

FIGURE 11–5

Deltas. *A,* photographed from *Gemini IV.* The relatively smooth, curved surface of the Nile Delta can be seen. The Mediterranean Sea is to the left, and the northern Red Sea is in the upper right. (Photo courtesy of NASA.) *B,* digitate structure of the Mississippi River delta results from low-energy environment. Location of present main channel and older main channels (Atchafalaya River and Bayou Lafourche) show how river shifts position with time.

A.

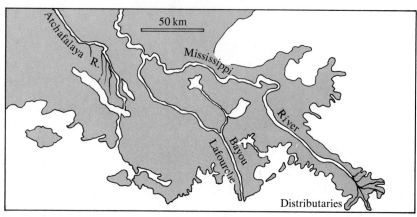

B.

Winds blow sand inland during dry periods to produce *dunes,* which are stabilized by dune grasses that can withstand salt spray and burial by sand. Dunes are the estuary's primary protection against excessive flooding during storm-driven high tides. Numerous passes exist through the dunes, particularly along the southeast Atlantic coast, where dunes are less well developed than to the north.

Behind the dunes the *barrier flat* forms as the result of deposition of sand driven through the passes during storms. These flats are quickly colonized by grasses. If for some reason the frequency of overwash by storms decreases, the grasses will successively be replaced by thickets, woodlands, and forests.

Salt marshes typically lie inland of the barrier flat. They are divided into the *low marsh,* extending from about mean sea level to the high neap tide line, and the *high marsh,* extending to the highest spring tide line. The low marsh is by far the most productive part of the salt marsh. New marsh is formed as overwash carries sediment into the lagoon, filling portions so they are intermittently exposed by the tides. Marshes may be poorly developed on parts of the island great distances from flood tide inlets. Their development is greatly restricted behind barrier islands where artificial dune enhancement and inlet filling prevent overwashing and flooding.

Due to the gradual rise in sea level relative to the eastern North American coast, most barrier islands are migrating landward, a fact clearly visible to those who build structures on these islands. Evidence for such migration can be seen in peat deposits cored beneath the barrier islands. Since the only barrier island environment in which peat forms is the salt marsh, the island must have moved inland over previous marsh development.

A possible cycle of salt marsh formation and destruction by landward migration of barrier islands is presented in figure 11–4*C*.

Deltas Some rivers carry more sediment than can be distributed by the longshore current. Such rivers develop a *delta deposit* as sediment settles out at the river mouth. One of the largest such features is that produced by the Mississippi River. Deltas are fertile, flat areas that are subject to periodic flooding.

Delta formation begins when a river has filled its estuary with sediment. Once the delta forms, it grows through the distribution of sediment by *distributaries,* branching channels that radiate out over the delta. Distributaries lengthen as they deposit sediment and produce fingerlike extensions to the delta. When the fingers get too long, they become choked with sediment. At this point, a flood may easily cause a shift in the distributary course and provide sediment to the low-lying areas between the fingers.

In areas where stream depositional processes are dominant over coastal erosion and transportation processes, the "bird-foot" Mississippi-type delta results. Where erosion and transportation processes exert a significant influence, the shoreline of the delta will be smoothed to a gentle curve like that of the Nile Delta (figure 11–5). The Nile Delta is presently eroding due to the entrapment of sediment behind the Aswan High Dam, which was completed in 1964. Erosion is only one of the negative effects of the construction of this dam, and this may have made possible the recent discovery of the ruins of ancient Alexandria beneath the Mediterranean waters at the edge of the delta. The problem of reduced availability of sediment may become very significant along the eastern end of the delta, where subsidence has occurred at a rate of 0.5 cm/year (0.2 in./year) for 7500 years.

CHANGING LEVELS OF THE SHORELINE

Along some coasts there are flat platforms backed by cliffs called *marine terraces, stranded beach deposits,* and other evidence of marine processes many meters above the present shoreline. These features characterize shorelines of emergence that have reached their present position relative to the existing shoreline either by an uplift of the continent, the lowering of sea level, or a combination of the two.

In other areas one may find *drowned beaches* and *submerged dune topography* beneath the water overlying the continental shelf. These features along with the drowned river mouths along the present shoreline indicate that past shorelines have been submerged. This submergence must have been caused by a subsidence of the continent, a rise in sea level, or a combination of the two.

Attempts to determine the causes of the change in relative level of the ocean and the continent have not met with great success. Whether the shoreline has become submerged because of a rising sea level or a subsiding continent in a particular region cannot be determined by examining the coastal features in that area, since both processes produce the same end results (figure 11–6).

Tectonic and Isostatic Movements

Changes in sea level relative to the continent may occur because of movement of the land—tectonic movement. Such movement includes large-scale uplift or subsidence of large portions of continents or ocean basins or more localized deformation of the continental crust involving folding, faulting, and tilting. The earth's crust also responds isostatically to the accumulation of heavy loads of ice, water, sediment, or lava.

There is evidence that during the last 2.5 to 3 million years at least four major accumulations of glacial ice developed in high latitudes. Although Antarctica is still covered by a very large glacial accumulation, much of the ice cover that once existed in northern Asia, Europe, and North America has disappeared. The most recent period of melting began about 18,000 years ago. Accumulations of ice that were up to 3 km (1.9 mi) in thickness have disappeared from northern Canada and Scandanavia. While these areas were beneath the thick ice sheet, they were pushed down and are still in the process of recovery after the melting of the ice. Some geologists believe that by the time the isostatic recovery is finished, the floor of Hudson Bay, which is now about 150 m (492 ft) deep will be above sea level. There is evidence in the Gulf of Bothnia of 275 m (900 ft) of isostatic recovery during the last 18,000 years. Generally, tectonic changes in the level of the shoreline are confined to a segment of the shoreline of a given continent.

FIGURE 11–6
Evidence of Changing Levels of the Shoreline. Terraces resulting from the exposure of ancient sea cliffs and wave-cut benches above present sea level mark ancient shorelines as do drowned beaches that lie below sea level.

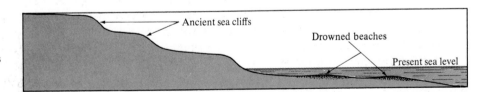

Ancient sea cliffs

Drowned beaches

Present sea level

Eustatic Movements

Changes in the level of the shoreline that can be measured on a worldwide basis because they are caused by the increase or decrease of water volume in the ocean or the capacity of the ocean basin are termed *eustatic*. This term refers to the highly idealized situation in which all the continents remain static while the sea rises or falls. Such changes in sea level could be created by the formation or the destruction of large inland lakes. Probably more important is the locking into or the release from continental glaciers of the earth's water during glacial and interglacial stages. It is also possible that the changes in the capacity of the ocean basin could result from changes in the average rate of sea-floor spreading. High rates of spreading would produce broader oceanic rises and cause a rise in sea level.

During the Pleistocene epoch, when the previously mentioned glacial advances were occurring, the amount of water in the ocean basin fluctuated considerably. Since the climate was colder during ice advances, we might account for some of the lowering of the shoreline by the contraction of the ocean volume as its temperature decreased. It has been calculated that for every 1°C (1.8°F) decrease in the mean temperature of the ocean water, sea level would drop 2 m (6.6 ft). Temperature indications derived from study of fossils from Pleistocene ocean sediments indicate the ocean temperature may have been as much as 5°C (9°F) lower than at present. Therefore, contraction of the ocean water may have lowered sea level by about 10 m (33 ft).

Although it is difficult to say with certainty what the range of shoreline fluctuation had been during the Pleistocene, there is cause to believe that the shoreline has been at least 120 m (394 ft) below the present shoreline. It is also estimated that if all the remaining glacial ice on earth were to melt, sea level would rise by another 60 m (197 ft). This would give a minimum possible range of sea level during the Pleistocene of 180 m (590 ft), most of which must be explained through the capture and release of the earth's water by glaciers. Such changes in sea level are termed *glacioeustatic oscillations*.

During the last 18,000 years the ocean volume has been increasing due to the expansion of the water resulting from warmer temperature and the melting of polar sea ice and glacial ice. Since the combination of tectonic and eustatic changes in sea level may be very complex and difficult to identify, it is hard to classify coastal regions as purely emergent or submergent. Most coastal areas show evidence of having experienced both submergence and emergence in the recent past.

In 1988, the World Meteorological Organization published a report stating that the *greenhouse effect* is being enhanced by the increased amounts of carbon dioxide, methane, and other gases that are entering the atmosphere. The report estimates that the average temperature of the earth will increase by from 0.25° to 0.8°C (0.45° to 1.44°F) per decade. This temperature increase could melt enough glacial ice to raise sea level from 20 to 152 cm (8 to 60 in.) in 50 years. Figure 11–7 shows the change in global mean sea level that has been recorded from 1880 to 1980.

The report said that even if we stopped emitting the gases into the atmosphere, temperatures and sea level would rise for another 100 years because of the gases already added to the atmosphere. Many coastal cities are reported to be in danger of flooding, and some low-lying nations such as Bangladesh and the Netherlands

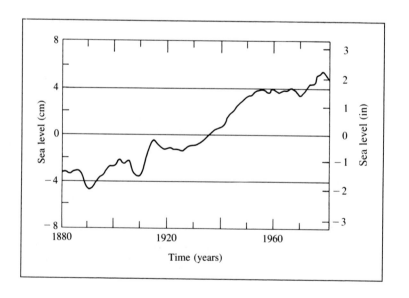

FIGURE 11–7

Sea-level Change from 1880 to 1980. Tide-gauge data averaged over 5-year periods show that global mean sea level has increased about 10 cm (4 in.) over the last 100 years. More recent data that remove the effect of isostatic recovery, however, indicate that for the last 50 years global sea level has risen 2.4 ± 0.90 mm/year. That rate is 2.4 times the rate shown in this figure. (After V. Gornits, S. Lebedeff, and J. Hausen. 1982. *Science* 215: 1611–14.

may suffer severe damage, especially during large storms.

If such changes do occur, we will have to add human activities as a significant cause of changing sea level. We can only hope to moderate these predicted events by modifying our industrial and transportation policies and by putting an end to deforestation. Should we not reverse the present trend, humans will become the most effective cause of change in climate and sea level. The predicted rate of change is as great or greater than that of any changes known to have been caused by natural processes.

THE UNITED STATES COASTS

The recent rise in sea level discussed in the previous section may well have been caused primarily by climatic changes brought about by human activities. However, other lines of evidence indicate that prior to this change there had been no significant change in sea level due to glacioeustatic oscillations during the last 3000 years. If this is so, we must look to tectonic processes to explain the changes in the shoreline relative to the continents during the recent past. To understand the underlying tectonic cause of emergence or submergence, we must return to the concept of global plate tectonics.

Atlantic-type Margins Considering only the coasts of North America, we find that the Atlantic coast is subsiding and the Pacific coast is emerging. Beginning with the breakup of Pangaea and the presently forming Atlantic Ocean, this emergence-submergence pattern can be readily understood. It was previously discussed that the lithosphere thickened, the ocean deepened, and heat flow decreased as the lithospheric plates cooled with age—with increasing distance from the spreading centers. Due to this process, the ocean floor of the North American Atlantic coast is thought to have subsided about 3 km (1.9 mi) over the last 150 million years.

Erosion of the continent produced sediment that has accumulated to a maximum thickness of 15 km (9.3 mi) along the Atlantic coast. This thickness of sediment has been made possible by the plastic asthenosphere allowing the lithosphere to flex downward as the sediment burden is increased. It is this thick sedimentary wedge underlying the continental shelf, slope, and rise that has been exploited for its large petroleum reserves in the Gulf of Mexico. Recent attention to its exploration off the Atlantic coast has resulted in cautiously optimistic projections of large petroleum reserves there.

As the upwelling initially split Pangaea, the granitic, continental-crustal rocks near the split were heated sufficiently to stretch. This produced a gradually thinning continental or transitional crust that gave way to oceanic crust 100 to 150 km (62–93 mi) from the present shore beneath the continental rise (figure 11–8). Thus, the thick wedge of sediment underlying the continental shelf, slope, and rise is deposited on marginal continental crust, transitional crust, and oceanic crust.

This describes the general conditions producing the subsidence of the Atlantic and Gulf coasts. Although the rate of thermal subsidence decreases with time and the rate of subsidence due to sediment loading depends on sediment supply, subsidence will continue as long as the existing spreading and depositional regimes prevail.

Pacific-type Margins. In contrast to the *passive* and subsiding Atlantic-type margin, the Pacific-type or *active* margin is the scene of intense tectonic activity. Along such margins, the eruption of volcanoes and shocks of earthquakes are common. The thick, broad sediment wedge characteristic of the Atlantic coast is not well developed. Instead of being the product of the creation of a new ocean basin, the Pacific-type margin is the scene of lithospheric plate destruction by subduction.

The characteristic alignment of mountain ranges parallel to the continental margin can be seen along such coasts. In all cases of continental margins associated with plate convergence, the compressive forces produce uplift. Along the California coast, the process has been further complicated with the development of the San Andreas Fault, a transform fault. This has made possible the local subsidence of the depositional basins in the Los Angeles area. But on the broader scale, emergence along Pacific-type margins is the characteristic condition.

A discussion of the Gulf of Mexico, Atlantic, and Pacific coasts will combine the effects of submergence and emergence that result from the plate tectonic interactions with the conditions of erosion and deposition along these coasts. The sum of these processes along these coastal regions should explain why the features found along specific segments of the U.S. coast have developed.

The Gulf Coast

The Louisiana-Texas coastline is dominated by the Mississippi River delta, which is being deposited in a microtidal and generally low-energy environment. The tidal range is normally less than 0.5 m (1.6 ft), and with the exception of the hurricane season, wave action is generally low. The sediments that have been deposited in the Gulf Coast region indicate that this area has been generally subsiding for approximately 150 million years. The sedimentary section that can be found beneath the coast is essentially continuous from the present back to the Cretaceous. For these sediments to have been de-

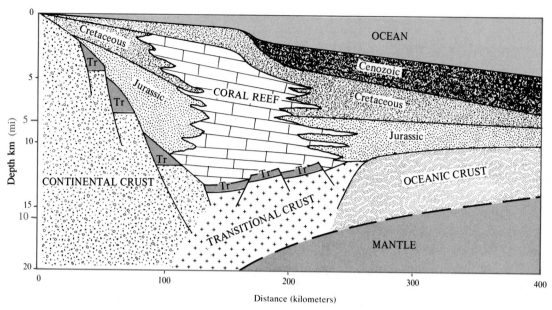

FIGURE 11–8

Atlantic-type Margin Crustal and Sediment Units. As Pangaea was rifted apart by upwelling magma, the continental crust was heated enough to flow plastically and thinned at the edge of the continent, which produced a transitional crust that gave way to oceanic crust. The Triassic sediments filling the initial rift valleys are overlain by Jurassic, Cretaceous, and Cenozoic sediments that reach thicknesses of 10 to 15 km (6 to 9 mi) as the lithosphere flexes down in the asthenosphere under the increasing burden of the overlying sediment.

posited, preserved, and overlain by younger sediment, at least a low rate of subsidence was required. Even though the Gulf coast is subsiding, the shoreline is building seaward in some regions. The tremendous volume of sediment that is being carried by the Mississippi River and deposited in its delta has extended the delta completely across the continental shelf (figure 11–9).

Along the Louisiana Gulf Coast a nip-and-tuck battle is going on between the forces of deposition, attempting to advance the coastal region, and submergence, threatening to make the coastline retreat inland. Locally, one process may be winning for a short period of time. Areas some distance from the delta cease to receive new sediment, continue to subside, and the shoreline moves inland. In other areas, the shoreline is pushed seaward by active deposition. In the city of New Orleans, it obviously would not be desirable to allow the Mississippi River to continue to add sediment. Without this deposition, the city has subsided so that much of it is well below sea level. Therefore, a constant battle must be waged to keep the area sufficiently diked to keep out the ocean and the river.

A chain of barrier islands extends along the coast east of the Mississippi Delta to the Florida panhandle and west of the delta to the Mexican border. These deposits, which are over 100 km (62 mi) long and more than 2 km (1.2 mi) wide, are separated from the mainland

by shallow lagoons and indicate the subsidence (figure 11–10).

The Atlantic Coast

Moving east to the Florida coastline, we encounter a gradually subsiding area with very low relief. The bedrock of most of the state of Florida is a resistant limestone that was alternately inundated and totally exposed numerous times during the Pleistocene. The area of the greatest ecological significance lies at the southern tip of the state—the Everglades. This coastal region is also a relatively low energy area. It is protected by the Florida Keys from large waves and receives high energy assault only during local thunderstorms and hurricanes, as is generally the case for the Mississippi Delta region.

Drainage of surface water across the Everglades from northern Florida once occurred as a thin sheet less than 1 m (3.3 ft) thick that flowed south along the gentle slopes of the Everglades to the ocean. This flow was not restricted to river channels but covered essentially the breadth of the state. Within this unique environment developed an assemblage of plant and animal life that is presently threatened.

To control flooding, the U.S. Army Corps of Engineers restricted the sheet runoff to flood control canals that have left portions of the Everglades short of water.

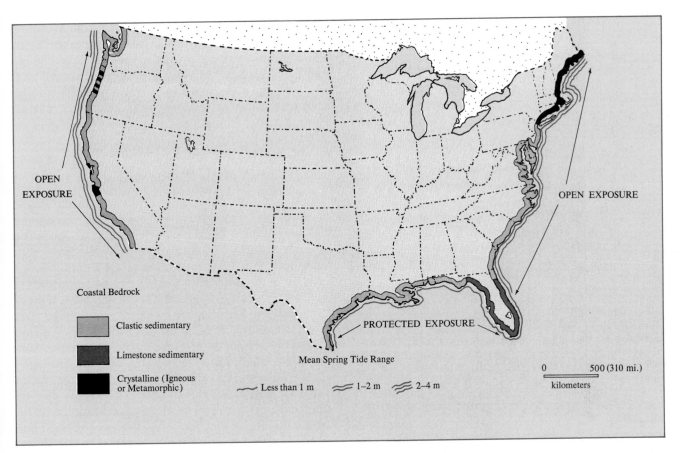

FIGURE 11–9
Distribution of Conditions Related to Coastal Erosion.

This shortage of water has endangered much of the plant and animal life. Also resulting from this controlled flow of water is a threat that was previously seldom known in the Everglades. Fire has become a serious problem.

Although isostatic rebound resulting from the removal of the continental glacier that once covered the continent north of New York has counteracted the tectonic subsidence, evidence indicates a gradually retreating shore to the south. South of New York, sea level is rising at an average rate of 0.3 m (1 ft) per century relative to the subsiding land.

The coastal rocks from Florida through New Jersey are predominantly poorly consolidated sedimentary rocks formed in the recent geologic past. These rocks do not provide great resistance to erosion and are sources of sand that is deposited as barrier islands and other depositional features common along this coast.

The increase in the volume of water in the ocean with the melting of the glaciers at the close of the Pleistocene has caused a significant "drowning" of river valleys that once flowed into an Atlantic Ocean that was a considerable distance seaward of the present shoreline. Delaware Bay, Chesapeake Bay, and the irregular land-

FIGURE 11–10
Galveston Bay. *Apollo 9* view from an altitude of 191 km (119 mi) of the Texas coast from Freeport to Sabine Lake. The barrier islands, Line Island, and Galveston Island extend respectively, to the north and south of Galveston Bay. (Courtesy of NASA.)

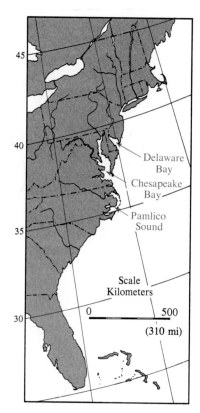

FIGURE 11–11
Drowned River Valleys along the Atlantic Coast.

much of the mainland coast is shielded from the direct attack of nor'easters by barrier islands, there are about 30 of these storms each year that generate large enough waves to erode barrier island beaches and frontal dunes.

The granite of the Maine coast has resisted erosion; however, zones of relatively weaker material have allowed erosion to produce a rugged and irregular coastline.

The Pacific Coast

Unlike the Gulf and Atlantic Coast regions, the Pacific Coast of the United States is characterized by a continental margin that is rising rapidly relative to sea level. The folding, faulting, and volcanic activity associated with this uplift can be observed along the entire western margin of the continent.

Throughout the Pacific Coast, the shoreline is characterized by cliffs that are being actively eroded. Old cliffs and wave-cut terraces that have been elevated above the present sea level during the past million years or so are manifested as terraces that are well developed in the Palos Verdes Hills in the Los Angeles area (figure 11–12). Here 13 uplifted wave-cut terraces can be counted, and the highest one is approximately 400 m (1300 ft) above the sea level. Because of the rapid coastal uplift, it is difficult to determine the effect of sea-level fluctuation due to the melting and forming of glaciers.

Rocks outcropping along the coast are, for the most part, relatively weak marine sediments that are geologically young. Locally, more resistant granite and metamorphic and volcanic rocks may form the bedrock along the coast. The effects of glaciation characterized by the fjords along the Canadian coast extend southward into the Washington coastal region and produce the Strait of Juan de Fuca and Puget Sound.

The Pacific Coast of the United States is characterized by high-energy conditions. The waves are normally about 1 m (3.3 ft) in height. Frequently, the wave height will increase to 2 m (6.6 ft), and a few times a year 6-m (20-ft) waves hammer the coast. In addition to the Pacific storms that generate these large waves, *tsunamis* periodically strike the coastal region. The strong winter waves that generally approach the coast from a west-northwesterly direction produce a strong longshore current that carries millions of tons of sediment in the longshore drift.

During the winter, many beaches have the sand eroded from the foreshore area due to the high energy waves. The exposed beaches, which are composed primarily of pebbles and boulders during the winter months, regain their sand as lower energy summer

ward margin of Pamlico Sound are of this origin. These estuaries are in the process of being filled by sediment carried to the coastal region by the rivers that drain the inland areas (figure 11–11).

Because these sediments remain for the most part in the estuaries, there is little sediment made available to the longshore currents by river runoff. This condition has produced locally severe coastal erosion problems. The wave energy released in these coastal areas is greater than in the Gulf of Mexico, since the area is exposed to the wave action from the open Atlantic Ocean. High-energy conditions develop during storms and when hurricanes strike the coastal area.

From New York north there is considerable evidence that glaciers directly affected the coastal region. This area is subject to even higher energy conditions than the coastal region farther south, particularly during the fall and winter. The "nor'easters" affect the coast from Cape Hatteras north. The high energy of these storms is manifested in up to 6-m (20-ft) waves with a 1-m (3.3-ft) rise in sea level that follows the low pressure as it moves northward. Such high energy conditions may be expected to cause considerable erosion along the coastal region that are predominantly depositional. Although

A.

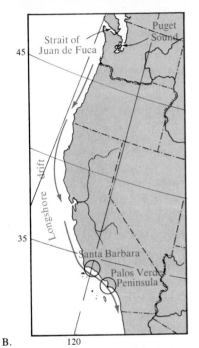

B.

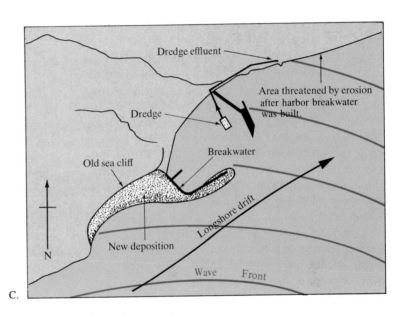

C.

FIGURE 11–12

Features of Special Interest along the Pacific Coast. *A,* Marine Terraces of Palos Verdes Hills, California. (Photo courtesy of John Shelton.) *B,* location of Palos Verdes Peninsula, Santa Barbara Harbor, Strait of Juan de Fuca and Puget Sound. *C,* Santa Barbara Harbor. Construction of a breakwater to the west of Santa Barbara Harbor interfered with the eastward moving longshore drift, creating a broad beach. Sand being deposited against the breakwater was no longer available to replace sand being removed by wave erosion to the east. As the beach extended around the breakwater into the harbor, dredging operations were initiated to keep the harbor open and to put the sand back into the longshore drift. This helped reduce the coastal erosion east of the harbor. (U.S. Coast and Geodetic Survey Chart 5161.)

waves beat more gently against the shore. Damming of many of the rivers along the Pacific Coast for flood control has reduced the amount of sediment for the longshore transport. Due to this sediment deficiency, erosion along the coast is becoming a significant problem as the sand-starved beaches disappear.

EFFECT OF ARTIFICIAL STRUCTURES

Jetties that have been constructed to protect harbor entrances from wave action are the most common barriers to the longshore transport of sediment. Many examples of the destructive results associated with the construction of such features along the shore can be cited. Along the Southern California coast, cases of excessive beach erosion related to the development of local harbor facilities are especially numerous. The longshore drift in California is predominantly from north to south, and the construction of a jetty results in the entrapment of sediment on the north side, causing increased erosion on the south side. A jetty constructed as a breakwater to protect harbor waters at Santa Barbara had a very significant effect upon the equilibrium established in this coastal region. The barrier represented by the breakwater on the west side of the harbor caused an accumulation of sand that was moving in an easterly direction along this portion of the California coast, which trends generally east-west. The beach to the west of the harbor continued to grow until finally the sand moved around the breakwater and began to fill in the harbor (figure 11–12*C*).

While deposition was occurring at an abnormal rate to the west, erosion was proceeding at an alarming rate east of the harbor. The energy available east of the harbor was no greater than it had been previously, but the sand that had formerly moved down the coast to replace sand removed was no longer available due to its entrapment behind the breakwater. To compensate for this deficiency of sand down-current from the harbor, dredging had to be initiated within the harbor. The dredging keeps the harbor from filling in, and the sand is pumped down the coast so it can reenter the longshore drift and replenish the eroded beach. The dredging operation has stabilized the situation, but at a considerable expense. It seems obvious that any time human beings interfere with natural processes in the coastal region, they will have to provide the energy needed to replace that which they have misdirected through modification of the shore environment.

Along many coasts, small jettylike structures called *groins* have been constructed at intervals along the beach (figure 11–13). They have helped beach development by causing sand to accumulate on the north side of these structures. Being shorter than jetties built to protect harbors, the groins eventually allow the sand to migrate around their ends. An equilibrium may be reached which allows sufficient sand transport along the coast before excessive erosion occurs down-current from the last groin. Although some serious erosional problems in these regions have developed, construction of a series of groins usually results in much less beach erosion than the construction of one large jetty.

One of the most destructive of structures built to "stabilize" eroding shores is the *seawall.* Instead of extending out into the longshore current as do groins and jet-

FIGURE 11–13
Groins. As sand drifts along the shore of Miami Beach, it accumulates on the up-current side of the groins to aid in increasing the width of the beach. Arrow indicates direction of longshore transport of sediment. (Photo courtesy of U.S. Army Corps of Engineers.)

ties, they are built parallel to the shore to protect developments landward of them from the action of ocean waves. Once waves begin breaking against the seawalls, turbulence generated by the abrupt release of wave energy quickly erodes the sediment on their seaward side, causing them to collapse into the surf (figure 11–14). Where they have been used to protect property on barrier islands, the seaward slope of the island beach is steepened and the rate of erosion increased.

Although most state laws say beaches belong to the public, government actions allow the owners of shore property to destroy beaches in an effort to provide short term protection to their property. To add to the public insult, taxpayers subsidize property owners in many shore developments by rebuilding their storm-damaged properties at great public expense.

An interesting possible alternative to the solid structures used to protect local shore areas from ocean waves is the tethered-float breakwater developed at Scripps Institution of Oceanography (figure 11–15). Tested successfully in San Diego Harbor, these inverted pendulums protect against coastal wave erosion by extracting energy from passing waves without stopping the transport of sediment. If they prove durable enough, they would certainly be a welcome substitute to traditional breakwaters.

FIGURE 11–14

Seawalls. When a seawall is built on a beach to protect beachfront property, the first large storm will remove the beach from the seaward side of the wall and steepen its seaward slope. Eventually, the wall is undermined and falls into the sea, and the property is lost as the oversteepened beach slope advances landward in an effort to reestablish its natural slope angle.

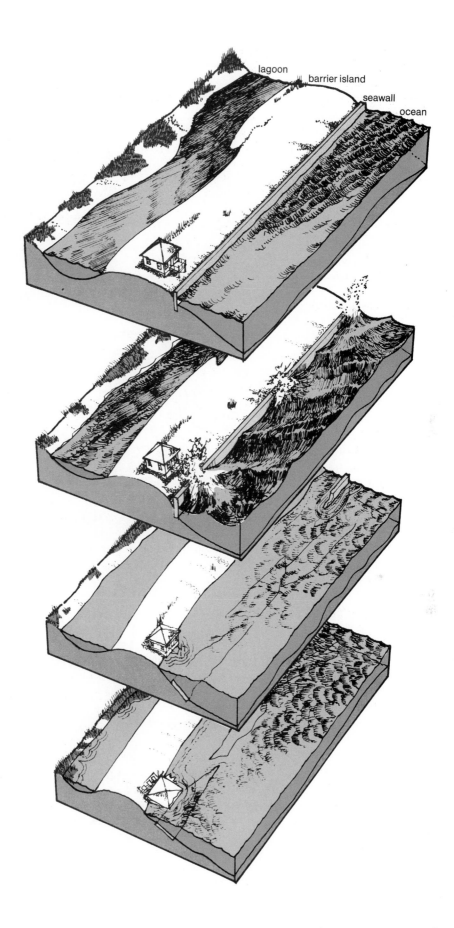

FIGURE 11–15

Tethered-Float Breakwaters. *A,* deployment of tethered-float breakwater at San Diego Harbor. (Photograph courtesy of Scripps Institute of Oceanography, University of California, San Diego.) *B,* artist's sketch of a tethered-breakwater system incorporating 1.5-m (4.8-ft) spheres of steel placed 1.5 m apart. The spehres are anchored to the ocean floor and held just beneath the ocean surface. This system has the advantage over traditional breakwater structures in that it removes energy from waves but does not interfere with the transport of sediment along the shore. (Courtesy of Scripps Institute of Oceanography, University of California, San Diego.)

A.

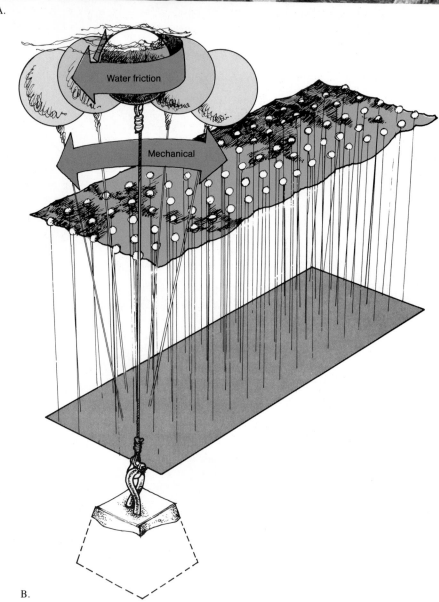

B.

SUMMARY

The region of contact between the oceans and the continents is marked by the **shore**, lying between the lowest low tides and the highest elevation on the continents affected by storm waves. The **coast** extends inland from the shore as far as land features related to marine processes can be found. The shore is divided into the **foreshore**, extending from low tide to high tide, and the **backshore**, extending beyond the high tide line to the **coastline** that separates the shore from the coast. Seaward of the low tide line are the **nearshore zone**, extending to the **breaker line**, and the **offshore zone** beyond.

Wave erosion of the shore produces a **wave-cut cliff** that constantly retreats, leaving behind such features as **sea caves**, **sea arches**, and **stacks**. The rate of wave erosion increases with increased **exposure** of the shore to the open ocean, **decreasing tidal range**, and decreasing **strength of bedrock**.

As waves break at an angle to the shore, a **longshore current** is set up that produces a **longshore drift** of sediment along the shore. Deposition of sediment transported by the longshore current may produce such features as **beaches**; **spits**, deposits attached at one end to land; **tombolos**, sand ridges connecting an island to another island or the mainland; and **barrier islands**, long offshore deposits lying parallel to the shore. (From the ocean side to the lagoon side, barrier islands commonly exhibit the following divisions: **ocean beach, dunes, barrier flat**, and **salt marsh**.) **Deltas** form at the mouths of rivers that carry more sediment to the ocean than can be distributed by the longshore current.

A drop in sea level may be indicated by **marine terraces** and **stranded beaches** well above the present shoreline, and a rise in sea level may be indicated by submerged **relict beaches**, wave-cut cliffs, and **drowned river valleys**. Such changes in sea level may result from tectonic processes causing local movement of the land mass or from **eustatic processes** changing the amount of water in the oceans or the capacity of ocean basins, thus causing worldwide changes in sea level. Melting of continental ice caps during the last 18,000 years has caused a **glacioeustatic** rise in sea level of about 140 m (460 ft).

The U.S. **Gulf Coast** is undergoing tectonic subsidence; therefore, the shoreline tends to move inland. This tendency is reversed only in areas where deposition of deltas, particularly the Mississippi River delta, is occurring. Although the tidal range is small and the bedrock easily eroded, coastal erosion is minimal because the shore is not exposed to the open ocean. Tidal range increases as we travel north along the **Atlantic Coast**, and the bedrock becomes more resistant to erosion. Although this coast is slowly retreating in the south, it may still be undergoing isostatic uplift in the north. Open exposure to attack by severe storms results in some significant erosion problems. The **Pacific Coast** is definitely rising tectonically, while tidal ranges from 1 to 2 m (3.3–6.6 ft) and relatively soft bedrock along much of the openly exposed shore cause erosion to dominate over deposition.

Artificial structures along the shore, such as **jetties** to protect harbors and **groins** used to widen beaches, will trap sediment on the upcurrent side, but erosion may then become a problem downcurrent. **Seawalls** built parallel to the shore hasten the rate of shore erosion.

QUESTIONS AND EXERCISES

1. To help you reinforce your knowledge of the shore, construct and label your own diagram similar to that in figure 11–1.

2. Discuss the formation of such erosional features as sea cliffs, sea caves, sea arches, and stacks.

3. List and discuss three factors in the rate of wave erosion.

4. What variables affect the velocity of the longshore current?

5. What is the longshore drift, and how is it related to the longshore current?

6. Discuss the composition of beaches and the relationship of beach slope to particle size.

7. List and define the following depositional features: spit, tombolo, bay barrier, and barrier island.

8. How do peat deposits develop beneath barrier islands?

9. Discuss why some rivers have deltas and others do not. Also include the factors that determine whether a bird foot or a smoothly curved Nile-type delta will form.

10. Compare the causes and effects of tectonic versus eustatic changes in sea level.

11. List the two basic processes by which coasts advance seaward, and list their counterparts that lead to coastal retreat.

12. Describe the tectonic and depositional processes causing subsidence along Atlantic-type margins.

13. How does the Pacific-type margin differ from the Atlantic-type margin?

14. Discuss the Gulf Coast, Atlantic Coast, and Pacific Coast by describing the conditions and features of emergence-submergence and erosion-deposition that are characteristic of each.

15. Describe the effect on erosion-deposition caused by putting an artificial structure such as a breakwater or jetty across the longshore current and drift.

REFERENCES

Bird, E. C. F. 1969. *Coasts,* vol. 4 of *An introduction to systematic geomorphology.* Cambridge, Mass.: The M.I.T. Press.

Burk, K. 1979. The edges of the ocean: an introduction. *Oceanus* 22:3, 2–9.

Coates, R., ed. 1973. *Coastal geomorphology.* Publications in Geormorphology. Binghamton, N.Y.: State University of New York.

Ingmanson, D. E., and Wallace, W. J. 1973. *Oceanology: an introduction.* Belmont, Calif.: Wadsworth.

Kuhn, G. G., and Shepard, F. P. 1984. *Sea cliffs, beaches, and coastal valleys of San Diego County: Some amazing histories and some horrifying implications.* Berkeley: University of California Press.

Leatherman, S. P. 1983. Barrier dynamics and landward migration with Holocene sea level rise. *Nature* 301:5899, 415–17.

Peltier, W. R., and Tushingham, 1989. Global sea level rise and the greenhouse effect: might they be connected? *Science* 244:4906, 806–10.

Remanathan V. 1988. The greenhouse theory of climate change: A test by an inadvertent global experiment. *Science* 240:4850, 293–99.

Shepard F. P. 1977. *Geological oceanography.* New York: Crane, Russak & Company.

Stanley, D. J. 1988. Subsidence in the northeastern Nile Delta: Rapid rates, possible causes, and consequences. *Science* 240:4851, 497–500.

SUGGESTED READING

Sea Frontiers

Carr, A. P. 1974. The ever-changing sea level. 20:2, 77–83.
A discussion of the causes of sea level change is very well presented.

Emiliani, C. 1976. The great flood. 22:5, 256–70.
An interesting discussion of the possible relationship of the rise in sea level resulting from the melting of glaciers 11,000 to 8000 years ago and ancient accounts of a great flood.

Feazel, C. T. 1987. The rise and fall of Neptune's kingdom. 33:1, 4–11.
Short-term and long-term fluctuations in sea level are discussed.

Fulton, K. 1981. Coastal retreat. 27:2, 82–88.
A discussion of the need to consider coastal erosion in the planning of coastal development.

Grasso, A. 1974. Capitola Beach. 20:3, 146–51.
The destruction of the beach of Capitola, California, shortly after the construction of a harbor by the U.S. Army Corps of Engineers at Santa Cruz to the north.

Mahoney, H. R. 1979. Imperiled sea frontier—barrier beaches of the East Coast. 25:6, 329–37.
The natural alteration of barrier beaches is considered.

Mooney, M. J. 1987. Forest in the dunes. 33:2, 120–27.
Centering on the dunes and Sunken Forest of Fire Island, the mobility of barrier islands is considered.

Schumberth, C. J. 1971. Long Island's ocean beaches. 17:6, 350–62.
A very informative article on the nature of barrier islands. The specific problems observed on the Long Island barriers serve as examples.

Westgate, J. W. 1983. Beachfront roulette. 29:2, 104–9.
The problems related to the development of barrier islands are discussed.

Scientific American

Bascom, W. 1960. Beaches. 203:2, 80–97.
A comprehensive consideration of the relationship of beach processes, of large and small scale, to release of energy by waves.

Dolan, R., and Lins, H. 1987. Beaches and barrier islands. 257:1, 68–77.
A discussion dealing with the ultimate futility of trying to develop and protect beaches and barrier islands.

Fairbridge, R. W. 1960. The changing level of the sea. 202:5, 70–79.
A discussion of what is known of the causes of changing level of the sea, which seem to be related mostly to the formation and melting of glaciers and changes in the ocean floor.

12

THE COASTAL OCEAN

GENERAL CONDITIONS

The primary difference between the coastal ocean and the open ocean is that of depth. Generally, rates of change in the nature of water with respect to distance and time are much greater in the shallower coastal ocean. Because of the shallowness of the coastal ocean, river runoff and tidal currents have a very significant effect on the nature of coastal water.

Salinity

River runoff has the direct effect of reducing salinity of the surface layer in areas where mixing is not significant and throughout the water column where mixing does occur. In areas where the precipitation on the landmass is predominantly rain, the runoff of the rivers will be at the maximum during the season of maximum precipitation. However, if the runoff is fed to a great extent by the melting of snow and ice, the season of maximum runoff will always occur during the summer. In general, salinity will be lower in coastal regions than in the open ocean due to the runoff of fresh water from the continents (figure 12–1).

Counteracting the effect of runoff in some coastal regions is the presence of prevailing offshore winds that usually have lost most of their moisture over the continent. These winds evaporate considerable quantities of water as they move across the surface of the coastal ocean. The increase in the evaporation rate in these areas tends to increase the surface salinity.

Temperature

In coastal regions of the ocean where the water is relatively shallow, very great ranges in temperature may occur on a yearly basis. Sea ice forms in many of the high-latitude coastal areas in which temperatures are determined by the freezing point of the water, which is generally above $-2°C$ ($28.4°F$). Maximum surface temperature in low-latitude coastal water may approach $38°C$ ($100°F$) in areas where the coastal water is somewhat restricted in its circulation with the open

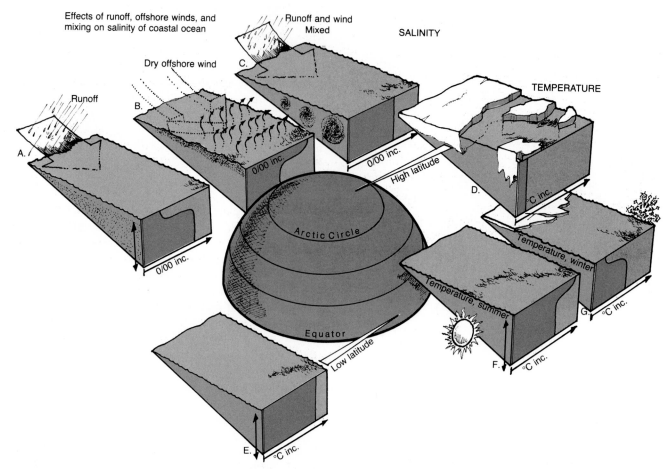

FIGURE 12–1

Temperature and Salinity Changes in the Coastal Ocean. *A*, freshwater runoff is not mixed into water column and forms surface layer of low salinity. A well-developed halocline occurs. *B*, warm, dry offshore winds cause a high rate of evaporation that may offset the effects of runoff, producing a halocline with a gradient that is the reverse of that seen in *A*. *C*, runoff is mixed with deeper water, producing an isohaline water column of generally lower salinity than the open ocean. *D*, in high latitudes where sea ice is forming or thawing throughout the year, the temperature of coastal water will remain uniformly near the freezing point. *E*, water in shallow, low-latitude coastal regions protected from free circulation with the open ocean may develop a high-temperature isothermal condition. *F*, in the mid-latitudes, coastal water will undergo a significant warming during summer. A strong seasonal thermocline may develop. *G*, winter may produce a layer of low-temperature water at the surface. The thermocline shown may develop; however, cooling may cause the surface water to sink because of increased density. This would result in a well-mixed isothermal water column. Mixing due to strong winds may drive the thermoclines shown in *F* and *G* deeper and may even cause mixing of the entire water column, producing a isothermal condition.

ocean. The seasonal change in temperature can most easily be detected in the coastal regions of the mid-latitudes, where surface temperatures are at a minimum in the winter and reach maximum values in the late summer.

Figure 12–1 shows how strong thermoclines may develop in areas where mixing does not occur. Very high temperature surface water may form a relatively thin layer. Mixing reduces the surface temperature by distributing the heat through a greater vertical column of water, thus pushing the thermocline deeper and making it less pronounced. One major factor in mixing coastal water is the effect of tidal currents, which can have a considerable influence on the vertical mixing of shallow water near the coast. Prevailing winds can also have a significant effect on surface temperatures if they blow from the continent. These air masses we previously mentioned in regard to salinity will usually be of relatively high temperature during the summer and cause an increase in the surface temperature of the ocean water. They will be of much lower temperature than the ocean surface during the winter and will cause a heat loss and cooling of the surface water near shore.

Coast Currents

Geostrophic Currents. The geostrophic effect that was discussed in association with the current gyres in chapter 8 also develops in the coastal ocean. The causes are basically the same. Wind blowing parallel to the coast in a direction that would cause water to pile up along the shore under the influence of the Coriolis effect produces a situation in which that water must eventually, under the influence of gravity, run back down the slope toward the open ocean. As the water runs down the slope away from the shore, the Coriolis effect causes it to veer to the north on the western coast and to the south on the eastern coast of continents in the Northern Hemisphere.

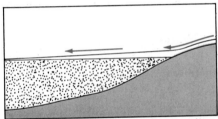

A. During the winter rainy season, fresh runoff water produces a seaward slope away from low-salinity surface water near the shore. A surface flow of low-salinity water from the shore toward the open ocean occurs.

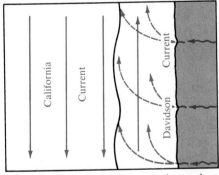

B. As the flow of surface water is acted on by the Coriolis effect, it veers right. This creates a north-flowing current (Davidson Current) along the coast of Washington and Oregon between the shore and the southbound California Current.

FIGURE 12–2
Davidson Current.

Another condition that produces geostrophic flow along the margins of a continent is the runoff of large quantities of fresh water that, due to its lower density, piles up on top of the oceanic mass in the coastal region. Gravity forces the water to begin to run out to sea down the slope of the mass of fresh water, as shown in figure 12–2. These variable currents, which depend upon the wind and the amount of runoff for their strength, are bounded on the ocean side by the more steady boundary currents of the open-ocean gyres.

These local geostrophic currents frequently flow in the opposite direction of the boundary current, as is the case with the Davidson Current that develops along the coast of Washington and Oregon during the winter. A very great amount of precipitation occurs in the Pacific Northwest during the winter months, and the winds that generally blow out of the southwest are the strongest during these same months. The combined effect is to produce a condition that develops a relatively strong northward-flowing geostrophic current between the southward-flowing California Current, which is part of the open-ocean circulation, and the continent of North America.

ESTUARIES

Estuaries (figure 12–3) are semienclosed coastal bodies of water in which the ocean water is significantly diluted by fresh water from land runoff. The mouths of large rivers throughout the world form the most economically significant estuaries, since many serve as ports and centers of ocean commerce. Many estuaries support important commercial fisheries as well, and the environmental changes that are occurring as a result of the commercial importance of the estuaries will be a major concern as the future development of these areas is planned. Many bays, inlets, gulfs, and sounds may be considered estuaries on the basis of their structure.

Origin of Estuaries

Essentially all estuaries in existence today owe their origin to the fact that in the last 18,000 years sea level has been raised approximately 130 m (426 ft) due to the melting of much of the major continental glaciers that covered portions of North America, Europe, and Asia during the Pleistocene epoch of geologic time, more commonly referred to as the *Ice Age.* Four major classes of estuaries can be identified on the basis of their origin (figure 12–3).

1. *Coastal plain estuaries* were formed as the rising sea level caused the oceans to invade the existing river

FIGURE 12–3
Classification of Estuaries Based on Origin.

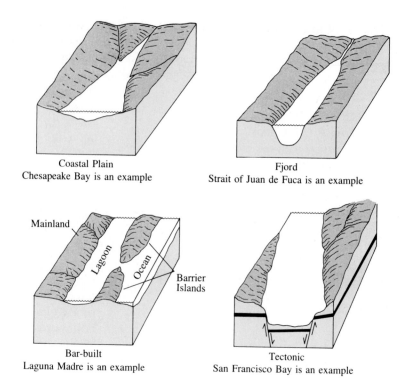

Coastal Plain
Chesapeake Bay is an example

Fjord
Strait of Juan de Fuca is an example

Bar-built
Laguna Madre is an example

Tectonic
San Francisco Bay is an example

valleys. These estuaries are sometimes referred to as *drowned river valleys.*

2. *Fjords* are glaciated valleys that are U-shaped with steep walls. They usually have a glacial deposit forming a sill near the ocean entrance.

3. *Bar-built estuaries* are shallow estuaries separated from the open ocean by bars composed of sand deposited parallel to the coast by wave action.

4. *Tectonic estuaries* are produced by faulting or folding, which causes a restricted down-dropped area into which rivers flow.

Water Mixing in Estuaries

Generally, the freshwater runoff that flows into an estuary moves as an upper layer of low-density water across the estuary toward the open ocean. An inflow from the ocean takes place below the upper layer, and mixing takes place at the contact between these water masses. Stommel has classified estuaries on the basis of the degree of mixing as determined by the distribution of water properties into the following categories, which are shown in figure 12–4.

Longitudinal sections showing salinity distribution ($^{0}/_{00}$)

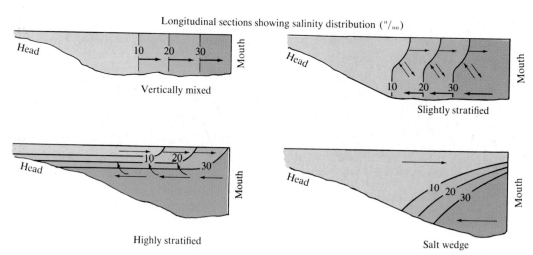

Vertically mixed

Slightly stratified

Highly stratified

Salt wedge

FIGURE 12–4
Classification of Estuaries Based on Degree of Mixing.

processes. They are of two types: *salt marshes* and *mangrove swamps*. Both are intermittently submerged by ocean water and are characterized by oxygen-poor mud and peat deposits. Marshes, characteristically inhabited by a variety of grasses, are known to occur from the equator to latitudes as high as 65°. Mangrove trees are restricted to latitudes below 30° (figure 12–6). Once mangroves colonize an area, they can normally outgrow and replace marsh grasses. With our strong desire to live near the oceans, marsh and mangrove wetlands have suffered from systematic filling and development for housing, industry, and agriculture.

Research is showing that wetlands have a very high economic value when left alone. Salt marshes are believed to serve as nursery grounds for over half the species of commercially important fishes in the southeast-

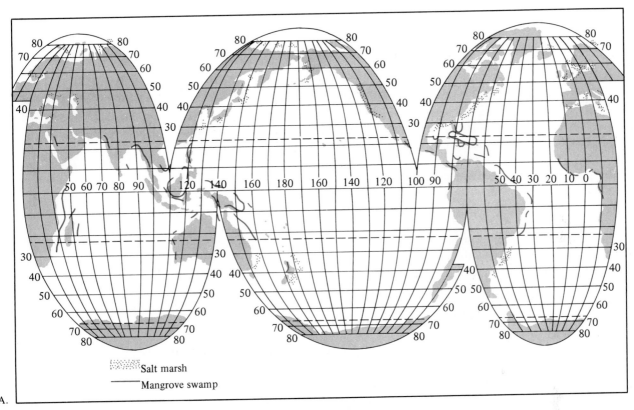

Salt marsh
Mangrove swamp

A.

B.

C.

FIGURE 12–6

World Distribution of Salt Marshes and Mangrove Swamps. *A,* note dominance of mangrove swamps in low latitudes and salt marshes in high latitudes. (Base map courtesy of National Ocean Survey.) *B,* salt marsh. (Photograph by Michael DiSpezio.) *C,* mangrove swamp. (Photograph by T. W. Vaughan, U.S. Geological Survey.)

ern United States. Other fishes such as flounder and bluefish use them for feeding and overwintering, and fisheries of oysters, scallops, clams, and fishes such as eels and smelt are located directly in the marshes. More than half of the nation's wetlands have been lost. Of the original 215 million acres of wetlands that once existed in the conterminous United States, only about 90 million acres remain. To help prevent the loss of remaining wetlands the Environmental Protection Agency established an Office of Wetlands Protection (OWP) in 1986. At that time wetlands were being lost to development at a rate of 300,000 acres per year. The OWP will actively pursue enforcement of regulations against wetlands pollution and identification of the most valuable wetlands that can be protected or restored.

A very important characteristic of wetlands is their ability to remove inorganic nitrogen compounds and metals from groundwater polluted by land sources. Most of the removal is probably achieved through adsorption on clay-sized particles. Some of the nitrogen compounds trapped in the sediment are decomposed by denitrifying bacteria, releasing the nitrogen to the atmosphere as nitrogen gas. Many of the remaining compounds are used for plant production in this environment, which is one of the most productive in the world. With the death of the plants, the organic nitrogen compounds either are incorporated into the sediment and converted to peat or are broken up and become food for bacteria, fungi, or detritus-feeding shell and fin fish.

Even though marshlands are considered to provide an annual return of over $5500 per acre based on their natural role in supporting fisheries and removing pollutants from groundwater, preserving them will surely be difficult. In many areas, one acre of marshland may sell for over $100,000 to industrial developers.

LAGOONS

Landward of barrier island salt marshes lie protected, shallow bodies of water called *lagoons*. Because of very restricted circulation between lagoons and the ocean, three distinct zones can usually be identified within a lagoon. There will usually be a *freshwater zone* near the mouths of rivers that flow into the lagoon, a *transitional zone* of brackish water (water with a salinity between that of fresh water and ocean water), and a *saltwater zone* close to the entrance where the maximum tidal effects within the lagoon can be observed. Moving away from the saltwater zone near the mouth of the lagoon, the tidal effects diminish and are usually undetectable in the freshwater region that is well protected from tidal effects.

The salinity of lagoons is also determined by factors other than position relative to the mouth of a lagoon. In latitudes with seasonal variations in temperature and precipitation, ocean water will flow through the entrance during a warm, dry summer to compensate for the volume of water that is lost through evaporation. This results in an increased salinity within the lagoon. Lagoons may actually become hypersaline in arid regions where the inflow of seawater is not sufficient to keep pace with evaporation that is taking place at the surface of the lagoon. During the rainy season the lagoon will become much less saline as the amount of freshwater runoff increases.

Laguna Madre

A hypersaline lagoon along the coast of Texas between Corpus Christi and the mouth of the Rio Grande is Laguna Madre, which is a long, narrow body of water protected from the open ocean by Padre Island, a 160-km (100-mi) long offshore island. Much of the lagoon, which probably formed about 6000 years ago as sea level was approaching its present height, is less than 1 m (3.3 ft) in depth. The tidal range of the Gulf of Mexico in this area is about 0.5 m (1.6 ft), and the inlets, one of which is shown in figure 12–7, at each end of the barrier island are quite small. There is, therefore, very little

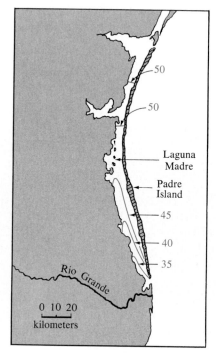

FIGURE 12–7
Laguna Madre.

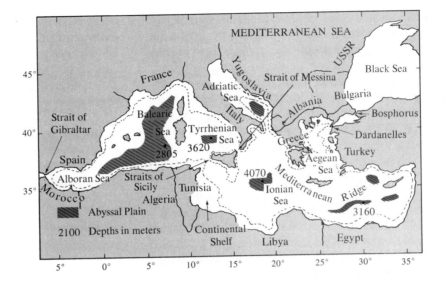

FIGURE 12–8
General Bathymetry and Circulation of the Mediterranean Sea.
(From *Encyclopedia of oceanography,* ed. Rhodes Fairbridge. 1966. Reprinted by permission of Dowden, Hutchinson, & Ross, Inc., Stroudsburg, Pa.)

tidal interchange between the lagoon and the open ocean.

The shallowness of the water in the lagoon makes possible a very great seasonal range of temperature and salinity in this semiarid region. The water temperatures are high in the summer, 30°C (86°F), and may fall below 5°C (41°F) in the winter. Salinities range from 2‰ to over 100‰, because of infrequent local storms that provide large volumes of fresh water in a short period of time and high evaporation that keeps the salinity generally well above 50‰. Because even the salt-tolerant marsh grasses cannot withstand such high salinities, the marsh is replaced by an open sand beach on Padre Island. At the inlets during periods of hypersalinity, the inflow from the ocean is as a surface wedge over the saltier and more dense water of the lagoon. In turn, the outflow from the lagoon occurs as a subsurface flow, just the opposite of the circulation described for estuaries.

MEDITERRANEAN SEA

Although the Mediterranean Sea is not in its entirety coastal, we will discuss it here because the relationship between climate and the pattern of circulation between it and the Atlantic Ocean is classically represented here. The Mediterranean Sea is an east-west–trending body of water that has a very irregular coastline dividing it into subseas with separate circulation patterns. Bounded by Europe and Asia Minor on the north and east and Africa to the south, it is surrounded by land except for a narrow connection with the Atlantic Ocean through the Strait of Gibraltar and the very narrow Bosporus to the Black Sea (figure 12–8).

Atlantic water enters the Mediterranean through the Strait of Gibraltar as a surface flow that is carried in to replace water evaporated at a high rate in the very arid eastern end of the Mediterranean Sea. The water level in the eastern Mediterranean is generally 15 cm (6 in.)

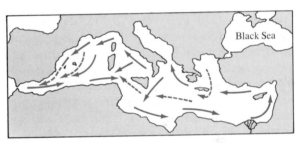

Surface flow ⟶
Intermediate flow (200–600 m; 650–2,000 ft) ----⟶

A. SURFACE AND INTERMEDIATE CIRCULATION

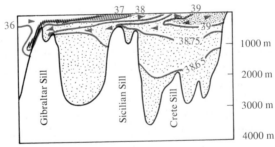

B. VERTICAL DISTRIBUTION OF SALINITY (‰)
Most of the mass of Mediterranean Water has salinity ranging between 38 ‰ and 39 ‰. Maximum salinities in excess of 39 ‰ are found in the surface waters at the east end of the sea. After sinking and moving as Intermediate Water toward the Atlantic Ocean, salinity is reduced to about 37.3 ‰ as it flows over the sill at Gibraltar.

FIGURE 12–9
Mediterranean Circulation.

lower than at the Strait of Gibraltar. This surface flow follows the northern coast of Africa throughout the length of the Mediterranean and spreads north across the sea (figure 12–9).

The remaining Atlantic water continues eastward to Cyprus where, during the winter, it sinks to form the Mediterranean Intermediate Water with a temperature of 15°C (59°F) and a salinity of 39.1‰. This water flows west along the North African coast at a depth of 200 to 600 m (650–2000 ft) and passes into the North Atlantic as a subsurface flow through the Strait of Gibraltar. By the time it passes through the strait, its temperature has dropped to 13°C (55°F) and the salinity is 37.3‰. It is still much denser than water at this depth in the Atlantic Ocean, so it moves down the continental slope until it reaches a depth of approximately 1000 m (3280 ft), where it encounters Atlantic water of the same density. It is at this level that it spreads out in all directions into the Atlantic water and becomes a detectable water mass over a broad area because of its high salinity.

The circulation between the Mediterranean Sea and the Atlantic Ocean is typical of closed restricted basins in areas where evaporation exceeds precipitation. In such situations the restricted basins will always lose water at a very high rate from the surface through evaporation, and this water will have to be replaced by surface inflow from the open ocean. The evaporation of the water that flows in from the open ocean increases its salinity to very high values, causing eventual sinking and the return to the open ocean as a subsurface flow. You will recall that during periods of hypersalinity, Laguna Madre has a pattern of circulation between it and the Gulf of Mexico that represents Mediterranean circulation on a smaller scale. Since the Mediterranean Sea is the classic area for this type of circulation, it is given the name *Mediterranean circulation*. You will recogizne it as being opposite the pattern we observed to be characteristic of estuaries, where the surface flow goes into the open ocean and subsurface flow of marine water is into the estuary. In areas where *estuarine circulation* develops between the marginal body of water and the ocean, the availability of fresh water from runoff and precipitation is greater than the rate of water loss to evaporation.

SUMMARY

Because of the shallow depth of the coastal ocean, river runoff, tidal currents, and seasonal changes in solar radiation, physical characteristics of the water such as temperature and salinity vary over a greater range than in the open ocean. Coastal **geostrophic currents** are produced as a result of runoff of fresh water and winds blowing in the proper direction along the coast.

Estuaries are semienclosed bodies of water where fresh water runoff from the land and ocean water mix. They are classified on the basis of origin as **coastal plain, fjord, bar-built**, and **tectonic**. Mixing patterns range from that of a **vertically mixed** shallow estuary through **slightly stratified** and **highly stratified** to the **salt wedge** characteristic of large-volume river mouths. The typical pattern of **estuarine circulation** consists of a surface flow of low-salinity water toward the mouth and a subsurface flow of marine water toward the head.

Chesapeake Bay is a classic example of a coastal plain estuary with most of its stream inflow entering along its western margin. This fact—along with the Coriolis effect acting on the estuarine circulation—makes the eastern side of the bay saltier. Possibly as a result of increased nutrient inflow resulting from human activities, greater amounts of the deep water in the estuary are undergoing greater periods of summer anoxia.

Wetlands are of two basic types: **salt marsh** and **tropical mangrove swamp**. These biologically productive regions are alternately covered and exposed by the tide. Marshes are also important in their ability to adsorb land-derived pollutants before they reach the ocean. Despite their ecological importance, wetlands are fast disappearing from our coasts as they are encroached upon by human activities.

Long offshore deposits called **barrier islands** protect marshes and **lagoons**. Some lagoons have very restricted circulation with the ocean and may exhibit a great range of salinity and temperature conditions as a result of seasonal change.

The **Mediterranean Sea** has a classical circulation characteristic of restricted bodies of water in areas where evaporation greatly exceeds precipitation. It is the reverse of estuarine circulation and is called **Mediterranean circulation**.

QUESTIONS AND EXERCISES

1. For coastal oceans where deep mixing does not occur, discuss the effect offshore winds and freshwater runoff will have on salinity distribution. How will the winter and summer seasons affect the temperature distribution in the water column?

2. How does coastal runoff of low-salinity water produce a longshore geostrophic current?

3. Describe the difference between vertically mixed and salt wedge estuaries in terms of salinity distribution, depth, and volume of river flow. Which displays the more classical estuarine circulation pattern?

4. Discuss the factors that may explain why the surface salinity of Chesapeake Bay is greater on its east side, and periods of summer anoxia in the deep water mass are increasing in severity with time.

5. Name the two types of wetland environments and the latitude ranges where each will likely develop. How do wetlands contribute to the biology of the oceans and the cleansing of polluted river water?

6. What factors lead to a wide seasonal range of salinity in Laguna Madre?

7. Describe the circulation between the Atlantic Ocean and the Mediterranean Sea, and explain how and why it differs from estuarine circulation.

REFERENCES

Fairbridge, R. W., ed. 1966. *Encyclopedia of oceanography*. New York: Van Nostrand Reinhold.

Godfrey, P. J. 1976. Barrier beaches of the East Coast. *Oceanus* 19:5, 27–40.

Molinari, R. L., Baig, S., Behringer, D. W., Maul, G. A., and Legeckis, R. 1977. Winter intrusions of the Loop Current. *Science* 198:505–6.

Officer, C. B., Biggs, R. B., Taft, J. L., Cronin, L. E., Tyler, M. A., and Boynton, W. R. 1984. Chesapeake Bay anoxia: Origin, development, and significance. *Science* 223:4631, 22–27.

Officer, C. B. 1976. Physical oceanography of estuaries. *Oceanus* 19:5, 3–9.

Pickard, G. L. 1975. *Descriptive physical oceanography: An introduction* 2nd ed. New York: Macmillan.

Valiela, I., and Vince, S. 1976. Green borders of the sea. *Oceanus* 19:5, 10–17.

SUGGESTED READING

Sea Frontiers

Baird, T. M. 1983. Life in the high marsh. 29:6, 335–41.
The ecology of the salt marsh is discussed.

Baker, R. D. 1972. Dangerous shore currents. 18:3, 138–43.
A discussion of the hazards to swimmers of various types of currents found near the shore.

de Castro, G. 1974. The Baltic—to be or not to be? 20:5, 269–73.
A review of the pollution threat to life in the Baltic Sea and what has been done to solve the problem.

Edwards, L. 1982. Oyster reefs: Valuable to more than oysters. 28:1, 23–25.
The value of oyster reefs to various estuarine life forms is detailed.

Heidorn, K. C. 1975. Land and sea breezes. 21:6, 340–43.
A discussion of the cause of land and sea breezes.

Osing, O. 1974. The meeting of the seas. 20:1, 21–24.
A description of the conditions at the northern tip of Denmark, where the North Sea and the Baltic Sea meet.

Sefton, N. 1981. Middle world of the mangrove. 27:5, 267–73.
The life forms associated with Cayman Island mangrove swamps are described.

Smith, F. 1982. When the Mediterranean went dry. 28:2, 66–73.
The life of global plate tectonics in causing the Mediterranean Sea to dry up some 5 million years ago.

Scientific American

Degens, E. T., and Ross, D. A. 1970. The Red Sea hot brines. 222:4, 32–53.
Describes the hot brine pools in the Red Sea. The source of brines and the economic potential resulting from metal enrichment associated with the brine deposits are considered.

Hsu, K. H. 1972. When the Mediterranean dried up. 227:6, 26–45.
Evidence is presented that shows the Mediterranean Sea was a dry basin 6 million years ago.

Ramage, C. S. 1986. El Niño. 254:6, 97–105.
An overview of how an atmospheric pressure seesaw is related to the El Niño phenomenon and affects climate on a worldwide basis.

13
THE MARINE HABITAT

The Crown-of-Thorns sea star, which threatened to destroy many coral reefs in the 1960s, being attacked by one of its few predators, the Pacific triton. Photograph by Christopher Newbert.

In Chapter 2 we discussed some evidence for believing that life originated in the oceans. Primary among the considerations was the simple fact that the ocean constitutes the major part of the water environment. Considering the need organisms have for water, we can see that life would be totally impossible on the continents were it not for the hydrologic cycle that provides water that runs across and percolates through the surface and is temporarily stored in lakes and rivers before it returns to the oceans. Although availability of water is seldom a problem, the success of individual organisms living in the oceans is dependent upon their ability to adjust to the many physical variables that we will consider in this chapter.

GENERAL CONDITIONS

The fact that all the major divisions of the animal kingdom, *phyla*, have members that live in the marine environment attests to the ocean's being a far more fit biological realm than the continents. In fact, all phyla are thought to have originated in the ocean, and five are exclusively marine. Although it may seem inconsistent at this point, the fact that only the simple one-celled plants inhabit the open oceans, whereas the more complex and highly developed plants are restricted to its margins and the continents, is further evidence pointing to the unique fitness of the oceans as a biological home.

The ocean environment is far more stable than the terrestrial environment, and as a result of this stability, the organisms that live in the oceans have generally not developed highly specialized regulatory systems to combat sudden changes that might occur within their environment. They are, therefore, affected to various degrees by quite small changes in concentrations of various salts in solution, temperature, turbidity, and other environmental variables.

Water constitutes over 80 percent of the mass of *protoplasm*, the substance of living matter. Over 65 percent of the weight of a human being and 95 percent of that of the jellyfish is accounted for by the presence

TABLE 13–1
Water Content of Organisms.

Organism	Water Content (%)
Human	65
Herring	67
Lobster	79
Jellyfish	95

of water (table 13–1). Water carries dissolved within it the gases and minerals needed by organisms and is itself one of the raw materials required by plants in the production of food through photosynthesis. Land plants and animals have developed very complex systems and devices to distribute water throughout their bodies and prevent it from being lost. The threat of atmospheric *desiccation* (drying out) does not exist for the inhabitants of the open ocean, as they are abundantly supplied with water.

Support

Another necessity of plants and animals is support. Land plants have vast root systems that fasten the plant securely to the earth. In the animal kingdom there are a number of support systems used, but each requires some combination of appendages that must totally support the land animal's weight.

In the ocean this is not the case. The water of the oceans, as it so lavishly bathes the organisms with their needed gases and nutrients, serves as their basic support as well. Organisms that live in the open ocean depend primarily upon buoyancy and frictional resistance to sinking to maintain them in their desired position. This is not to say that they do not have problems maintaining their position within the ocean; some of the special adaptations that are developed toward increasing their efficiency in this respect will be discussed in this and succeeding chapters.

Effects of Salinity

Marine animals are significantly affected by relatively small changes in their environmental conditions. One of these that is highly important is change of salinity. Some oysters that live in an estuarine environment at the mouths of rivers are capable of withstanding a considerable range of salinity. During floods the salinity will be extremely low. On a daily basis, as the tides force ocean water into the river mouth and draw it out again, the salinity changes considerably. The oysters, and most other organisms that inhabit coastal regions, have a tolerance for a wide range of salinity conditions. They are

euryhaline. By contrast, other marine organisms, particularly those that inhabit the open ocean, can withstand only very small changes in salinity; these organisms are *stenohaline*.

Certain plants and animals cause important alterations in the amount of dissolved material in ocean water by extracting minerals to construct hard parts of their bodies that serve as protective coverings. The compounds primarily used for this purpose are silica (SiO_2) and calcium carbonate ($CaCO_3$). The silica is used by the most important plant population in the ocean, the diatoms, and the microscopic animals called radiolaria. Calcium carbonate is used by the foraminifera, most members of the phylum Mollusca, corals, and some algae that secrete a calcium carbonate skeletal structure.

Cell membranes are semipermeable and will allow the passage of molecules of some substances while

A. DIFFUSION

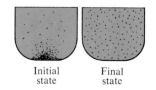

Initial state   Final state

If a substance soluble in water is placed in a pile on the bottom of a beaker of water, it will eventually become evenly distributed through random molecular motion. Diffusion means molecules of a substance move from areas of high concentration of the substance to areas of low concentration.

B.

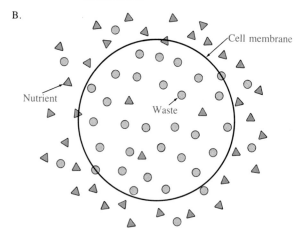

Cell membrane

Nutrient

Waste

Nutrients are in high concentration outside the cell and diffuse into the cell through the cell membrane. Waste is in higher concentration inside the cell and diffuses out of the cell through the cell membrane.

FIGURE 13–1
Diffusion.

OSMOSIS

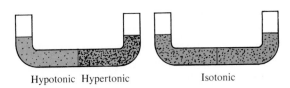

Hypotonic Hypertonic   Isotonic

Separating two water solutions of different salinities by a
membrane that will allow the passage of water molecules
but not the molecules of dissolved substances sets the
stage for osmosis to occur. Water molecules will achieve
a net motion by diffusion through the membrane into the
more concentrated (hypertonic) solution from the less
concentrated (hypotonic) solution. Osmosis will **not**
produce a net movement of water through the membrane
when the salinity of the two solutions is the same. They
are then isotonic.

FIGURE 13–2
Osmosis.

screening out others. By *diffusion,* cells may take in the
nutrient substances they need from the surrounding
fluid medium that contains them in high concentration.
The nutrient compounds, to which the cell wall is per-
meable, pass from the surrounding medium into the
cell, where they are in lesser concentrations (figure
13–1).

The waste materials that a cell must dispose of after
it has utilized the energy held in these nutrients is
passed out of the cell by the same method. As the concentration of waste materials becomes greater within the
cell than within the fluid medium surrounding the cell,
these materials will pass out of this area of high concentration within the cell into the surrounding fluid. The
waste products will then be carried away by the circulating fluid that services the cells in higher animals or
by the surrounding water medium that bathes the simple one-celled organisms of the oceans.

Osmotic Pressure When aqueous solutions of unequal
salinity are separated by a water-permeable membrane,
there will be diffusion of water molecules through the
membrane into the more concentrated solution. This
process is called *osmosis* (figure 13–2). Such solutions
can be compared in terms of their osmotic pressures.
The *osmotic pressure* of a solution is that which must be
applied to keep pure water that is separated from it by
a water-permeable membrane from passing through the
membrane and diluting the solution. The higher the salinity of the solution, the higher its osmotic pressure.
We can compare the osmotic pressure of solutions by
comparing their concentrations of dissolved solids—
their salinity. This comparison is readily done by determining their freezing point, since the addition of dissolved solids to an aqueous solution lowers its freezing
point (figure 13–3). This effect is due to the interference
of the dissolved particles with the formation of ice crystals. The lower the freezing point, the higher the osmotic pressure of a solution.

If the salinity of body fluid and the external medium
are equal, they are said to be *isotonic* and have equal
osmotic pressure. No net transfer of water will occur
through the membrane. If the external fluid medium
has a lower osmotic pressure and lower salinity than the
body fluid within the cells of an organism, water will
pass through the cell walls into the cells. This organism
is *hypertonic* relative to the external medium. Should
the salinity or osmotic pressure within the cells of an
organism be less than that of the external medium, water from the cells will pass through the cell membranes
into the external medium. This organism is described as
hypotonic relative to the external medium.

In an oversimplification, we will consider osmosis as
simply a diffusion process. If we consider the relative
concentration of water molecules on either side of the
semipermeable membrane, we will see that the net
transfer of water molecules is from the side where the

FIGURE 13–3
**Osmotic Pressure and Freezing
vs. Salinity.** Osmotic pressure increases and the freezing point decreases with increased salinity.

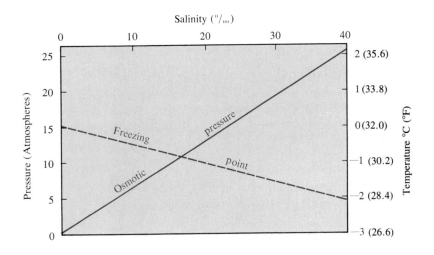

greatest concentration of water molecules exists through the membrane to the side that contains a lesser concentration of water molecules. The processes of water molecules moving through the semipermeable membrane (osmosis), nutrient molecules moving from the external medium where they are more concentrated into the cell where they will be used to maintain the cell, and the movement of molecules of waste materials from the cell into the surrounding medium that carries them away are all going on at the same time.

We must keep in mind that during this process molecules of all the substances in the system are passing through the membrane in both directions. Nevertheless, there will be a net transport of molecules of a given substance from the side on which they are most highly concentrated to the side where the concentration is less.

In the case of the marine invertebrates, the body fluids and the external medium in which they are immersed are in a nearly isotonic state. Since no significant difference exists, they do not have to develop special mechanisms to maintain their body fluids at a proper concentration. Thus, they have an advantage over their freshwater relatives, whose body fluids are hypertonic in relation to the low-salinity external medium (table 13–2).

Marine fish have body fluids slightly more than one-third as saline as ocean water. This level may be the result of their having evolved in low-salinity coastal waters. They are, therefore, hypotonic in relation to the surrounding medium. This difference produces a problem in that the saltwater fish, without some means of regulation, would be continually losing water from its body fluids to the surrounding ocean environment. This loss is counteracted by the fact that marine fish drink ocean water and excrete the salts through "chloride cells" located in the gills. They also help maintain their body water by discharging a very small amount of very highly concentrated urine (figure 13–4). To summarize the process by which marine fish maintain low osmotic pressure in their body fluids: They drink large quantities of water, extract the salts of that water, and dispose of the salts by excreting them. They conserve body water by excreting a highly concentrated urine solution.

Freshwater fish are hypertonic in relation to the very dilute external medium in which they live. The osmotic pressure of the body fluids of such fish may be 20 to 30 times greater than that of the fresh water that surrounds them. Their problem is to avoid taking into their body cells large quantities of water that would eventually rupture the cell walls. In order to meet this problem, the freshwater fish does not drink water. Its cells have the capacity to absorb salt. The water that may be gained as a result of body fluids being greatly hypertonic is disposed of with large volumes of very dilute urine (figure 13–4). To summarize the means by which the freshwater fish avoid having their cells ruptured by the intake of water through osmosis: They do not drink, they have cells that can absorb salt to maintain their osmotic pressure, and they dispose of large quantities of water through copious dilute urine.

We should all be aware that we cannot drink salt water when at sea because the osmotic pressure of our body fluids is about one-fourth that of ocean water. To drink ocean water would cause a drastic loss of water through the wall of the digestive tract by osmosis, causing eventual dehydration.

Availability of Nutrients

The distribution of life throughout the ocean's breadth and depth is basically dependent upon the availability of plant nutrients. In areas where the conditions are right

TABLE 13–2

Osmotic Pressure of Body Fluids in Various Organisms. A comparison of the freezing points of the body fluids of some marine and freshwater organisms and the waters in which they live. The lower the freezing point of the fluid, the higher the salinity and osmotic pressure. Marine invertebrates and sharks are essentially isotonic, while most marine vertebrates are hypotonic. Essentially all freshwater forms are hypertonic.

Marine Animals			Freshwater Animals		
	Freezing Points (°C)			Freezing Points (°C)	
Animals	Body Fluid	Water	Animals	Body Fluid	Water
INVERTEBRATES					
Annelid worm	−1.70	−1.70	Mussel	−0.15	−0.02
Mussel	−2.75	−2.12	Water flea	−0.20	−0.02
Octopus	−2.15	−2.13	Crayfish	−0.80	−0.02
Lobster	−1.80	−1.80			
Crab	−1.87	−1.87			
VERTEBRATES					
Cod	−0.74	−1.90	Eel	−0.62	−0.02
Shark	−1.89	−1.90	Carp	−0.50	−0.02
Turtle	−0.50	−1.90	Water snake	−0.50	−0.02
Seal	−0.50	−1.90	Manatee	−0.50	−0.02

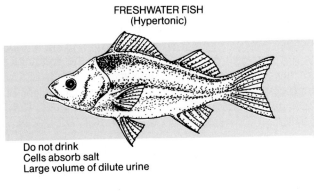

FRESHWATER FISH
(Hypertonic)

Do not drink
Cells absorb salt
Large volume of dilute urine

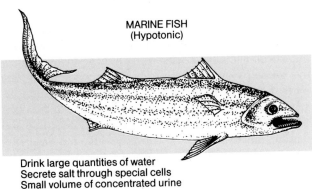

MARINE FISH
(Hypotonic)

Drink large quantities of water
Secrete salt through special cells
Small volume of concentrated urine

FIGURE 13–4
Freshwater and Marine Fish.

for the supply of large quantities of nutrient materials, marine populations will reach their greatest concentration. Where are these areas in which we find the greatest concentration of biomass? To answer this question, we must consider the sources of nutrients.

Previously, we discussed the role played by water in eroding the continents, carrying the eroded material to the oceans, and depositing it as sediment at the margins of the continents. During this process, the water dissolves inorganic compounds in the form of nitrate and phosphate compounds that serve as the basic nutrient supply for plants. Through the process of photosynthesis, plants combine these with carbon dioxide and water to produce the carbohydrates, proteins, and fats that store the energy upon which the rest of the biological community depends.

If the continents are the major sources of these inorganic nutrients, then we might expect that the greatest concentrations of marine life will be found at their margins. This is the case. As we go to the open sea from the margins of the continents, we will travel through waters containing progressively lower concentrations of marine life. This situation is basically due to the vastness of the world's oceans and the great distance between the open ocean and the coastal regions where the nutrients are supplied.

Availability of Solar Radiation

To make use of these available nutrients, the plants must meet another requirement. Photosynthesis cannot proceed without the appropriate energy provided by solar radiation. The radiation from the sun penetrates the atmosphere quite readily, so plants on the continents would seem to have an advantage over those in the ocean, as there is hardly ever any shortage of solar radiation to drive the photosynthetic reactions for these land-based plants.

By contrast, the ocean water does represent a significant barrier to the penetration of solar radiation, although in the clearest water solar energy may be detected to depths in excess of about 1 km (0.62 mi). However, the amount of solar energy reaching these depths is extremely small and would not be sufficient to support the photosynthetic process. Photosynthesis in the ocean is restricted to a very thin layer, approximately the upper 100 m (328 ft), which is the depth to which sufficient light energy to carry on photosynthesis can penetrate. Near the coast this zone is much thinner as the water contains more suspended material that restricts the light penetration.

Considering these two factors necessary for photosynthesis, the supply of nutrients and the presence of solar radiation, we find that in the open ocean away from the continental margins the column of water having solar energy available extends deeper, but that this column contains only a small concentration of nutrients. Conversely, in the coastal regions where the water is more turbid, light penetrates to much shallower depths, but the nutrient supply is quite rich. Since the coastal zone is by far the most productive zone, we can see manifested in this comparison the overriding importance of nutrient availability to life in the oceans.

Margins of the Continents

We have mentioned that the stable condition of the ocean environment makes it ideal for the continuation of life processes, but we are now indicating that the richest concentration of marine organisms is found in the very margins of the oceans, where the conditions are the least stable. When we consider the physical conditions that would appear at first to be deterrents to the establishment of life, we will see that many are present in the coastal regions.

The water depths are shallow, allowing seasonal variations in temperature and salinity that are much greater than would be found in the open ocean. A varying thickness of water column exists in the nearshore region as a result of the tide movements that periodically cover and uncover a thin strip along the margins of the continents, where there is a constant beating of the surf. The surf condition represents a sudden release of energy that has been carried for great distances across the open ocean. During this transit, little energy was lost and little effect on the marine environment was observed as a consequence of its transfer.

The development of such a great concentration of biomass in the environment containing so many factors that would seem to inhibit the development of life serves to highlight the importance of the evolutionary process in developing new species by natural selection. Over the eras of geologic time, new life forms have developed to fit into all biological niches. Many of these forms have adapted to live under conditions that would seem deleterious to the conservation of life. This fact attests to the great range of physical conditions within which life can exist so long as those basic requirements for the production of the food supply are met.

We find that along continental margins there are those areas where life is more abundant than others. What are the characteristics of these areas that result in such an uneven distribution of life along the continental margins? Again, we need consider only those basic requirements for the production of food for the answer. If we measure the various properties of the water along the coast of the continents and compare those measurements from place to place, we will find that the areas containing the greatest concentrations of biomass are those where the water temperatures are lower. Due to the low temperature, it is possible for the water to maintain greater amounts of the important gases, oxygen and carbon dioxide, in solution than would be found in warmer waters. Of particular importance is the increased availability of carbon dioxide, and we see again an example of the greater availability of the basic requirements of plants affecting the distribution of life in the oceans.

Upwelling In certain areas of the coastal margins we find an additional factor that enhances the conditions for life—upwelling. Upwelling occurs along the western margins of continents where surface currents are moving toward the equator. It consists of a flow of subsurface water to the surface that grows directly out of the relationship existing between the direction of the current flow and the continental margin.

In the case of northward-flowing currents along the western margin of the Southern Hemisphere continents or southward-flowing currents along the western margin of Northern Hemisphere continents, the Ekman transport tends to move the surface water away from the coast. As the surface water moves away from the continents, water surfaces from depths of 200 to 1000 m (656–3280 ft) to replace it. This water that surfaces from depths below those where photosynthesis occurs is rich in plant nutrients because there are no plants in these deeper waters to utilize them. This constant replenishing of nutrients at the surface enhances the conditions for life in these areas. Upwelling water is usually of low temperature, which produces the additional benefit of having a high capacity for dissolved gases (figure 13–5).

Water Color and Life in the Oceans

It is usually possible to visually determine areas of high and low organic production in the ocean from the color of the water. Coastal waters are almost always greenish in color because they contain more large particulate matter that disperses solar radiation in such a way that the wavelengths most scattered are those that represent the greenish or yellowish light. This condition is also partly the result of the presence of yellow-green microscopic marine plants in these coastal waters. In the open ocean, where particulate matter is relatively scarce and marine life exists in low concentration, the water appears blue due to the size of the water molecules and the scattering they effect on the solar radiation, which is similar to the process that produces the apparent blueness of the skies.

Green color in water usually indicates the presence of a lush biological population such as might correspond to the jungle environments on the continents. The deep indigo blue of the open oceans, particularly between the tropics, usually indicates an area that lacks abundant life and could be considered a biological desert.

Size

Earlier in this chapter we referred to the fact that marine plants are quite simple in comparison to the specialized forms we find throughout the continents. How can we explain the fact that so many marine plants have not followed the path of aggregation (becoming multicellular)?

Assuming the availability of nutrients, the major requirements of marine plants are that they maintain themselves within the upper layers of ocean water where solar radiation is available to carry on photosynthesis and that they take in nutrients from the surrounding waters and expel waste materials as efficiently as possible. These requirements lead us to a consideration of size and shape. The ease with which marine plants with no means of locomotion will maintain their position in the upper layers of the ocean is closely related to the ratio that exists between surface area and body mass. The greater the surface area per unit of body mass, the easier it will be for an organism to maintain

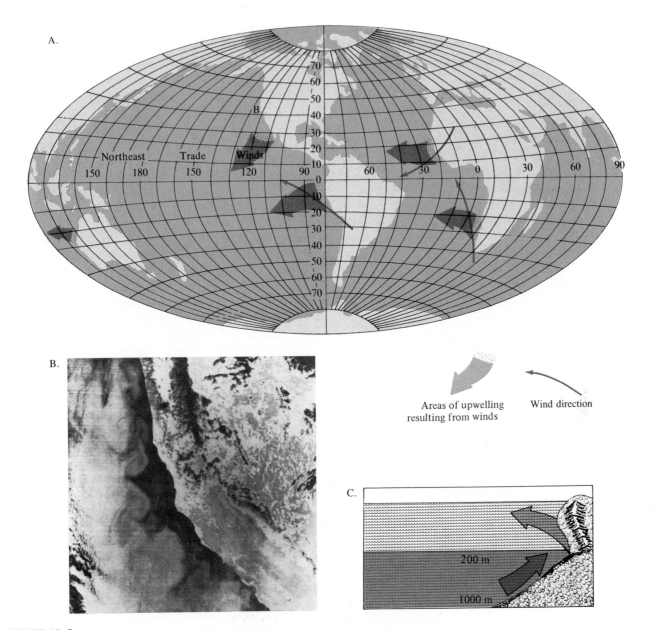

FIGURE 13–5
Upwelling. *A,* the northeast trade winds drive currents south along the western margins of continents in the Northern Hemisphere, while the southeast trade winds drive them north in the Southern Hemisphere. The Coriolis effect, turning the surface water to the right in the Northern Hemisphere and to the left in the Southern Hemisphere, moves it away from the western margins of the continents. (Base map courtesy of National Ocean Survey.) *B,* on this infrared satellite image of the California and Oregon coastal water, cold upwelling water near the shore appears light, while warmer water is darker. See *B* in Part *A* for the location. (Photo courtesy of NOAA.) *C,* as the surface water moves away from the continent, water from depths of 200–1000 m rises to replace it. Since this water comes from below the photosynthetic zone, it is rich in nutrients.

its position because there is more surface per unit of body mass in frictional contact with the surrounding water. Greater surface area per unit of body mass amounts to a greater resistance to sinking per unit of body mass. An examination of figure 13–6 will help you understand that this ratio of surface area per unit of mass increases with decrease in size. It can readily be seen that it would benefit the one-celled plants making up the bulk of marine plant life to be as small as possible. They are, in fact, microscopic.

The efficiency with which the plants take in the nutrients from the surrounding water and expel their waste through their cell membranes is also related to the amount of surface area per unit of mass. The greater the surface area per unit of mass, the easier it is for the organism to carry on these functions. Since both processes are dependent on diffusion, there would be a point at which increased size would reduce the surface area to body mass ratio to the point that the cell could not be serviced and would die. It is a direct result of this condition that we find cells in all plants and animals to be microscopic, regardless of the size of the organism. Each cell must take in nutrients and dispose of wastes by diffusion.

Examining diatoms carefully, we see that it is quite common for them, as well as other members of the microscopic marine community, to have long needlelike extensions from their siliceous *tests,* or protective coverings. These extensions serve to increase further the ratio of surface area to the mass of the organism.

Small plants and animals use another contrivance to increase their ability to stay in the upper layers of the ocean: They produce a drop of oil, which tends to decrease their overall density and increase their buoyancy.

With all these adaptations made in an attempt to increase the ability to float, the organisms still have a density in excess of the density of water and therefore tend to sink, if ever so slowly. This is not a serious handicap, since they are so small and are carried quite readily along with the water movements. In these surface layers, there is considerable turbulent action associated with the mixing that results from wind action on the surface. This turbulence tends to dominate the movement of these small organisms to the point that they are able to spend adequate time bathed in the solar radiation to carry on their role of producing the energy needed by the other members of the marine biological community.

Viscosity

We are familiar with the fact that liquids flow with varying degrees of ease, and we know that syrup flows less readily than water. This condition is indicative of the higher viscosity of syrup as compared to water. *Viscosity,* defined as internal resistance to flow, is a characteristic of all fluids.

The viscosity of ocean water is affected by two variables—temperature and salinity. Viscosity of ocean water increases with an increase in salinity but is affected even more by temperature, a lowering of which increases viscosity. As a result of this temperature relationship, we find that floating plants and animals in colder waters have less need for extensions to aid them in floating than those that occupy warmer waters. It has been observed that members of the same species of floating crustaceans will be very ornate with featherlike appendages where they occupy warmer waters, while these appendages are missing in the colder, more viscous environments. It appears then that high viscosity would benefit the floating members of the marine biocommunity because it would make it easier for them to maintain their position in the surface layers.

With increasing size of organisms and a change in the mode of life, viscosity ceases to enhance the ability of

CUBE *A*
Linear dimension—1 cm

→|1 cm|←

Area—6 cm²
Volume—1 cm³

Ratio of surface area to volume:

$$\frac{6 \text{ cm}^2}{1 \text{ cm}^3}$$

Cube *A* has 6 units of surface area per unit of volume.

CUBE *B*
Linear dimension—10 cm

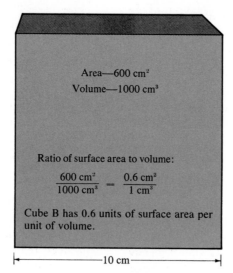

Area—600 cm²
Volume—1000 cm³

Ratio of surface area to volume:

$$\frac{600 \text{ cm}^2}{1000 \text{ cm}^3} = \frac{0.6 \text{ cm}^2}{1 \text{ cm}^3}$$

Cube B has 0.6 units of surface area per unit of volume.

|←——————10 cm——————→|

Cube *A* has ten times as much surface area per unit of volume as cube *B*.

If both cubes are composed of the same substance, their masses are proportional to their volumes, since mass = volume × density.

If cubes *A* and *B* were plankton, cube *A* would have ten times as much frictional resistance to sinking per unit of mass as cube *B*. It could stay afloat by exerting far less energy than cube *B*.

Cube *A*, were it a planktonic alga, could also take in nutrients and dispose of waste through the cell wall ten times as efficiently as an alga with the dimensions of cube *B*.

FIGURE 13–6
The Importance of Size.

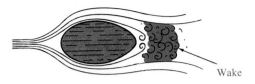
A body with a low degree of streamlining. Its broad structure causes too much resistance area in front and too much wake region behind.

Teardrop shape (fusiform) front is well rounded and the body tapers gradually to the rear, producing a small wake region.

FIGURE 13–7

Streamlining. Due to the viscosity of water, any body moving rapidly through it must be designed properly to produce as little stress as possible as it displaces the water through which it moves. After the water has moved past the body, it must fill in behind the body with as little eddy action as possible.

an organism to pursue its lifestyle and becomes an obstacle instead. This is particularly true of the large organisms that swim freely in the open ocean. They must pursue prey and displace water to move ahead. The more rapidly they swim, the greater the stress that is created on the organism. Not only must water be displaced ahead of the animal, but water must move in behind it to occupy the space that it has vacated. The latter is one of the more important considerations in *streamlining*.

The familiar shape of the free-swimming fish and the mammals manifests the type of adaptations that must be achieved in streamlining an organism to move with a minimum of stress through the water. We see that a common shape achieved to meet this need is a laterally flattened body presenting a small cross section at the anterior end with a gradually tapering posterior. This form, which is characteristic of the bony fishes, reduces stress that results from movement through the water and makes it possible for this movement to be achieved with the exertion of a minimum of energy (figure 13–7).

Temperature

Of all the conditions we can measure in the ocean, there are probably no physical characteristics we have discussed that are more important to the life in the oceans than temperature. It has been stated that the marine environment is much more stable than that of the land. A comparison of temperature ranges that exist in the sea

with those that are found on land will serve to exemplify this condition.

The minimum temperature observed in the open sea is never much less than $-2°C$ (28.4°F), and the maximum seldom exceeds 27°C (80.6°F). The continental temperatures range from a low of $-88°C$ ($-127°F$) to a high of about 58°C (136°F). The temperature found on the continent varies considerably on a seasonal and daily basis. Daily variations at the ocean surface rarely exceed more than 0.2°C (0.4°F) or 0.3°C (0.5°F), although they may be as great as 2° or 3°C (3.6 or 5.4°F) in shallower coastal waters. Annual temperature variations are also small. They range from 2°C (3.6°F) at the equator to 8°C (14.4°F) at 35° to 45° latitude and decrease again in the higher latitudes. Annual variations of temperature in shallow coastal areas may be as high as 15°C (27°F).

The temperature variations we are discussing as characteristic of surface waters are reduced in magnitude with an increase in water depth. In the deep oceans daily or seasonal variation in temperature becomes a consideration of little or no importance. Throughout the deeper parts of the ocean the temperature remains uniformly low. At the bottom of the ocean basin, where the depth is in excess of 1.5 km (0.9 mi), temperatures hover around the 3°C (37.4°F) mark regardless of the latitude.

We previously discussed the fact that a decrease in temperature will increase density, viscosity, and the capacity of water to hold gases in solutions. All these changes have significant effects upon the organisms inhabiting the ocean. A direct consequence of the increased capacity of water to contain dissolved gases is seen in the vast plant communities that develop in the high latitudes during the summer seasons when solar energy is available to carry on photosynthesis. One of the major factors in this phenomenon is the abundance of dissolved gases—carbon dioxide for the use of plants in carrying on photosynthesis and oxygen needed by the animals that feed upon the plants.

It has been observed that floating organisms are larger individually in colder waters than in warmer waters of the tropics, although the tropical populations appear to be characterized by a larger number of species. However, the total biomass of floating organisms in the colder high-latitude planktonic environments greatly exceeds that of the warmer tropics. The fact that organisms in the tropics are smaller than those observed in the higher latitudes could quite possibly be related to the lower viscosity found in the lower latitude waters. Being smaller, these tropical species can expose more surface area per unit of body mass, and they are also characterized by ornate plumage to further increase surface area per unit of body mass (figure 13–8). These plumose ad-

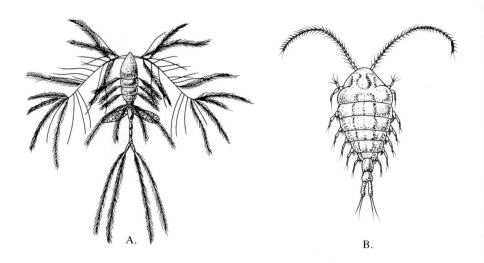

FIGURE 13–8
Water Temperature and Appendages. *A,* copepod (*Oithona*) that displays the ornate plumage characteristic of warm water varieties. *B,* copepod (*Calanus*) that displays the less ornate appendages found on temperate and cold water forms. (After Sverdrup, Johnson, and Fleming.)

A.

B.

aptations are strikingly absent in the larger cold-water species.

Increases in temperature increase the rate of biological activity, which more than doubles with an increase in the temperature of 10°C (18°F). Tropical organisms apparently grow faster, have a shorter life expectancy, and reproduce earlier and more frequently than their counterparts in the colder waters.

There are some species of animals that can successfully live only in cooler waters and others that can do so only in warmer waters. Many of these can withstand only a very small change in temperature and are called *stenothermal.* Other varieties apparently are little affected by temperature change and can withstand changes over a large range. These are classified as *eurythermal.* Stenothermal organisms are found predominantly in the open ocean and at the greater depths, where it is very unlikely that large ranges of temperature will occur. The eurythermal organisms are more characteristic of the shallow coastal waters, where the largest ranges of temperature are found, and the surface waters of the open ocean.

DIVISIONS OF THE MARINE ENVIRONMENT

We can readily divide the marine environment into two basic units: the ocean water itself is the *pelagic environment,* and the ocean bottom constitutes the *benthic environment* (figure 13–9*A*).

Pelagic Environment

The pelagic environment is further divided into two provinces. The *neritic province,* which extends from the shore seaward, includes all water overlying an ocean bottom less than 200 m (660 ft) in depth. Beyond this depth the *oceanic province* is found.

The oceanic province, which includes water with a very great range in depth from the surface to the bottom of the deepest ocean trenches, is further subdivided on the basis of water depth. The oceanic province is subdivided into the *epipelagic zone* from the surface to a depth of 200 m (660 ft), the *mesopelagic zone* from 200 to 1000 m (660–3280 ft), the *bathypelagic zone* from 1000 to 4000 m (3280–13,000 ft), and the *abyssopelagic* zone, which includes all the deepest parts of the ocean below 4000 m (13,000 ft) depth.

An important factor in determining the distribution of life in the oceanic province and boundaries of biozones within it is the availability of light. The *euphotic region* extends from the surface to a depth where there is enough light to support photosynthesis, rarely to more than a depth of 100 m (328 ft). Below this zone, there are small but measurable quantities of light within the *disphotic zone* to a depth of about 1000 m (3280 ft). Below this depth there is no light in the *aphotic zone.*

The upper epipelagic zone is the only oceanic biozone in which there is sufficient light to support photosynthesis. The boundary between it and the mesopelagic zone (200 m or 660 ft) is also the approximate depth at which the level of dissolved oxygen begins to decrease significantly. This decrease in oxygen results because no plants are found beneath about 150 m (490 ft), and the dead organic tissue (detritus) descending from the biologically productive upper waters is undergoing decomposition by bacterial oxidation (figure 13–9*B*). Nutrient content of the water also increases abruptly below 200 m (600 ft) (figure 13–9). This depth also serves as the approximate bottom of the mixed layer, seasonal thermocline, and surface water mass.

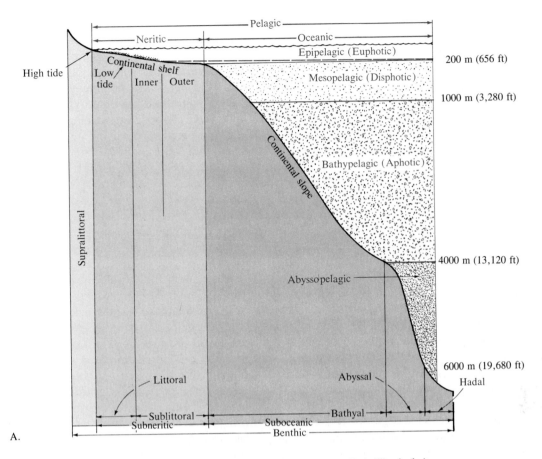

A.

FIGURE 13–9

Biozones and Distribution of Oxygen (O_2) and Plant Nutrient Phosphate (PO_4) in the Water Column, Mid- to Low Latitudes.
Oxygen is abundant in surface water due to mixing with the atmosphere and plant photosynthesis. Nutrient content is low in surface water due to uptake by plants. Oxygen content decreases and nutrient content increases abruptly below the euphotic zone. An oxygen minimum and nutrient maximum are recorded at or near the base of the mesopelagic zone. Nutrient levels remain high to the bottom, while oxygen content increases with depth as deep- and bottom-water masses carry oxygen into the deep ocean.

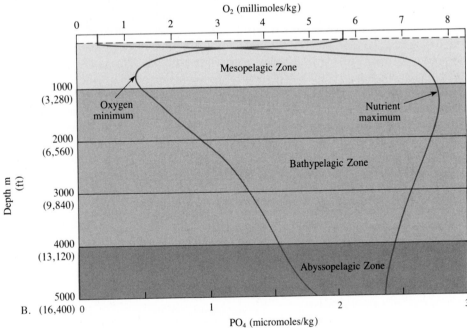

Within the mesopelagic zone, a dissolved oxygen minimum occurs at a depth of about 700 to 1000 m (2300–3280 ft). The intermediate-water masses that move horizontally in this depth range often possess the highest levels of plant nutrient content in the ocean. The base of the permanent thermocline and the boundary between the disphotic and aphotic zones characteristically occur at the 1000-m (3280-ft) boundary between the mesopelagic and bathypelagic zones.

Within the mesopelagic zone, we see evidence that animals sense the presence of light in the phenomenon known as the *deep scattering layer* (DSL). During sonar equipment tests by the U.S. Navy early in World War II, a sound-reflecting surface that changed depth on a daily basis was observed. The variation indicated a reflecting mass at a depth of 100 to 200 m (660–1320 ft) during the night that sank to depths as great as 900 m (2950 ft) during the day. After considerable investigation it was determined that this echo, the DSL, was produced by masses of migrating marine life that moved closer to the surface at night and then to a greater depth during the day. DSL appeared to be a response to the changing intensity of solar radiation (figure 13–10).

The use of plankton nets and submersibles has indicated that the DSL contains layers of siphonophores, copepods, euphausiids, cephalopods, and small fish. Tests that have been done using fish with swim bladders suggest that they may be the primary cause of the reflection, since a very small concentration of such fish is sufficient to produce such a phenomenon. Fish, however, are predators and if their presence is responsible for the echoes, the organisms on which they prey are probably responsible for the cyclic nature of their movement. The smaller organisms responding to a changing intensity of light could trigger the vertical movements, while the presence of fish that prey upon them could be primarily responsible for the echoing phenomenon associated with the migration.

The mesopelagic zone is also inhabited by species of fish that have unusually large and sensitive eyes, capable of detecting light levels 100 times less than those which the human eye can detect. Another important inhabitant of this zone is the bioluminescent group, especially shrimp, squid, and fish. Approximately 80 percent of the inhabitants carry light-producing *photophores*. These are glandular cells containing luminous bacteria surrounded by dark pigments. Some contain lenses to amplify the radiation. This cold light is produced by a chemical process involving the compound luciferin. The molecules of luciferin are excited and emit photons of light in the presence of the enzyme luciferase and oxygen. Only a 1 percent loss of energy is required to produce the illumination. This system is similar to that of the firefly.

The aphotic bathypelagic zone and abyssopelagic zone below the mesopelagic zone represent in excess of 75 percent of the living space in the oceanic province. In this region of total darkness many totally blind fish exist. Very bizarre small predaceous species make up the total fish population. Many species of shrimp that normally feed on detritus become predators at these depths, where the food supply has been greatly reduced from what is available in the shallower waters. The animals that live in the aphotic bathypelagic and abyssopelagic zones feed mostly upon one another and have developed impressive warning devices and unusual apparatus to make them more efficient predators. They are characterized by small expandable bodies, extremely large mouths relative to body size, and very efficient sets of teeth (figure 13–11).

Aside from the depth of light penetration, the primary physical criteria used in defining the boundaries of the bathypelagic zone are the top and bottom of deep-water masses. Oxygen content increases with depth, as these water masses carry oxygen from the cold surface waters where they formed to the deep ocean.

FIGURE 13–10
Deep Scattering Layer. The deep scattering layer, which scatters and reflects sonar signals well above the bottom, may be caused by euphausids that grow to lengths greater than 2 cm (0.8 in.) and lantern fish (myctophid) that reach lengths of 7 cm (2.8 in.). They are predators that feed on smaller planktonic organisms migrating vertically in the water column on a daily cycle.

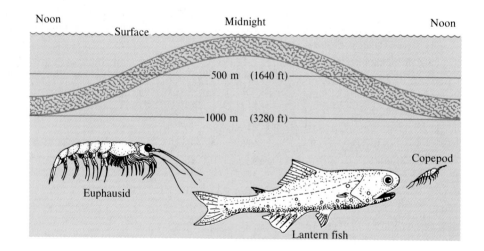

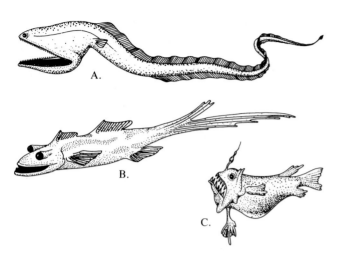

FIGURE 13–11
Deep-Sea Fish. *A, Eurypharynx pelecanoides.* Length, 50–60 cm (20–24 in.). *B, Gigantura chuni.* Length, to 12 cm (5 in.). *C, Lino-phryne bicornis.* Length, 5–8 cm (2–3 in.).

The abyssopelagic zone is the realm of the bottom-water masses that commonly move in the opposite direction of the overlying deep water masses of the bathypelagic zone.

Benthic Environment

The benthic environment can be subdivided into two larger units, which are then subdivided on the basis of various criteria (figure 13–9). These are the *subneritic* province, which extends from the spring high tide shoreline to a depth of 200 m (660 ft) (approximately the continental shelf), and the *suboceanic* province, which includes all the benthic environment below the 200 m (660 ft) depth.

A transitional region above the high tide line is the *supralittoral.* The supralittoral zone, commonly called the *spray zone,* is covered with water only during periods of extremely high tides and when tsunamis or large storm waves break on the shore.

The intertidal zone is classified as the *littoral zone,* and from low tide to 200 m (660 ft), the subneritic is called the *sublittoral* or shallow subtidal zone. The *inner sublittoral* includes that part of the sublittoral to a depth of approximately 50 m (160 ft). This seaward limit will vary considerably because it is determined by that depth at which we find no plants growing attached to the ocean bottom. Its extent is determined primarily by the amount of solar radiation that penetrates the surface water. In areas of high turbidity the inner sublittoral will have its seaward limit at shallower depths due to the

decreased penetration of solar radiation, while in areas where the water is unusually clear it may stand well below the 50 m (160 ft) mark. The *outer sublittoral* includes that portion of the sublittoral zone from the seaward limit of the inner sublittoral to a depth of 200 m (660 ft) or the shelf break, the seaward edge of the continental shelf.

With increased depth in the suboceanic system, we find the *bathyal* zone extending from a depth of 200 to 4000 m (660–13,100 ft) and corresponding generally to that geomorphic province found beyond the continental shelf—the continental slope. From a depth of 4000 to 6000 m (13,100–19,700 ft) stretches the *abyssal* zone, which includes in excess of 80 percent of the benthic environment. A restricted environment, including all depths below 6000 m (19,700 ft), found only in the trenches along the margins of continents is the *hadal* zone.

Once we have reached the outer sublittoral benthic region, we will see no more attached plants as we go from there into the deep ocean basin. All the photosynthesis seaward of the inner sublittoral zone is carried on by the microscopic algae floating in the neritic province above the outer sublittoral zone and in the epipelagic zone of the oceanic province. The *base,* or seaward limit, of the inner sublittoral can be said to coincide with the base of the *euphotic* zone, which is that zone of water near the ocean surface into which enough light penetrates to support photosynthesis. The seaward limit of the outer sublittoral approximately coincides with the shelf break. Currents are in general stronger across the outer sublittoral bottom than along the upper continental slope of the bathyal zone. Sediments are also coarser on the continental shelf than on the continental slope.

Changes in the sediment type of the continental slope to a depth of about 4000 m (13,100 ft) are subtle. Throughout much of the ocean, the calcium carbonate compensation depth occurs at about this depth, so sediments in the bathyal zone may contain considerable calcium carbonate, while those below contain little or none. In general, the deposits of neritic sediment found around the continents begin to be replaced by oceanic sediment below 4000 m (13,100 ft).

The ocean floor representing the abyssal zone is covered by soft oceanic sediment, primarily abyssal clay. The tracks and burrows of animals that live in this sediment are frequently recorded in bottom photographs (figure 13–12). The abyssal zone represents over 60 percent of the surface area of the benthic environment, or almost 43 percent of the earth's surface.

The hadal zone below 6000 m (19,700 ft) is primarily ocean trenches. Isolation in these deep, linear depressions allows the development of faunal assemblages found only in these trenches.

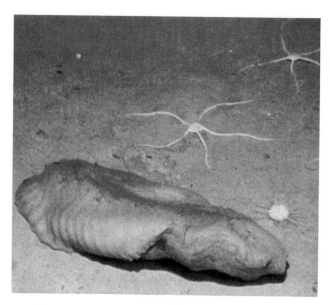

FIGURE 13–12
Benthic organisms. Deep-sea sea cucumber, brittle stars, and sea urchin. (Photograph by Fred Grassle.)

DISTRIBUTION OF LIFE IN THE OCEANS

It is difficult to describe the degree to which the sea is inhabited because of its immense space and the paucity of our knowledge of it. We do know, however, that some populations fluctuate greatly each season, and this fact increases the difficulty of describing with numbers the extent to which the marine environment is populated. Although it is perhaps meaningless to estimate the number of individual organisms in the ocean, we may derive some means of comparing the marine and terrestrial environment by comparing the number of marine species to land species.

Well over 1.2 million species of animals are known, and only about 20 percent of these live in the ocean. Many biologists believe there may be from 3 to 10 million unnamed and undescribed animals living on earth. A large number of these may well inhabit the oceans. Yet, the following theory may explain why we can expect fewer species of marine animals to exist than terrestrial animals.

If the ocean is such a prime habitat for life and if life originated in this environment, why do we now see such a small percentage of the world's animal species living there? This lesser number may well result because the marine environment is more stable than the terrestrial environment. The relatively uniform conditions of the open ocean do not produce pressures for adaptation. Also, once we get below the surface layers of the ocean, the temperatures are not only stable but also relatively low. Chemical reactions are retarded by this lower temperature; this, in turn, may reduce the tendency for variation to occur.

If we are to consider briefly the great variety of species of organisms found on the continents, we can assume that this development was the product of an environment less stable than is found in the ocean, one possessing many opportunities for natural selection to produce new species to inhabit new niches within this terrestrial environment. At least 75 percent of all land animals are insects that have evolved species capable of inhabiting very restricted environmental niches. If we ignore the insects, the sea does possess in excess of 65 percent of the remaining animal species living in the marine and terrestrial environments

In describing the distribution of life within the ocean, we can get a general idea of the relative concentrations in the various marine environments if we consider that of the 200,000 animal species inhabiting the marine environment, only 4000 (about 2 percent) live in the pelagic environment. The remainder inhabit the ocean floor (figure 13–13).

Before further discussion of the organisms of the oceans, we need to define some terms to describe these organisms on the basis of the portion of the ocean they inhabit and the means by which they move.

Plankton

The *plankton* include all those organisms that drift with ocean currents. This does not mean that all are without the ability to move. Many plankters have this capacity

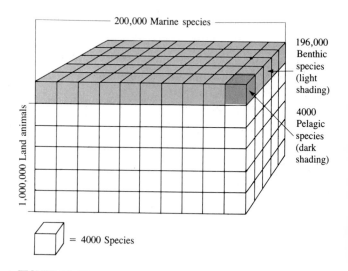

FIGURE 13–13
Distribution of Animal Species on Earth. The cube represents the 1.2 million species of animals living on the earth. The colored division indicates the proportion (20 percent) that live in the oceans. Of the marine species, 98 percent live on or in the ocean floor.

FIGURE 13–14

Phytoplankton and Zooplankton.
Maximum dimension in parentheses.
Phytoplankton *A*, and *B*, coccolitho-
phoridae (15μm, or 0.0006 in.). *C–L*,
diatoms (80 μm, or 0.0032 in.): *C,
Corethron; D, Asteromphalus; E, Rhi-
zosolenia; F, Coscinodiscus; G, Bid-
dulphia favus; H, Chaetoceras; I, Lic-
mophora; J, Thalassiorsira; K,
Biddulphia mobiliensis; L, Eucampia.
M–S,* dinoflagellates (100 μm, or
0.004 in.): *M, Ceratium recticulatum;
N, Goniaulax scrippsae; O, Gymnodi-
nium; P, Goniaulax triacantha; Q,
Dynophysis; R, Ceratium bucephalum;
S, Peridinium.*

Zooplankton *A,* fish egg (1 mm, or
0.04 in.); *B,* fish larva (5 cm, or 2 in.);
C, radiolaria (0.5 mm, or 0.02 in.); *D,*
foraminifera(1 mm, or 0.04 in.); *E,* jellyfish (30 m, or 98 ft); *F,* arrow
worms (3 cm, or 1.2 in.); *G* and *H,*
copepods (5 mm, or 0.2 in.); *I,* salp
(10 cm, or 4 in.); *J,* doliolum; *K,* jellyfish (30 m, or 98 ft); *L,* worm larva (1
mm, or 0.04 in.); *M,* fish larva (5 cm,
or 2 in.); *N,* tintinnid (50 μm, or
0.002 in.); *O,* foraminifera (1 mm, or
0.04 in.); *P,* dinoflagellate (*Noctiluca*)
(1 mm, or 0.04 in.).

PHYTOPLANKTON

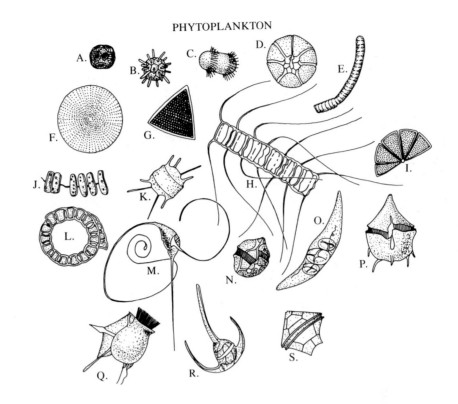

ZOOPLANKTON

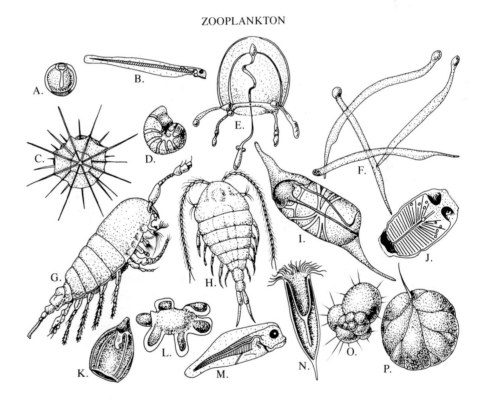

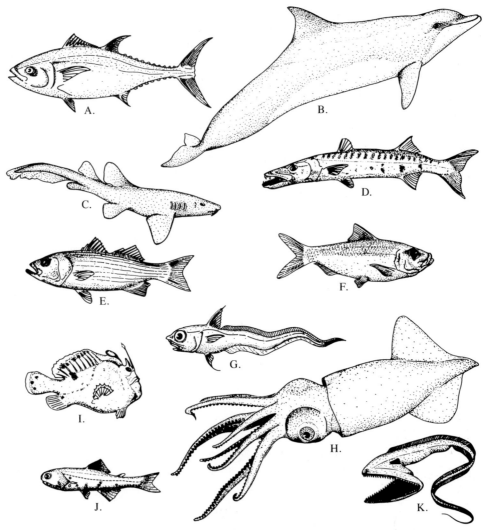

FIGURE 13–15

Nekton. Not drawn to scale; typical maximum dimension is in parentheses. *A,* bluefin tuna (2 m, or 6.6 ft); *B,* bottlenosed dolphin (4 m, or 13 ft); *C,* nurse shark (3 m, or 10 ft); *D,* barracuda (1 m, or 3.3 ft); *E,* striped bass (0.5 m, or 1.6 ft); *F,* sardine (15 cm, or 6 in); *G,* deep-ocean fish (8 cm, or 3 in.); *H,* squid (1 m, or 3.3 ft); *I,* angler fish (5 cm, or 2 in.); *J,* lantern fish (8 cm, or 3 in.); *K,* gulper (l5 cm, or 6 in.).

but either are able to move only weakly or are so restricted to vertical movement that they cannot determine their horizontal position within the ocean. The plants that follow this life style in the upper layers of the ocean are called *phytoplankton,* and the animals are *zooplankton* (figure 13–14). It has recently been discovered that free-living bacteria are much more abundant in the plankton community than they were previously thought to be. Having an average dimension of only 0.5 μm (0.00002 in.), the *bacterioplankton* were missed in earlier studies because of their small size.

In considering the plankton in more detail, we will try to develop some basic understanding of this portion of marine life that many of us have never seen, or imag-

ined. The fact is, most of the biomass of the earth is found adrift in the oceans as plankton. The volume of the earth's space inhabited by animals that either drift or swim greatly exceeds that occupied by all animals that live on land or the ocean's bottom.

The plankton range in size from larger animals and plants such as jellyfish, salps, and Sargassum (*macroplankton,* 2 to 20 cm or 0.8–8 in.) to bacteria that are too small to be filtered from the water by a silk net and must be removed by other types of microfilters (*picoplankton,* 0.2 to 2 μm or 0.000008–0.00008 in.).

An additional scheme for classifying plankton is based on the portion of their life cycle spent within the plankton community. Those organisms such as planktonic diatoms and copepods which spend their entire life in the plankton are *holoplankton*. Many of the organisms we normally consider as *nekton* or *benthos* because they spend their adult life in one or the other of these living modes actually spend a portion of their life cycle as plankton. Many of the nekton and very nearly all the benthos during their larval stages make the plankton community their home. These organisms, which as adults sink to the bottom to become members of the benthos or begin to swim freely as members of the nekton, are *meroplankton*.

Nekton

The *nekton* include all those animals capable of moving independently of the ocean currents that are not only capable of determining their own positions within relatively small areas of the ocean but are also capable, in many cases, of long migrations. Included in the nekton are most adult fish and squid, marine mammals, and marine reptiles (figure 13–15).

The freely moving nekton, with some exceptions, are not able to move at will throughout the breadth of the ocean. By the presence of nonvisible barriers they are effectively limited in their lateral range. Gradual changes in temperature, salinity, viscosity, and availability of nutrients create impenetrable barriers. The death of large numbers of fish has frequently been caused by temporary lateral shifts of masses of ocean water. Vertical range is normally determined by pressure for all fish that contain air bladders and mammals that belong to the nekton.

Fish appear to be everywhere but are normally considered to be more abundant near continents and islands and in colder waters. Some fish, such as the salmon, ascend rivers to spawn. Many eels do just the reverse, growing to maturity in fresh water, then descending the streams to breed in the great depths of the ocean.

FIGURE 13–16
Benthos—Some Intertidal and Shallow Subtidal Forms. Not drawn to scale; typical maximum dimension is in parentheses. *A,* sand dollar (8 cm, or 3 in.); *B,* clam (30 cm, or 12 in.); *C,* crab (30 cm, or 12 in.); *D,* abalone (30 cm, or 12 in.); *E,* sea urchins (15 cm, or 6 in.); *F,* sea anemones (30 cm, or 12 in); *G,* brittle star (20 cm, or 8 in.); *H,* sponge (30 cm, or 12 in.); *I,* acorn barnacles (2.5 cm, or 1 in.); *J,* snail (2 cm, or 0.8 in.); *K,* mussels (25 cm, or 10 in.); *L,* gooseneck barnacles (8 cm, or 3 in.); *M,* sea star (30 cm, or 12 in.); *N,* brain coral (50 cm, or 20 in.); *O,* sea cucumber (30 cm, or 12 in.); *P,* lamp shell (10 cm, or 4 in.); *Q,* sea lily (10 cm, or 4 in.); *R,* sea squirt (10 cm, or 4 in.).

Benthos

The *benthos* live on or in the ocean bottom. There are those called *infauna* that live buried in the sand, shells, or mud. Those that live attached to rocks or move over the surface of the ocean bottom are the *epifauna*. Benthos such as some shrimp and demersal flounder that live in the bottom but move with relative ease through the water above the ocean floor are the *nektobenthos* (figure 13–16).

The littoral and inner sublittoral are the only zones in which we find the macroscopic algae attached to the bottom because they are the only benthic zones to which sufficient light can penetrate. There is a great diversity of physical and nutritive conditions in these zones. Animal species have developed in great numbers within this nearshore benthic community as a result of the variations existing within the habitat. As one moves across the bottom from the littoral into the deeper benthic environments, one commonly observes that an inverse relationship exists between the distance from shore and the number of benthic individuals and biomass that can be found. The number of species per unit of surface area may remain relatively constant throughout the benthic environment.

Throughout most of the benthic environment, animals live in a region of perpetual darkness where photosynthetic production cannot occur. They must feed on each other or on whatever outside nutrients fall from the productive zone near the surface. The deep-sea bottom is an environment characterized by coldness, stillness, and darkness. Under these conditions, it would be expected that life would move at a relatively slow pace. For those animals that move around on the bottom, streamlining, which is very important to the nekton, is of little importance.

The organisms that live in the deep sea normally have quite a wide range of distribution because the physical conditions for life do not vary greatly, even over great distances, on the deep-ocean floor. A few species appear to be extremely tolerant of pressure changes in that members of the same species may be found in the littoral province and at depths of several kilometers.

The 1977 discovery of the first hydrothermal vent biocommunity in the Galapagos Rift has shown us that high concentrations of deep-ocean benthos are possible. It appears that the primary limiting factor for life on the deep ocean floor is the availability of food. At these vents, food in the form of chemosynthetic bacteria is abundant. Thus, the size of individuals and the total biomass in the hydrothermal communities far exceeds that previously known for the deep-ocean benthos.

SUMMARY

The relatively stable marine environment is thought to have given rise to all **phyla** of organisms. Those organisms that have established themselves in the terrestrial realm have had to develop complex systems for support and for acquiring and retaining water.

Although the relative proportions of the constituents of salinity in the ocean and in the body fluids of the organisms are often very nearly the same, the salinity of one may differ greatly from the other. If the body fluids of an organism and ocean water are separated by a membrane that allows water molecules to pass, problems requiring **osmoregulation** may develop. Marine invertebrates and sharks are essentially **isotonic**, having body fluids with a salinity similar to that of ocean water. They rarely face osmoregulatory problems. Most marine vertebrates are **hypotonic**, having body fluids with a salinity lower than that of ocean water, and tend to lose water through **osmosis**, the passing of water molecules from a region in which they are in higher concentration through a semipermeable membrane into a region where they are in lower concentration. Freshwater organisms are essentially all **hypertonic**, having body fluids much higher in salinity than the water in which they live, so they must compensate for a tendency to take water into their cells through osmosis.

For life to flourish in any environment, there must be a sufficient food supply. The basic producers of food are plants, so the requirements of plants must be met if food is to be plentiful. The availability of **nutrients** and **solar radiation** make plant life possible. Since solar radiation is available only in the surface water of the ocean, plant life is restricted to a thin layer of surface water usually no more than 100 m (330 ft) deep. Nutrients derived ultimately from the continents are much more abundant near continental features. Although much is yet to be learned about the distribution of life in the oceans, it appears that the **biomass** concentration of the oceans decreases away from the continents and with increased depth. The color of the oceans ranges from green in highly productive regions to blue in areas of low productivity.

The plants that must stay in surface water to receive sunlight and the small animals that feed on plants do not have effective means of locomotion. They depend, therefore, on their small **size** and other adaptations to give them a high ratio of surface area per unit of body mass, which results in a greater frictional resistance to sinking. Large animals that swim freely face an altogether different problem and generally have **streamlined bodies** to reduce frictional resistance to motion.

Compared to life in colder regions, organisms living in warm water tend to be individually smaller, comprise a greater number of species, and constitute a much smaller total biomass. Warm water organisms also tend to live shorter lives and reproduce earlier and more frequently than their cold water counterparts.

The marine environment is divided into two basic units—the **pelagic** (water) and the **benthic** (bottom) environments. These regions, which are further divided primarily on the basis of depth, are inhabited by organisms we can classify into three categories on the basis of life style. These categories are the **plankton**, or free-floating forms with little power of locomotion; the **nekton**, or free swimmers; and the **benthos**, or bottom-dwellers.

QUESTIONS AND EXERCISES

1. Discuss the major differences between marine plants and land plants, and explain the need for greater complexity of land plants.

2. Define the terms euryhaline, stenohaline, eurythermal, and stenothermal. Where in the marine environment will organisms displaying a well-developed degree of each characteristic be found?

3. Describe the relationships existing among osmotic pressure, salinity, and freezing point of a solution.

4. What is the problem requiring osmotic regulation that is faced by a hypotonic fish in the ocean? How have these animals adapted to meet this problem?

5. An important variable in determining the distribution of life in the oceans is the availability of nutrients. What are the relationships among the continents, nutrients, and the concentration of life in the oceans?

6. Another important determinant of plant productivity is the availability of solar radiation. Why is biological productivity relatively low in the tropical open ocean where the penetration of sunlight is greatest?

7. Discuss the characteristics of the coastal ocean where unusually high concentrations of marine life are found.

8. What factors create the color difference between coastal waters and the less productive open-ocean water?

9. Compare the ability to resist sinking of an organism with an average linear dimension of 1 cm (0.4 in.) to that of an organism with an average linear dimension of 5 cm (2 in.). Discuss some adaptations other than size used by organisms to increase their resistance to sinking.

10. Changes in water temperature significantly affect the density, viscosity of water, and ability of water to hold gases in solution. Discuss how decreased water temperature changes these variables and may affect marine life.

11. Describe how higher water temperatures in the tropics may account for the greater number of species in these regions compared to low-temperature, high-latitude areas.

12. Construct a table listing the subdivisions of the benthic and pelagic environments and the physical factors used in assigning their boundaries.

13. Describe the vertical distribution of oxygen and nutrients in the oceanic province, and discuss the factors that are responsible for this distribution.

14. Discuss the probable cause and composition of the deep scattering layer.

15. List the relative number of species of animals found in the terrestrial, pelagic, and benthic environments, and discuss the factors that may account for this distribution.

16. Describe the lifestyles of plankton, nekton, and benthos. Why is it proper to consider that plankton account for a relatively larger percentage of the biomass of the oceans than the benthos and nekton?

17. List the subdivisions of plankton and benthos and the criteria used for assigning individual species to each.

REFERENCES

Borgese, E. M., and Ginsburg, N. 1988. *Ocean yearbook 7.* Chicago: The University of Chicago Press.

Coker, R. E. 1962. *This great and wide sea: an introduction to oceanography and marine biology.* New York: Harper and Row.

Hedpeth, J., and Hinton, S. 1961. *Common seashore life of southern California.* Healdsburg, Calif.: Naturegraph.

Isaacs, J. D. 1969. The nature of oceanic life. *Scientific American* 221:65–79.

May, R. M. 1988. How many species are there on Earth? *Science* 241:4872, 1441–48.

Roughgarden, J., Gaines, S., and Passingham, H. 1988. Recruitment dynamics in complex life cycles. *Science* 241:4872, 1460–66.

Sieburth, J. M. N. 1979. *Sea microbes*. New York: Oxford University Press.

Sumich, J. L. 1976. *An introduction to the biology of marine life*. Dubuque, Ia: Wm. C. Brown.

Sverdrup, H., Johnson, M., and Fleming, R. 1942. *The oceans*. Englewood Cliffs, N.J.: Prentice Hall.

Thorson, G. 1971. *Life in the sea*. New York: McGraw-Hill.

SUGGESTED READING

Sea Frontiers

Burton, R. 1977. Antarctica: Rich around the edges. 23:5, 287–95.
The high level of biological productivity around the continent of Antarctica is the topic.

Gruber, M. 1970. Patterns of marine life. 16:4, 194–205.
Many varieties of life in the ocean are discussed in terms of how their form and size fit them for life in a particular environmental niche.

Hammer, R. M. 1974. Pelagic adaptations. 16:1, 2–12.
A comprehensive discussion of the adaptations of pelagic organisms to reduce the energy required to maintain their position in the open ocean.

Patterson, S. 1975. To be seen or not to be seen. 21:1, 14–20.
A discussion of the possible role of color in the protection and behavior of tropical fishes.

Perrine D. 1987. The strange case of the freshwater marine fishes. 33:2, 114–19.
Explains how marine crevalle jacks are able to inhabit the fresh waters of Crystal River, Florida.

Schellenger, K. 1974. Marine life of the Galapagos. 20:6, 322–32.
A discussion of the unique life forms of the Galapagos Islands, 950 km (590 mi) from South America.

Thresher, R. 1975. A place to live. 21:5, 258–67.
An interesting discussion of how bottom-dwelling animals compete for space on the ocean floor.

Williams, L. B., and Williams E. H., Jr. l988. Coral reef bleaching: Current crisis, future warning. 34:2, 80–87.
Corals and related reef animals underwent "bleaching" along the Central American coast in 1983 and in the Caribbean Sea in l987.

Scientific American

Denton, E. 1960. The buoyancy of marine animals. 203:1, 118–29.
The means by which some marine animals reduce the energy expenditure required to live in the ocean water far above the ocean floor are discussed.

Eastman, J. T., and DeVries, A. L. 1986. Antarctic fishes. 255:5, 106–14.
Explains how one group of fish survived when the Antarctic turned cold.

Horn M. H., and Gibson, R. N. 1988. Intertidal fishes. 258:1, 64- 71.
Intertidal fishes have undergone remarkable adaptation to survive this physically harsh environment.

Isaacs, J. D., and Schwartzlose, R. A. 1975. Active animals of the deep-sea floor. 233:4, 84–91.
A surprisingly large population of large fishes on the deep-sea floor is suggested by automatic cameras dropped to the ocean bottom.

Isaacs, J. D. 1969. The nature of oceanic life. 221:3, 146–65.
A well-developed survey of the conditions for life in the ocean as they relate to the variety and distribution of marine life forms.

Palmer, J. D. 1975. Biological clock and the tidal zone. 232:2, 70–79.
This article investigates the mechanism of biological clocks set to the rhythm of the tides, which are found in organisms from diatoms to crabs.

Partridge, B. L. 1982. The structure and function of fish schools. 246:6, 114–23.
Schooling benefits and the means by which fish maintain contact with the school are considered.

Vogel, S. 1978. Organisms that capture currents. 239:2, 128–39.
The manner in which sponges use ocean currents is an important part of this discussion.

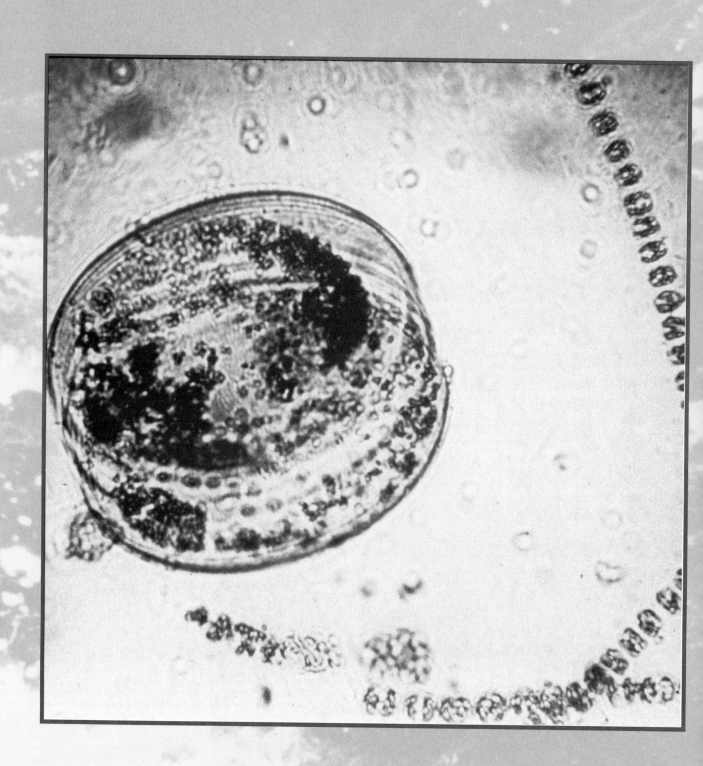

14

BIOLOGICAL PRODUCTIVITY—ENERGY TRANSFER

The major primary producers of the oceans, marine algae, capture most of the energy used to support the marine biological community. When we think of marine plants, most of us undoubtedly first consider the large macroscopic plants we see growing near shore in many coastal areas. These large plants, however, play only a minor part in the production of energy for the ocean population as a whole. Instead, marine organisms depend primarily on the small planktonic varieties of marine plant life that inhabit the upper sunlit water of the world's oceans. They are not so obvious. All are microscopic, but they are scattered throughout the breadth of the ocean surface layer and represent the largest biomass community in the marine environment—phytoplankton.

In this chapter we first consider the classification of organisms. We then discuss the composition of the marine algae, the ecological factors related to plant and bacterial productivity, and the passing of stored chemical energy on to other marine organisms within food webs.

TAXONOMIC CLASSIFICATION

All living things belong to one of the five kingdoms shown in figure 14–1 (see appendix V). The simplest of all organisms are classified as *Monera*. These organisms are single-celled and have their nuclear material spread throughout the cell. Included in this kingdom are the cyanobacteria (blue-green algae) and heterotrophic bacteria.

Representing a higher stage of evolutionary development is the kingdom *Protista*. It includes those organisms that are single-celled but have their nuclear material contained within a nuclear sheath. Here we find the algae (simple plants) and the single-celled animals, *Protozoa*.

Kingdom *Mycota*, the fungi, appear to be poorly represented in the oceans. Less than 1 percent of the known 50,000 species are found there. Fungi are found throughout the marine environment, but they are much more common in the intertidal zone in a mutu-

Coscinodiscus diatom (left) and chain diatom. Photograph by Michael DiSpezio.

FIGURE 14–1
The Five Kingdoms of Organisms.

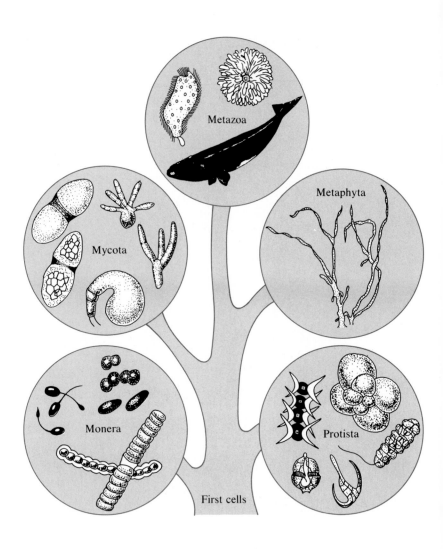

alistic relationship with cyanobacteria or green algae. In this relationship, called *lichen*, the fungus provides a protective covering that retains water during periods of exposure, while the algae provides food for the fungus through photosynthesis. Other fungi function primarily as decomposers in the marine ecosystem: They remineralize organic matter.

Though there is one more kingdom of marine plants—*Metaphyta*, the multicelled plants—they are mostly restricted to the shallow coastal margins of the ocean and are not a major component of the marine productive community. They do play a very important role as producers within restricted communities, such as mangrove swamps and salt marshes.

The *Metazoa* consists of multicelled animals. Metazoans range in complexity from the simple sponges to the vertebrates (animals with backbones).

The kingdoms are divided into increasingly specific groupings (*taxa*). The system of taxonomic classification introduced by Carl von Linné (Linnaeus) in 1758 includes the following major categories:

Kingdom

Phylum

Class

Order

Family

Genus
Species $\Big\}$ Species name

Every organism's scientific name includes its genus and species. For example, the toothed whale called the common dolphin is the species *Delphinus delphis*.

MACROSCOPIC PLANTS

As we discuss the marine plants, we will first consider those groups of plants with which we are most familiar. These are the attached forms of macroscopic algae and metaphyta found in shallow waters along the margins of the ocean. In classifying algae, we will use criteria that are based partly on the pigment content of the plants within the group (figure 14–2).

Phaeophyta (Brown Algae)

The phylum *Phaeophyta* contains the largest members of the marine plant community that are attached to the bottom wherever suitable substrate is available in the littoral and inner sublittoral zones. The dominant pigment in the brown algae is *fucoxanthin*, and the color may range from a very light brown to black. The brown algae occur primarily in temperate and cold water areas. Their sizes range from the small black encrusting patch of the *Ralfsia,* found in the upper and middle intertidal zones, where it is exposed to an extensive threat of desiccation and may become crisp and dry in the sun without dying, to the *Pelagophycus* (bull kelp), which may grow from small holdfasts in water depths in excess of 30 m (98 ft).

Chlorophyta (Green Algae)

The predominant forms of algae found in freshwater environments belong to the phylum *Chlorophyta,* green algae, and are not well represented in the ocean. Most species are intertidal, or grow in shallow waters of bays. Because of the pigment *chlorophyll* they range in color from yellow-green to a very dark green, but most are grass green in color. They grow only to moderate size, seldom exceeding 30 cm (12 in.) in the largest dimension. Forms range from that of finely branched filaments to flat thin sheets. The various species of the genus *Ulva* (sea lettuce), a thin membranous sheet two cell layers thick, may be found widely scattered throughout cold water areas, whereas the genus *Codium* (sponge weed), a dichotomously branched form more commonly found in warm waters, can be in excess of 6 m (20 ft) in length.

Rhodophyta (Red Algae)

Red algae belong to the phylum *Rhodophyta* and are the most abundant and widespread of the marine macroscopic algae. Over 4000 species are found, many of them attached, from the very highest intertidal levels to the outer edge of the inner sublittoral zone. They are very rare in fresh water. The red algae range in size from just visible to the naked eye to lengths of up to 3 m (10 ft). While Rhodophyta are found in both warm and cold water areas, the warm water varieties are relatively small. The characteristic pigment of the red algae is *phycoerythrin.* The color of the red algae will vary considerably depending on their depth in the intertidal or inner sublittoral zones. In the upper, well-lighted areas it may be green to black or purplish in color, and it changes through a brown to a pinkish red in the deeper water zones, where the light concentrations are lower. Although the bulk of marine plant productivity is believed to occur above water depths where the amount of light is reduced to 1 percent of that available at the surface (approximately 100 m, or 330 ft), a red alga has been observed growing at a depth of 268 m (880 ft) on a seamount near San Salvadore, Bahamas. Available light at this sighting was thought to be only 0.05 percent of the light available at the ocean's surface.

Spermatophyta (Seed-Bearing Plants)

The only metaphyta observed in the marine environment belong to the highest group of plants, the seed-bearing *Spermatophyta.* Two seed-bearing plants found in the marine environment are *Zostera* (eelgrass) and *Phyllospadix* (surf grass). Zostera, a grasslike plant with true roots, is found primarily in quiet waters of bays and estuaries from the low tide zone down to a depth of some 6 m (20 ft). *Phyllospadix* prefers the high-energy environment of an exposed rocky coast and can be found from the intertidal zones down to a depth of 15 m (50 ft). Both of these plants are considered to be important sources of the detrital food for the marine animals that inhabit their environment. Found in salt marshes are grasses belonging mostly to the genus *Spartina,* while mangrove swamps contain primarily the mangrove genera *Rhizophora* and *Avicennia.*

MICROSCOPIC PLANTS

The phyla discussed next include all the members of the important phytoplankton that produce in excess of 99 percent of the food supply for marine animals. They are primarily floating forms, although some live on the bottom in the nearshore environment.

Chrysophyta (Golden Algae)

Containing predominantly the yellow pigment *carotin,* these microscopic plants belong to the phylum *Chrysophyta* and store food in the form of *leucosin* (a carbohydrate) and oils.

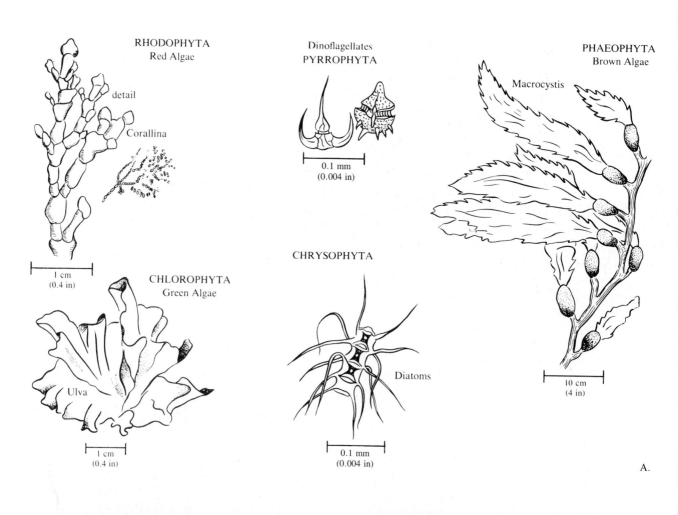

RHODOPHYTA
Red Algae

detail

Corallina

1 cm
(0.4 in)

Dinoflagellates
PYRROPHYTA

0.1 mm
(0.004 in)

PHAEOPHYTA
Brown Algae

Macrocystis

CHLOROPHYTA
Green Algae

Ulva

1 cm
(0.4 in)

CHRYSOPHYTA

Diatoms

0.1 mm
(0.004 in)

10 cm
(4 in)

A.

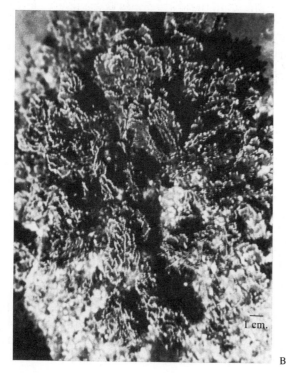

1 cm.

B.

1 cm.

C.

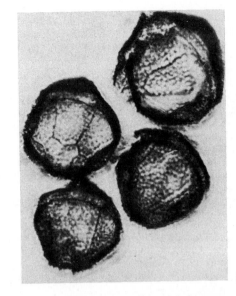

D.

E.

FIGURE 14–2

Algae. *A*, detailed drawing of several algae. (Line drawing by Phil David Weatherly.) *B*, red algae, *Corallina vancouveriensis*. *C*, brown algae, *Egregia laevigata*. Feather-boa kelp, common in the surf zone. Note the flat axis with featherlike fronds, some of which have air bladders at the base. *D*, green algae, *Codium fragile*. The prominent fingerlike plant is called *sponge weed*. (*B–D*, photos from John D. Roche, Jr.) *E*, photomicrograph of *Gonyaulax polyedra* magnified 1100 times. *Gonyaulax* is a large genus of phosphorescent marine dinoflagellates that, in great abundance, cause what is known along the shoreline as the "red tide." (Photo from Scripps Institute of Oceanography, University of California, San Diego.) *F*, Coccoliths, disc-shaped calcium carbonate, plates that cover coccolithophore cells.

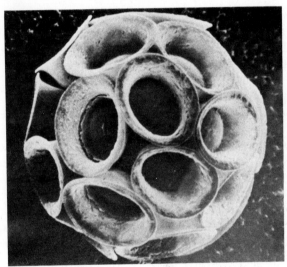

F.

Diatoms The diatoms are a class of plants contained in a shell, or *frustule*, composed of opaline silica ($SiO_2 \cdot n\ H_2O$). These silica housings are important geologically because they accumulate on the ocean bottom and produce a siliceous sediment, *diatomite*. Some raised deposits of diatomite on land are mined and used primarily in the manufacture of filtering devices.

The frustule of the diatom is similar in structure to a microscopic pill box. The top and the bottom of the box are called *valves*; the larger of the valves is the *epitheca* and the smaller valve is the *hypotheca*. The protoplasm of the plant is contained within this housing and exchanges nutrients and waste products with the surrounding water through slits or pores in the valves.

The reproduction of diatoms is by simple cell division. To achieve this division, the valves of the frustule must separate. Each valve then serves as one-half the housing for each newly formed cell. A new valve must form to complete the enclosure of each daughter cell, and each of these newly formed halves forms as a hypotheca.

As can be seen in figure 14–3, this process leads to the formation of smaller and smaller organisms. Eventually, the newly formed daughter cell approaches a size so small that if it were decreased further, abnormalities would result. At this point, a change in the process occurs, namely, the formation of an *auxospore*. The auxospore forms between the two separating valves within

FIGURE 14–3

Diatom Reproduction. The manner in which diatom frustules separate during cell division. As the epitheca and hypotheca of a diatom separate, they become the epitheca of each new cell. The new frustule is completed by the generation of a new hypotheca by each new diatom. Following the arrow through the formation of three generations of diatoms, it can be seen that some cells will be crowded into ever smaller frustules. When the size of the frustule becomes critically small, an auxospore forms to allow growth of a larger cell.

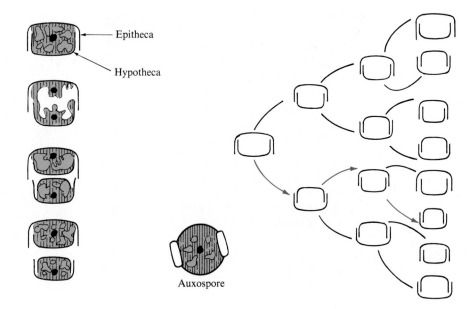

Epitheca

Hypotheca

Auxospore

a membrane that allows the auxospore to grow into a full-sized diatom so that the process of reproduction by the splitting of the frustule can commence again.

Coccolithophores Coccolithophores are for the most part flagellated organisms covered with small calcareous plates, *coccoliths,* made of calcium carbonate ($CaCO_3$). The name of the group means "bearers of coccoliths." The individual plates are about the size of a bacterium, and the entire organism is too small to be captured in plankton nets. Coccolithophores are included in the nanoplankton, with dimensions of less than 0.06 mm (0.002 in.). The coccolithophores contribute significantly to calcareous deposits in all the temperate and warmer oceans.

Pyrrophyta (Dinoflagellate Algae)

A group second in importance to the diatoms in marine productivity is the *dinoflagellates,* which belong to the phylum *Pyrrophyta.* Possessing *flagella* (whiplike structures) for locomotion, they have a slight capacity to move into more favorable areas for plant productivity. The dinoflagellates are not important geologically because many have no protective covering, and those that do, have coverings made of cellulose that is easily decomposed by bacterial action once the organism dies. Many of the 1100 species undergo structural changes as a result of changes that occur in the environment. Many are luminescent.

Red tides result from conditions where up to 2 million dinoflagellates may be found in 1 L (1 qt) of water. Mainly responsible for this red tide phenomenon are the genera *Ptychodiscus* and *Gonyaulax,* which produce water-soluble toxins. Gonyaulax toxin is not poisonous to shellfish but will concentrate in their tissue and is poisonous to humans who eat the shellfish. *Ptychodiscus* toxin kills fish and shellfish. April through September are particularly dangerous months in the Northern Hemisphere. In most areas there is a quarantine against the taking of shellfish that feed on these microscopic organisms and concentrate the poisons that they secrete to levels that are dangerous to humans. *Ptychodiscus breve* is an important contributor to Gulf of Mexico red tides. *Gonyaulax* is an important participant in the cooler waters of the New England and West Coast areas.

A potentially tragic epidemic of paralytic shellfish poisoning (PSP) from a *Gonyaulax tamarensis* red tide occurred in the coastal waters of Massachusetts in the fall of 1972. Fortunately, no deaths occurred, although 30 cases of poisoning were reported. Symptoms of PSP are similar to those of drunkenness, including incoherent speech, uncoordinated movement, dizziness, and nausea. Documented cases throughout the world include 300 deaths and 1750 nonfatal cases. There is no known antidote for the toxin, which attacks the human nervous system, but the critical period usually passes in 24 h.

PRIMARY PRODUCTIVITY

Primary productivity is defined as the amount of carbon fixed by autotrophic organisms through the synthesis of organic matter from inorganic compounds such as CO_2 and H_2O using energy derived from solar radiation or chemical reactions. The major process through which primary productivity occurs is thought to be photosynthesis. Although new knowledge of the role of chemosynthesis in supporting hydrothermal vent communities on oceanic spreading centers has emerged, chemosynthesis is much less significant in overall marine primary production than is photosynthesis. Because of the far greater importance of photosynthesis in the overall productivity of the oceans and our greater knowledge of the factors that affect marine photosynthesis, the following considerations will relate primarily to this process.

Photosynthetic Productivity

The total amount of organic matter produced by photosynthesis per unit of time represents the *gross primary production* of the oceans. Plants use some of this organic matter for their own maintenance through respiration. That which remains is the *net primary production,* which is manifested as growth and reproduction products. It is the net primary production that supports the heterotrophic marine populations—animals and bacteria.

In the 1920s, the *Gran Method* of measuring net primary productivity was developed based on the fact that oxygen is produced by photosynthesis in proportion to the amount of organic carbon synthesized. The method involves putting equal quantities of phytoplankton into a series of bottles, all of which contain the same amount of dissolved oxygen. The bottles are then arranged in pairs, one being transparent and the other totally opaque. These pairs are suspended on a hydrographic line through the euphotic zone, where they are left for a specific period of time. After the bottles are brought to the surface, the oxygen concentration is determined for each bottle.

Photosynthesis, which is confined to the transparent bottles, will add oxygen to the water, whereas respiration will reduce the oxygen content in both the transparent and opaque bottles. Increased oxygen concentration in the transparent bottles is proportional to the amount of photosynthesis that has occurred minus the oxygen consumed by plant respiration and thus represents the net increase in biomass of the plants within the transparent bottle. Decreased oxygen content within the opaque bottles corresponds to the respiration rate. For any depth, an assessment of gross production can be made by adding the oxygen gain in the clear bottles to the oxygen loss in the opaque bottles.

The depth at which the oxygen production and the oxygen consumption are equal is called the *oxygen compensation depth,* and it represents that light intensity below which plants do not survive. Since respiration goes on at all times, during the daylight hours plants must produce, through photosynthesis, biomass in excess of that which is consumed by respiration in any 24-h period if the total biomass of the community is to increase (figure 14–4). An analogy can be made with the common paycheck: Gross photosynthesis (gross pay earned) = oxygen change in

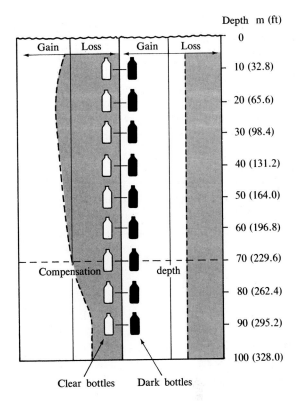

Depth m (ft)

FIGURE 14–4
Compensation Depth of Oxygen. Biological assay of phytoplankton production and consumption of oxygen (O_2) shows to what depth light penetrates with sufficient intensity to sustain photosynthesis at a level sufficient to increase phytoplankton mass. This occurs only above the compensation depth, the depth at which oxygen production by photosynthesis is equal to consumption through respiration. The change in oxygen content in the clear bottles represents a net photosynthesis resulting from gross photosynthesis less the consumption of oxygen by respiration. The loss of oxygen in the dark bottles gives a value for the amount of respiration that occurred in the clear bottles. This amount added to the net photosynthesis that occurred in the clear bottles gives a value for the gross photosynthesis in the clear bottles.

clear bottle (take home pay) + oxygen loss in dark bottle (income tax withheld).

During the 1950s, a method involving the use of radioactive carbon (^{14}C) was developed. It has been refined and is currently the most often used method for determining marine primary productivity. The procedure is similar to that described above, because it involves the use of a series of paired clear and opaque bottles. Each bottle contains identical phytoplankton samples, equal amounts of CO_2 containing carbon-14, and equal amounts of CO_2 containing stable carbon.

The phytoplankton sample is filtered from each bottle after the system has been suspended in the ocean for a sufficient period of time. The amount of beta radiation emitted by each sample is measured by a radiation counting device, and the rate of assimilation of carbon-14 is computed from these measurements. High levels of radioactive emission indicate high levels of productivity.

A third method used in studies of marine primary productivity is the measurement of *chlorophyll a* content of living phytoplankton samples taken from ocean surface waters. Although this method is not as precise as the measurement of net primary productivity using carbon-14, there is a direct relationship between the amount of chlorophyll *a* in a phytoplankton sample from a given volume of ocean water and gross primary productivity. The satellite sensor, Coastal Zone Color Scanner (CZCS), which failed in June, 1986, was used to sense the effect of phytoplankton pigment on the color of ocean water. Data from this system were used to prepare the map of pigment concentration used on the book cover and the maps on plates 6, 21, and 22.

Distribution of Productivity

The production of organic compounds such as carbohydrates, proteins, and fats by plants through the process of photosynthesis will vary considerably throughout the areal extent of the ocean as well as in time. Photosynthetic productivity in the oceans depends on (1) the availability of solar radiation and (2) the availability of nutrients. Both variables are related to the seasonal pattern that grows out of the earth's revolution around the sun while rotating on an axis that is tilted relative to the ecliptic.

Primary photosynthetic productivity of the oceans varies from about 0.1 g of carbon per square meter per day ($gC/m^2/d$) in the open ocean to over $10gC/m^2/d$ in highly productive coastal areas. This variability is primarily the result of the uneven distribution of nutrients throughout the photosynthetic zone. Some of the causes of this variability follow.

Temperature Stratification and Nutrient Supply

Solar radiation penetrates only the surface layer of the ocean, and infrared wavelengths that can be converted directly to heat are essentially 100 percent absorbed in the upper 2 m (6.5 ft) of the ocean. It is not surprising that the surface waters are warmed. This condition produces a thermocline that separates the surface waters, which are relatively warmer, from the deep water masses that are colder and more dense.

Thermoclines are relatively permanent in low latitude areas, while a seasonal thermocline may be present during the summer months in the midlatitudes. They are generally nonexistent in high latitude areas. In the following discussion of productivity, we will see that the

presence or absence of the thermocline will have an important effect on nutrient availability.

In considering the availability of solar radiation and nutrients, we can rather conveniently divide the oceans into polar, temperate, and tropical regions. We will concern ourselves primarily with the nature of the open ocean in all three regions. The polar region will be discussed first as the productivity considerations are simplest there.

Polar Region As an example of productivity on a seasonal basis in a polar sea, we will consider the Barents Sea off the northern coast of Europe, where there is peak diatom productivity during the month of May that tapers off through July (figure 14–5A). This brings about a peak period of zooplankton development, consisting primarily of small crustaceans of the genus *Calanus*. The zooplankton biomass reaches a peak in June and continues at a relatively high level until winter darkness begins in October. In this region above 70°N latitude, there is continuous darkness for about 3 months of winter and continuous illumination for a period of 3 months during summer.

In the Antarctic region, particularly at the southern end of the Atlantic Ocean, productivity patterns are similar to those found in the Barents Sea, except that the seasons are reversed, and productivity is somewhat greater. The most likely explanation for this greater productivity in Antarctic waters is the continual upwelling of water that has sunk in the North Atlantic. Moving southward as a deep-water mass, the North Atlantic Deep Water surfaces hundreds of years later carrying high concentrations of nutrients (figure 14–5B).

To illustrate the very great productivity that occurs during the short summer season in polar oceans, we can consider the growth rate of baby whales. The largest of all whales, the blue whale, migrates through the temperate and polar oceans at the times of maximum zooplankton productivity. As a result of this excellent timing, they are able to develop and support calves that during a gestation of 11 months reach lengths in excess of 7 m (23 ft) at birth. The mother suckles the calf for 6 months with a teat that actually pumps the youngster full of rich milk. By the time the calf is weaned, it is over 16 m (53 ft) in length and will in a period of 2 years reach a length approaching 23 m (75 ft). In 3½ years a 60-tn blue whale has developed. This phenomenal growth rate gives some indication of the enormous biomass of copepods and krill that these large mammals feed upon.

In polar waters there is little density stratification to prevent the mixing of deeper water with shallow water. Polar waters are relatively isothermal (figure 14–5C). In most polar areas the surface waters freely mix with the deeper nutrient-rich water. There can be, however,

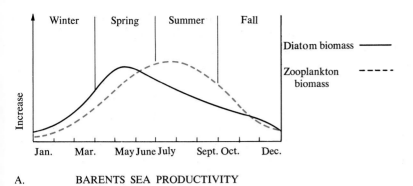

A.   BARENTS SEA PRODUCTIVITY

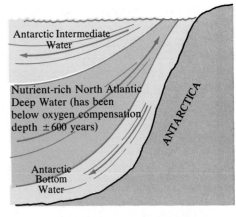

B.   ANTARCTIC UPWELLING

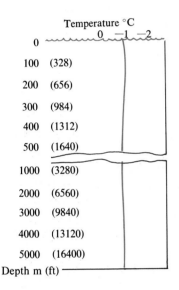

C.   Depth m (ft)

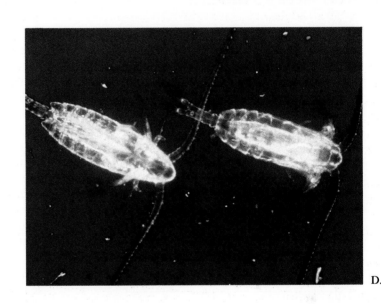

D.

FIGURE 14–5

Productivity in Polar Oceans. *A*, diatom mass increases rapidly in the spring when the sun rises high enough in the sky to cause deep penetration of sunlight. As soon as this diatom food supply develops, the zooplankton population begins feeding on it, and the zooplankton biomass peaks early in the summer. *B*, the continuous upwelling of North Atlantic Deep Water keeps the Antarctic waters rich in nutrients. When the summer sun provides sufficient radiation, there is a huge period of biological productivity. *C*, polar water shows little temperature stratification. Nearly isothermal conditions from the surface to the bottom are common. *D*, copepod, *Calanus*. (Photo courtesy of Scripps Institution of Oceanography, University of California, San Diego.)

some density stratification of water masses due to the summer melting of ice, which lays down a thin low-salinity layer that does not readily mix with the deeper waters.

There are usually high concentrations of phosphates and nitrates in the surface waters. Thus, plant productivity in the high latitudes is more commonly limited by the availability of solar energy than by the availability of nutrients. The productive season in these waters is relatively short but is characterized by an outstandingly high rate of production.

Tropical Region   In direct contrast with high productivity associated with the summer season in the polar seas, low productivity is the rule in the tropical regions of the open ocean. Light penetrates much deeper into the open tropical ocean than into the temperate and polar waters. This produces a very deep compensation depth. In the tropical ocean, however, a permanent thermocline produces a relatively permanent stratification of water masses and prevents mixing between the surface waters and the nutrient-rich deeper waters (figure 14–6A).

A. NORMAL TROPICAL REGIONS

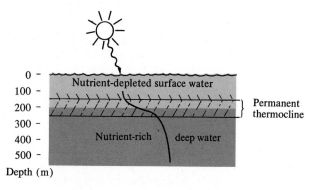

Deep penetration of sunlight produces a deep compensation depth with a good supply of solar radiation available for photosynthesis.

Thermocline serves as an effective barrier to mixing of surface and deep water. As plants use nutrients in surface layer, productivity is retarded because thermocline prevents replenishment from deeper water.

B. TROPICAL REGIONS OF HIGH PRODUCTIVITY
(local areas where nutrients are brought to the surface)

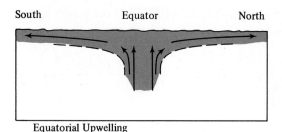

Equatorial Upwelling

Nutrient-enriched water from 200 to 300 m (656 to 984 ft) replaces diverging equatorial surface water.

Coastal Upwelling

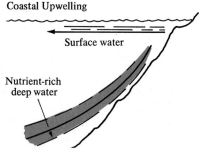

Surface water moved away from the shore is replaced by deep water that is rich in nutrients because it comes from below the euphotic layer.

FIGURE 14–6
Productivity in Tropical Oceans.

At about 20° latitude the concentrations of phosphate and nitrate are commonly less than $\frac{1}{100}$ of the concentrations of these nutrients in temperate oceans during winter. Nutrient-rich waters within the tropics lie for the most part below 150 m (492 ft), and the highest concentration of nutrients occurs between 500 and 1000 m (1640–3280 ft) depth.

There is generally a steady, low rate of primary productivity in tropical oceans. Although the rate of tropical productivity is low, when we compare the total annual productivity of tropical oceans with that of the more productive temperate oceans, we find that the tropical productivity is generally at least half of that found in the temperate regions on an annual basis.

Within tropical regions there are three environments where productivity is unusually high—regions of equatorial upwelling, coastal upwelling, and coral reefs (figure 14–6B). In areas where trade winds drive westerly equatorial currents on either side of the equator, surface water diverges as a result of the Ekman spiral. This surface water that moves off toward higher latitudes is replaced by nutrient-rich water that surfaces from depths of up to 200 m (660 ft). This condition of equatorial upwelling is probably best developed in the eastern Pacific Ocean. Where the prevailing winds blow toward the equator and along the western margin of continents, the surface waters are driven away from the coast. They are replaced by nutrient-rich waters from depths of 200 to 900 m (660–2950 ft). As a result of this upwelling of nutrient-rich water, there is a high rate of primary productivity in these areas, which supports large fisheries. Such conditions exist along the southern coast of California and the southwest coast of Peru in the Pacific Ocean. Upwelling also occurs along the northwest coast of Morocco and the southwest coast of Africa in the Atlantic Ocean. The relatively high productivity of coral reef environments is not related to the upwelling process and is discussed in Chapter 16.

Temperate Region We have discussed the general productivity picture in the polar regions, where productivity is limited primarily by the availability of solar radiation, and in the tropical low-latitude areas, where the limiting factor is the availability of nutrients. We will next consider the temperate regions, where an alternation of these factors controls productivity in a pattern that is somewhat more complex.

Productivity in temperate oceans is at a very low level during the winter months, although high concentrations of nutrients are available in the surface layers. In fact, the nutrient concentration is higher during the winter season than at any time throughout the year. The limiting factor on productivity during the winter season in the temperate ocean is the availability of solar energy.

Since the sun is at its lowest elevation above the horizon during this season, a higher percentage of solar energy is reflected and a smaller percentage absorbed into the surface waters. The compensation depth for basic producers such as diatoms is so shallow that it does not allow growth of the diatom population (figure 14–7).

As the sun rises higher in the sky during the spring season, the compensation depth deepens as the amount of solar radiation being absorbed by the surface water increases. Eventually, there is sufficient water volume included within the water column above the compensation depth to allow for the exponential growth of the diatom population. This expanding population puts a tremendous demand on the nutrient supply in the euphotic zone. In most Northern Hemisphere areas decreases in the diatom population will occur by May as a result of insufficient nutrient supply.

As the sun rises higher in the sky during the summer months, the surface waters in the temperate parts of the ocean are warmed, and the water becomes separated from the deeper water masses by a seasonal thermocline that may develop at depths of approximately 15 m (50 ft). As a result of this thermocline development, there is little or no exchange of water across this discontinuity, and the nutrients that are depleted from surface waters cannot be replaced by those available in the deep waters. Throughout the summer months the plant population will remain at a relatively low level but will again increase in some temperate areas during the autumn months.

The autumn increase is much less spectacular than that of the spring because of the decreasing availability of solar radiation resulting as the sun drops lower in the sky. This causes a decrease in surface temperature and the breakdown of the summer thermocline. A return of nutrients to the surface layer occurs as increased wind strength mixes it with the deeper water mass in which the nutrients have been trapped throughout the summer months. This bloom is very short-lived. The phytoplankton population begins to decrease rapidly. The limiting

FIGURE 14–7
Productivity in Temperate Oceans.

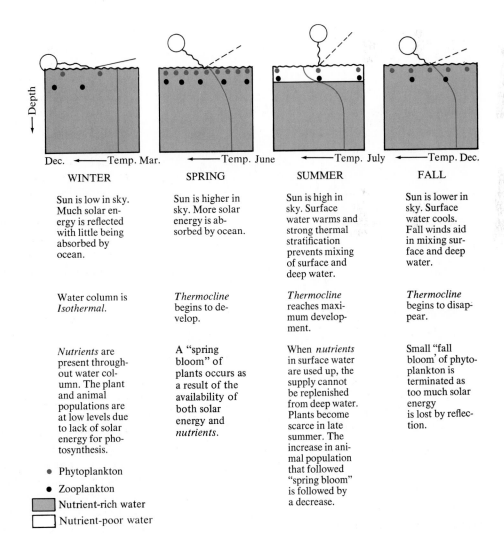

WINTER

Sun is low in sky. Much solar energy is reflected with little being absorbed by ocean.

Water column is *Isothermal.*

Nutrients are present throughout water column. The plant and animal populations are at low levels due to lack of solar energy for photosynthesis.

SPRING

Sun is higher in sky. More solar energy is absorbed by ocean.

Thermocline begins to develop.

A "spring bloom" of plants occurs as a result of the availability of both solar energy and *nutrients.*

SUMMER

Sun is high in sky. Surface water warms and strong thermal stratification prevents mixing of surface and deep water.

Thermocline reaches maximum development.

When *nutrients* in surface water are used up, the supply cannot be replenished from deep water. Plants become scarce in late summer. The increase in animal population that followed "spring bloom" is followed by a decrease.

FALL

Sun is lower in sky. Surface water cools. Fall winds aid in mixing surface and deep water.

Thermocline begins to disappear.

Small "fall bloom" of phytoplankton is terminated as too much solar energy is lost by reflection.

- Phytoplankton
- Zooplankton
- Nutrient-rich water
- Nutrient-poor water

factor in this case is the opposite of that which reduced the population of the spring phytoplankton. In the case of the spring bloom, solar radiation was readily available, and the decrease in nutrient supply was the limiting factor.

Additional Considerations Coastal waters with high nutrient levels are highly productive or *eutrophic*. Most of the open ocean has a low level of nutrients and productivity. This low productivity is referred to as an *oligotrophic* condition. Figure 14–8 shows the general patterns of ocean productivity based on photosynthetically fixed carbon, whereas the image on the book cover shows the concentration of phytoplankton pigment throughout the world ocean.

New measurements of photosynthetic productivity of oligotrophic waters in the North Atlantic and North Pacific oceans indicate they may be from 2 to 7 (or even more) times as productive as ^{14}C data indicate. Instead of using the small bottles used in the ^{14}C measurements and suspending them for a short time in the ocean,

some physical oceanographers have used a much larger "bottle." In their studies of the circulation within the subtropical gyres, physical oceanographers use bottles that are "capped" by the pycnoclines produced by the thermoclines beneath the warm layers of surface water within the subtropical gyres of the open oceans. These bottles contain (depending on the design of the study) from tens to thousands of cubic kilometers of water that record the average results of photosynthetic activity over periods of from a few months to decades.

The new studies analyze the effect of photosynthesis on the oxygen concentrations in (1) the euphotic zone and (2) beneath the euphotic zone. In the North Pacific Ocean, oxygen saturations in *subsurface oxygen maximums* (SOM) at depths between 50 and 100 m (165–330 ft) were from 110 percent to 120 percent at latitudes of from 30° to 40°N during summer months based on data obtained from 1962 through 1979. The excess oxygen (anything over 100 percent) may be due to trapped photosynthetically produced oxygen. The North Atlantic study determined the rate at which oxygen is used up beneath the euphotic zone (100

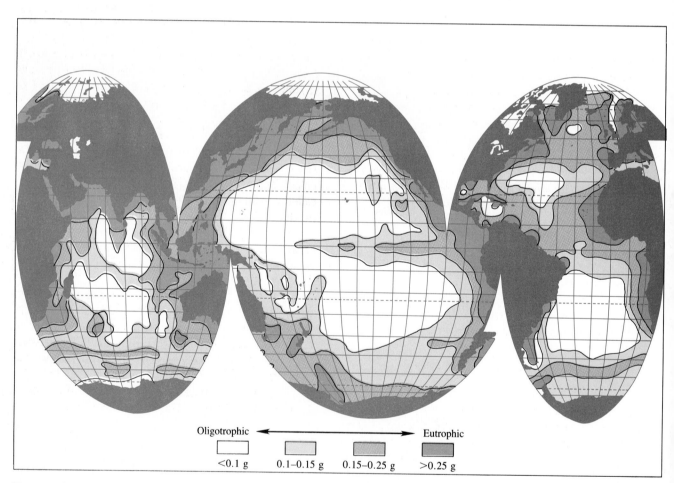

FIGURE 14–8

Distribution of Photosynthetic Primary Production (gC/m²/day).

FIGURE 14–9

Nitrogen Fixation by Aggregates of Diatoms and Cyanobacteria in Oligotrophic Ocean Waters. *A*, typical mat of *Rhizosolenia* about 5 cm long. It is composed of intertwining chains of *R. castracanei* (wider cells) and *R. imbricata* (narrower cells). Within these cells, symbiotic bacteria fix nitrogen for uptake by the plant cells. (Photo by James M. King.) *B*, an aggregation of the nitrogen-fixing cyanobacteria *Oscillatoria* spp. with an O_2 probe inserted into it. The probe has a maximum diameter of about 5 μm. (Photo courtesy of Hans W. Paerl, University of North Carolina.)

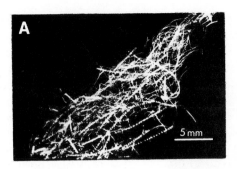

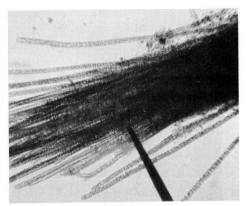

B.

m, or 330 ft) by decomposing organic matter falling toward the ocean floor. *Oxygen utilization rates* (OUR) of twice the accepted mean oligotrophic rate indicate there may be much more organic matter synthesized in the oligotrophic euphotic zone than is indicated by ^{14}C data.

The problem with the ^{14}C method may be that by using such a small sample and exposing it for such a short time, periods of unusually high productivity may be missed and not averaged into the results. Some interesting findings may help explain how some of the biological productivity of oligotrophic water may have been missed. Mats of diatoms belonging to the genus *Rhizosolenia* have been found floating in the North Pacific Gyre and the Sargasso Sea. Averaging 7 cm (2.8 in.) in length, these mats possess symbiotic bacteria that fix molecular nitrogen (N_2) into nitrate (NO_3) useable as a nutrient by the diatoms (figure 14–9). Also, bundles of a nitrogen-fixing cyanobacteria, *Oscillatoria* spp. have been verified in the Sargasso Sea. Such aggregations are readily broken up by net tows and missed by standard bottle casts.

Recent studies in the Pacific and Atlantic oceans indicate that a large percentage (60 percent in one study) of oligotrophic photosynthesis is achieved by picoplankton (0.2–2.0 μm; 0.000008–0.00008 in.) that pass through many filters used in productivity studies. Continued investigation of this problem may reveal the overall productivity of the oligotrophic ocean waters has been greatly underestimated.

Chemosynthetic Productivity

Another source of potentially significant biological productivity in the oceans is occurring in the rift valleys of the oceanic spreading centers at depths of over 2500 m (8200 ft), where there is no light for photosynthesis. In regions where ocean water seeps down fractures in the ocean crust to depths where it is heated by underlying magma chambers, hydrothermal springs exist. Once the water is heated, it rises to the ocean floor dissolving

minerals from the crustal rocks as it does so. Potentially significant deposits of minerals are associated with biotic communities. A significant component of these communities is the vestimentiferan worm population, a sample of which can be seen in figure 14–10. These 1-m (3-ft) tube worms and other chemosynthetic symbionts such as clams and mussels reach unusually large size in association with autotrophic bacteria that derive energy to produce their own food from the hydrogen sulfide (H_2S) gas dissolved in the water of the hydrothermal springs. By oxidizing the gas to form free sulfur (S) and sulfate (SO_4^{2-}), the bacteria release chemical energy that they use much as plants use solar energy to carry on photosynthesis. Similar communities have been discovered near cold water seeps at the base of the Florida Escarpment in the Gulf of Mexico. Since the bacterial synthesis of inorganic nutrients into organic molecules depends on the release of chemical energy, it is called *chemosynthesis*. The true significance of bacterial productivity on the deep ocean floor will not be fully understood until much more research on the phenomenon is conducted. It has the potential, however, of increasing our estimates of the biological productivity of the ocean.

Biochemists have recently discovered that bacteria can obtain the chemical energy needed for the synthesis of organic molecules through the oxidation of a large variety of compounds containing the metals iron, manganese, copper, nickel, and cobalt. These microorganisms may well be an important factor in the deposition of ore-quality deposits of the oxides of these metals on the ocean floor in the form of manganese nodules.

ENERGY TRANSFER

We have been discussing the general consideration related to availability of nutrients. Our attention will now

ENERGY TRANSFER | 269

A.

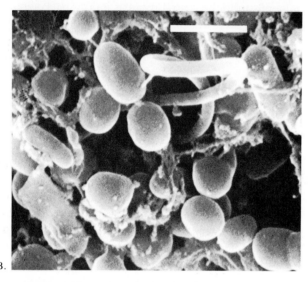

B.

FIGURE 14–10
Chemosynthetic Life of the Galapagos Rift. *A,* this photograph taken at the Galapagos Rift shows the large tube worms found there and at other deep-sea hydrothermal vents. These worms possess symbiont sulfur-oxidizing bacteria and chemosynthetically produce their own food by combining inorganic nutrients dissolved in the deep-ocean water. *B,* sample of sulfur-oxidizing bacteria from the Galapagos Rift with small particles of what is probably free sulfur attached to them. These and similar bacteria live symbiotically within the tissue of the tube worms, clams, and mussels found at the vents. They chemosynthetically produce food to support these organisms. Magnification 20,000×. Bar is 1 μm. (Courtesy of Woods Hole Oceanographic Institution.)

be turned to the cycling of specific important classes of nutrients and the flow of energy.

Marine Ecosystem

The term *biotic community* refers to the assemblage of organisms that live together within some definable area. An *ecosystem* includes the biotic community and abiotic environment with which it interacts in the exchange of energy and chemical substances. Within an ecosystem there are generally three basic categories of organisms—*producers, consumers,* and *decomposers.* Plants and some bacteria are the *autotrophic* producers and have the capacity to nourish themselves through chemosynthetic and photosynthetic processes. The consumers and the decomposers are *heterotrophic* organisms that depend on the organic compounds produced by the autotrophs for their food supply.

Animals may be divided into three categories: *herbivores,* which feed directly on the plants; *carnivores,* which feed only on other animals; and *omnivores,* which feed on both. As the role of bacteria in the marine ecosystem becomes better understood, a fourth category of animals, the *bacteriovores,* which feed on bacteria, may be identified as an important component of the marine ecosystem. The decomposers, such as bacteria, break down the organic compounds of dead plants and animals and animals' excretions while taking some of these decomposition products for their own energy requirements. They characteristically release simple inorganic salts that are used by the plants as nutrients.

Energy Flow

Before considering the biogeochemical cycles that involve the transfer of organic and inorganic matter, we will consider the flow of energy in general. Most energy is put into a biotic community through plants. From the plants, it follows a unidirectional path (although cycles of energy exchange occur at many intermediate biological levels) that leads to a continual degradation of energy culminating in *entropy,* or energy converted to a form where it is no longer available to do work. As can be seen in figure 14–11, which depicts the flow of en-

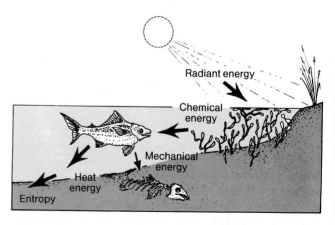

FIGURE 14–11
Energy Flow. Energy may enter the biological community as solar radiation, a high grade of energy. The energy in the system follows a path of degradation to lower forms of energy until it achieves a high degree of entropy and is no longer available to do work.

ergy through a plant-supported biotic community, energy enters the system as the high-grade radiant form, solar energy, which is absorbed by the plants. Photosynthesis converts it to a medium-grade, chemical energy, which is used for plant respiration and passes on to the animals to be used for growth and carrying on their various life-sustaining functions. Energy is expended by the animals as mechanical and heat energy, which are progressively lower forms of energy. Finally, it becomes biologically useless as entropy increases.

Composition of Organic Matter

Having discussed the noncyclic nature of flow of energy through the biotic community and observed that it is a unidirectional flow, let us now consider the biogeochemical cycles involving matter that is not lost to the biotic community but is cycled by being converted from one chemical form to another by the various members of the community.

Biological mass is made up of compounds involving a number of elements that are abundant on earth. All the elements that are naturally occurring can be assumed to be present in the oceans, at least in extremely small concentrations. However, some of these elements are in concentrations so small that we cannot detect their presence. Fewer than 60 have been identified by chemical analysis as being present in the oceans.

In view of the fact that living organisms are composed of carbohydrates, fats, and proteins, let us first concern ourselves with the elements that go into the makeup of these substances. About 20 elements in these basic organic compounds are considered to be essential to their production:

1. The major constituents of organic materials, those that individually make up at least 1 percent each of organic material on a dry weight basis, are, in the order of their abundance, carbon, oxygen, hydrogen, nitrogen, and phosphorus.
2. Occurring in concentrations from 500 ppm to 10,000 ppm by weight are the elements sulfur, chlorine, potassium, sodium, calcium, magnesium, iron, and copper.
3. A third group occurring in concentrations of less than 500 ppm of dry weight is composed of boron, manganese, zinc, silicon, cobalt, iodine, and fluorine.

Considering the availability of these elements as solutes in ocean water, we find that the second group, which we may refer to as the *secondary constituents* of organic composition, is in sufficient concentration to make it unlikely that members of this group would ever be limiting factors in plant productivity. However, the third group, the *tertiary constituents,* which occur in very low concentrations in organic compounds, are also found to be in very low concentrations in the marine environment. It seems probable that the tertiary constituents might create by their absence conditions that would limit productivity.

Returning to a consideration of the major constituents we find that carbon dioxide and water should certainly, on the basis of their concentrations, provide enough carbon, oxygen, and hydrogen to assure that these elements would never limit productivity. Nitrates and phosphates are the nutrients containing nitrogen and phosphorus that we need to consider in more detail since they do, under many conditions, limit marine productivity.

Biogeochemical Cycling

In biogeochemical cycles, elements will follow a pattern where an inorganic form is taken in by an autotrophic organism that synthesizes it into organic molecules. The food is passed through a food web that usually ends in bacterial decomposition of the organically produced compounds into the inorganic forms that may again be used in plant production.

A New Role for Bacteria Although it has been widely accepted that zooplankton are the primary grazers of phytoplankton, new evidence indicates that free-living bacteria may consume up to 50 percent of the production of phytoplankton. The bacteria are thought to consume the dissolved organic matter that is lost from the conventional food web by three processes:

1. Phytoplankton exudate. As phytoplankton age they lose some of their cytoplasm directly into the ocean.
2. Phytoplankton "munchates." As phytoplankton are eaten by zooplankton, cytoplasm is "spilled" into the ocean.
3. Zooplankton excretions. The liquid excretions of zooplankton are dissolved into ocean water.

Free-living heterotrophic bacteria absorb this dissolved organic matter and reenter the conventional food web primarily through the grazing of microscopic flagellates. Other larger zooplankton may be important in this process locally.

The details of the process by which free-living bacteria participate in the marine food web are still under investigation, but microbial ecologists studying the problem are convinced the evidence indicates bacteria have an important role in the transfer of energy through the marine ecosystem. Thus, bacteria may have more varied roles in the biogeochemical cycling of matter in the oceans than had previously been known.

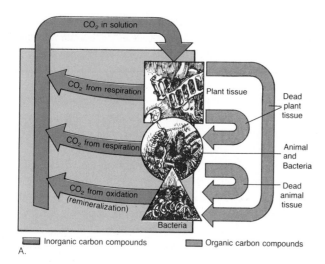

CO2 in solution

CO2 from respiration

Plant tissue

Dead plant tissue

CO2 from respiration

Animal and Bacteria

Dead animal tissue

CO2 from oxidation (remineralization)

Bacteria

Inorganic carbon compounds ▢ Organic carbon compounds

A.

FIGURE 14–12

Biogeochemical Cycling. The chemical components of organic matter enter the biological system through plant photosynthesis (or bacterial chemosynthesis). They are passed on to animal populations through feeding patterns. When the plants and animals die, their organic remains are converted to inorganic form by bacterial or other decomposition processes. In this form, they are again available for uptake by plants. *A.* Carbon Cycle. There is a large supply of inorganic carbon in the oceans, and it is not a limiting factor in biological productivity. Only about 1 percent of the carbon dissolved in the ocean is involved in plant productivity. *B.* Nitrogen Cycle. The total amount of nitrogen fixed into organic molecules at any given time may be 10 times as great as the yearly average of soluble nitrogen compounds dissolved in ocean water. Therefore, each nitrogen atom must be recycled biogeochemically about 10 times per year. Also, the decomposition of organic nitrogen compounds back into the preferred inorganic form of nitrogen, nitrate (NO_3), requires three steps of bacterial decomposition. As a result, some of this process may not be completed until the molecules sink beneath the euphotic zone, where they are unavailable to plants. Nitrogen is considered to be the nutrient that is most likely to limit biological productivity as a result of its depletion. *C.* Phosphorus Cycle. Although each phosphorus atom may need to be recycled up to four times per year to maintain biological productivity in the oceans, it is seldom depleted to the point of limiting biological productivity. Two factors that help get organic phosphorus back into a form useable as nutrients by plants rather quickly are autolytic breakdown and the single step of bacterial decomposition required.

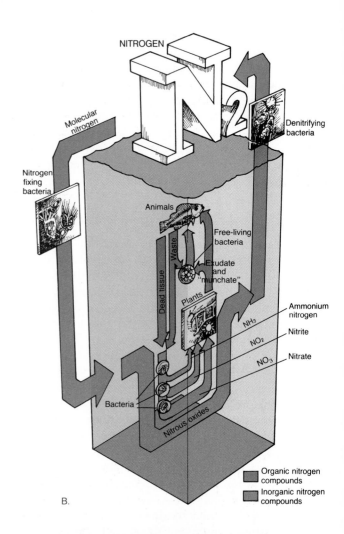

NITROGEN

Molecular nitrogen

Nitrogen fixing bacteria

Denitrifying bacteria

Animals

Dead tissue

Waste

Free-living bacteria

Exudate and "munchate"

Plants

Ammonium nitrogen

NH_3

NO_2 Nitrite

NO_3 Nitrate

Bacteria

Nitrous oxides

▢ Organic nitrogen compounds
▢ Inorganic nitrogen compounds

B.

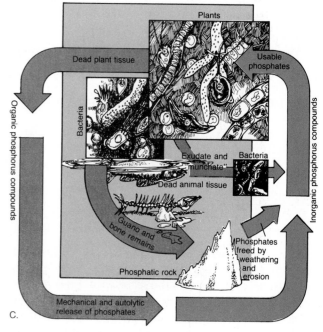

Plants

Dead plant tissue

Usable phosphates

Organic phosphorus compounds

Bacteria

Exudate and "munchate"

Bacteria

Dead animal tissue

Guano and bone remains

Phosphatic rock

Phosphates freed by weathering and erosion

Inorganic phosphorus compounds

Mechanical and autolytic release of phosphates

C.

Carbon, Nitrogen, and Phosphorus Although there is no concern that carbon concentrations in the marine environment are a limiting factor in marine productivity, we will examine the carbon cycle, since this element is the basic component of organic compounds. To appreciate the excess of carbon dioxide in the world's oceans as it is related to the requirements for plant productivity, we should know that only about 1 percent of the total carbon present in the oceans is involved in plant productivity.

Comparatively, the soluble nitrogen compounds involved in plant productivity may be 10 times the total nitrogen compound concentration that can be measured as a yearly average. This level implies that the soluble nitrogen compounds must be cycled completely up to 10 times per year. Available phosphates may be turned over up to 4 times per year.

Comparing the ratio of carbon to nitrogen to phosphorus in dry weights of diatoms, we find that proportions are 41:7:1. This ratio is also observed in the zooplankton that feed on the diatoms as well as in ocean water samples taken throughout the world. Thus, phytoplankton take up nutrients in the ratio in which they are available in the ocean water and pass them on to zooplankton in the same ratio. When these planktonic plants and animals die, carbon, nitrogen, and phosphorus are restored to the water in this ratio.

The Carbon Cycle This cycle involves the uptake of carbon dioxide by plants that use it in the photosynthetic process. Carbon dioxide is returned to the ocean water primarily through respiration of plants, animals, and bacteria and secondarily by autolytic breakdown of dissolved organic materials. *Autolytic decomposition* results from the action of enzymes present in the cells of organic tissue and does not require bacterial action (figure 14–12A).

The Nitrogen Cycle Nitrogen is primarily important in the production of *amino acids,* the building blocks of proteins that are synthesized by plants. These photosynthetic products are consumed by animals and free-living bacteria. They are then passed on to the saprobic (decomposing) bacteria, along with dead plant tissue, as dead animal tissue and excrement.

The saprobic bacteria gain energy from breaking down these compounds. This breakdown leads to the liberation of inorganic compounds, such as nitrates, that are the basic nutrient salts used by plants. Most nitrogen in the ocean is found as molecular nitrogen (N_2). The most common combined forms are nitrite (NO_2) and nitrate (NO_3). These are highly oxidized states of nitrogen. The most abundant reduced form is ammonia (NH_3).

Different bacteria involved in the nitrogen cycle make it somewhat complicated. Although most of the bacteria play the heterotrophic roles of consuming dissolved organic matter or converting organic compounds into inorganic salts, there are some that are able to fix molecular nitrogen into combined forms. These are *nitrogen-fixing bacteria.* *Denitrifying bacteria* make up another special group whose metabolism depends upon the breakdown of nitrates and the liberation of molecular nitrogen. The nitrogen cycle is graphically presented in figure 14–12B.

Autotrophic organisms of various types can use ammonium nitrogen and nitrites. However, the most important nutrient form of inorganic nitrogen, nitrate, can be utilized more efficiently by plants. Studies of nutrient cycles in various portions of the world's oceans indicate that the availability of nitrogen is clearly a limiting factor in productivity during summer months. This condition arises from the fact that the processes involved in converting particulate organic substance to nitrate salts through bacterial action may require up to 3 months. The conversion begins in the lower portions of the photosynthetic zone as the particulate matter is sinking toward the ocean bottom.

There are three basic stages in the conversion of particulate organic matter to nitrate salts that involve three distinct bacterial types. The first reduces the particulate organic nitrogen into ammonium nitrogen, which is then acted upon by the second bacterial community that converts it to nitrite. A third action is required to convert the nitrite to the prime organic salt species, nitrate. By the time this conversion is completed, the nitrogen compounds are usually below the euphotic zone and thus unavailable for photosynthesis. They cannot readily be returned to the euphotic zone during summer due to the strong thermostratification that exists throughout much of the ocean surface. If nitrates are to again become available to plants in these regions, their return must wait until the thermostratification disappears, allowing upwelling and mixing during the winter to effect the movement of the nutrients into the surface waters.

The Phosphorus Cycle The phosphorus cycle is simpler than the nitrogen cycle primarily because the bacterial action involved in breaking down the organic phosphorus compounds is simpler. This difference can be studied by comparing figure 14–12B and C. The rate at which the organic compounds can be decomposed into inorganic *orthophosphates*, which are the phosphorus compounds primarily used by plants, is much more rapid than that of nitrogen breakdown. This quicker rate is due to autolytic breakdown of phosphatic organic material by enzymes and the single step required in the bacterial breakdown. As a result of the greater rate of breakdown for organic phosphorus, much of it can be completed above the oxygen compensation depth. It is therefore made available to plants within the photosynthetic zone. Although the concentration of phosphorus in the oceans is only about one-seventh that of the concentration of nitrogen, the fact that the recycling can take place within the photosynthetic zone usually allows adequate quantities of inorganic phosphorus to be available for plant productivity. The lack of phosphorus is rarely a limiting factor of plant productivity.

The Silicon Cycle As previously discussed, frustules of diatoms are composed of silica (SiO_2). Although the availability of silica can be a limiting factor in the productivity of diatoms, it is rarely a limiting factor of total primary productivity, since not all phytoplankton require silica as a protective covering. There is very little likelihood that silicon will ever be a limiting nutrient in total productivity because it is very abundant in the rocks that make up the earth's crust, and the small fraction of the earth's crustal silica that will eventually find its way into the sea is very great in relation to the silicon needs of the diatom population. Silicon concentrations range from unmeasurable up to 400 mg/m^3. Fluctuations in concentration roughly coincide with those observed in nitrogen and phosphorus. However, the fluctuation in silica concentrations displays a much greater amplitude than fluctuations for nitrogen and phosphorus. This condition is probably due to the fact that silica does not undergo bacterial decay and is taken directly into solution after undergoing autolytic breakdown.

TROPHIC LEVELS AND BIOMASS PYRAMIDS
Trophic Levels

The transfer of chemical energy stored in the mass of the ocean's plant population to the animal community occurs, in part, through the feeding process. Zooplankton feed as herbivores on the diatoms and other microscopic marine plants, while larger animals feed on the macroscopic algae and "grasses" found growing attached to the ocean bottom near shore. The herbivores will be fed upon by larger animals, carnivores, who in turn will be fed upon by another population of larger carnivores, and so on. Each of these feeding levels is a *trophic level*. As was discussed earlier in this chapter and illustrated in figure 14–12, an amount of secondary production equal to that of the zooplankton may be achieved by free-living bacteria.

In general, the individual members of a feeding population will be larger but not too much larger than the individual member of the population on which they feed. Although we can consider this to be generally true, there are outstanding exceptions to this condition. The blue whale,

which is the largest animal known to have existed on earth, feeds upon the krill that grow to maximum lengths of 6 cm (2.4 in.).

We should recall that all consideration of energy transfer must be approached with the understanding that the transfer of energy from one population to another represents a continuous flow of energy. Small-scale recycling and conservation of this energy occurs and slows the process of conversion of potential (chemical) energy to kinetic energy, to heat energy, to be lost finally to entropy. However, despite the cycling of energy, all energy that enters the organic community is inevitably lost to entropy in the end.

Transfer Efficiency

In the transfer of energy between feeding, or trophic, levels we are greatly concerned with efficiency. There is a relatively high degree of variability in the efficiency of various plant species under various laboratory conditions. As an average, we consider the percentage of light energy absorbed by plants and ultimately synthesized into organic substances made available to herbivores to be about 2 percent.

The gross efficiency at any trophic level is the ratio of energy passed on to the next higher trophic level divided by the energy received from the trophic level below. When we examine the transfer of energy from feeding population to feeding population within the ocean, we can readily see that of the biomass representing the food intake of a given population only a portion will be passed on to the next feeding level.

Figure 14–13 shows that some of the energy taken in as food by herbivores is passed by the animal as feces and the rest is assimilated by the animal. Of that portion of the energy that is assimilated, much is quickly converted through respiration to kinetic energy for maintaining life, while what remains is available for growth and reproduction. Only a portion of this mass is passed on to the next trophic level through feeding. Figure 14–14 represents diagrammatically the passage of energy between trophic levels through an entire ecosystem from the solar energy assimilated by plants to the mass of the ultimate carnivore.

Many investigations have been conducted into the efficiency of energy transfer between trophic levels. There are many variables, such as the age of the animals involved, to be considered. Young animals display a higher growth efficiency than older animals. The availability of food can also alter efficiency. When food is plentiful, animals are observed to expend more energy in digestion and assimilation than when food is not readily available. Most efficiencies that have been determined range between 6 percent and 15 percent. It is well accepted that ecological efficiencies in natural eco-

not have any predators, which marks the end of the chain. In nature it is rare to find food chains comprising more than five trophic levels. Based on our previous considerations of the efficiency of energy transfer between trophic levels, we see that for a population within a food chain it may be beneficial to feed as close to the primary producing population as possible. This arrangement increases the biomass available for food and the number of individuals in the population to be fed upon.

An example of an animal population that is an important fishery and usually represents the third trophic level in a food chain is the herring off the coast of Newfoundland. Although some herring populations are involved in longer food chains, the Newfoundland herring feed primarily on a population of small crustaceans, copepods, that in turn feed upon diatoms (figure 14–15).

Food Webs It is, however, uncommon to see simple feeding relationships of the type described above in nature. More commonly the animals representing the last step in a linear food chain feed on a number of animals that have simple or complex feeding relationships, constituting a *food web*. The overall importance of food webs is not well understood, but one consequence for animals that feed through a web rather than a linear chain is their greater likelihood of survival if population extinctions or sharp decreases occur within the web at or below their feeding level. Those animals involved in food webs, such as the North Sea herring in figure 14–15, are less likely to suffer from the extinction of one of the populations upon which they depend for food than the Newfoundland herring that feed only on copepods. The extinction of the copepods in the latter case certainly would be expected to have a catastrophic effect on the herring population.

The Newfoundland herring population does, however, have an advantage over its relatives who feed through the broader-based food web. The Newfoundland herring more likely have a larger biomass to feed on since they are only two steps removed from the producers, while the North Sea herring represent the fourth level in some of the food chains within its web.

The ultimate effect of the transfer of energy between trophic levels can be seen in figure 14–16, which depicts the progressive decrease in numbers of individuals and total biomass at successive trophic levels resulting from decreased amounts of available energy.

SUMMARY

Organisms are divided into five **kingdoms**: **Monera**, single-celled organisms without a nucleus; **Protista**, single-celled organisms with a nucleus; **Mycota**, fungi; **Metaphyta**, many-celled plants; and **Metazoa**, many-celled animals. Taxonomic classification of organisms involves dividing the kingdoms into the increasingly specific groupings: **phylum**, **class**, **order**, **family**, **genus**, and **species**.

Protistan plants include the macroscopic algae **Phaeophyta** (kelp and *Sargassum*), **Chlorophyta** (green algae), and **Rhodophyta** (red algae). The microscopic algae include **Chrysophyta** (diatoms and coccolithophores) and **Pyrrophyta** (dinoflagellates). The more complex **Spermatophyta** are represented by a few genera of nearshore plants such as **Zostera** (eelgrass), **Phyllospadix** (surf grass), **Spartina** (marsh grass) and **mangrove trees**.

Photosynthetic productivity of the oceans is limited by the availability of solar radiation and of nutrients. The depth to which sufficient light penetrates to allow plants to produce only that amount of oxygen required for their respiration is the **oxygen compensation depth**. Plants cannot live successfully below this depth, which may occur at less than 20 m (65 ft) in turbid coastal waters or at a probable maximum of 150 m (492 ft) in the open ocean. Nutrients are most abundant in coastal areas due to runoff and upwelling. In high-latitude areas thermoclines are generally absent, so upwelling can readily occur, and productivity is commonly limited more by the availability of solar radiation than lack of nutrients. In low-latitude regions, where a strong thermocline may exist year-round, productivity is limited except in areas of upwelling. The lack of nutrients is generally the limiting factor. In temperate regions, where distinct seasonal patterns are developed, productivity peaks in the spring and fall and is limited by lack of solar radiation in the winter and lack of nutrients in the summer.

In addition to the primary productivity through plant photosynthesis, organic biomass is produced through bacterial **chemosynthesis**. Chemosynthesis, recently observed on oceanic spreading centers in association with hydrothermal springs, is based on the release of chemical energy by the oxidation of hydrogen sulfide.

Radiant energy captured by plants is converted to chemical energy and passed through the biotic community. It is expended as mechanical and heat energy and ultimately reaches a state of **entropy**, where it is biologically useless. There is, however, no loss of mass. The mass used as nutrients by plants is converted to **biomass**. Upon the death of organisms, the mass is decomposed to an inorganic form ready again for use as nutrients for plants. Of the **nutrients** required by plants, compounds of **nitrogen** are most likely to be depleted and restrict plant productivity. Since the total decomposition of organic nitrogen compounds to inorganic nutrients requires three stages of bacterial decomposition, these compounds may have sunk beneath the photosynthetic zone before decomposition was complete and therefore are unavailable to plants.

As energy is transferred from plants to **herbivore** and the various **carnivore** feeding levels, only about 10 percent of the mass taken in at one feeding level is passed on to the next. The ultimate effect of this decreased amount of energy that is passed between **trophic levels** higher in the food chain is a decrease in the number of individuals and total biomass of populations higher in the **food chain.**

QUESTIONS AND EXERCISES

1. List the five kingdoms of organisms and the fundamental criteria used in assigning organisms to them.

2. Compare the macroscopic algae in terms of pigment, maximum depth in which they grow, common species, and size.

3. The Chrysophyta contains two classes of important phytoplankton. Compare their composition and the structure of their hard parts as well as their geologic significance.

4. Discuss and compare the contributions of the Pyrrophyta genera *Ptychodiscus* and *Gonyaulax* to red tide development.

5. Define oxygen compensation depth, and explain the use of the dark and transparent bottle technique for its determination. Discuss how the quantity of oxygen produced by photosynthesis in each clear bottle (gross photosynthesis) is determined.

6. Compare the biological productivity of polar, temperate, and tropical regions of the oceans. Include a discussion of seasonal variables, thermal stratification of the water column, and the availability of nutrients and solar radiation.

7. Discuss chemosynthesis as a method of primary productivity. How does it differ from photosynthesis?

8. Describe the components of the marine ecosystem.

9. Describe the flow of energy through the biotic community, and include the forms to which solar radiation is converted. How does this flow differ from the manner in which mass is moved through the ecosystem?

10. What are the proportions by weight of carbon, nitrogen, and phosphorus in ocean water, phytoplankton, and zooplankton? Suggest how these amounts may support or refute the idea that life originated in the oceans.

11. Explain why nitrogen is much more likely than phosphorus to be a limiting factor in marine productivity.

12. How is the energy taken in by a feeding population lost so that only a small percentage is made available to the next feeding level? What is the average efficiency of energy transfer between trophic levels?

13. If a killer whale is a third-level carnivore, how much phytoplankton mass is required to add each gram of new mass to the whale? Assume 10 percent efficiency of energy transfer between trophic levels. Include a diagram.

14. Describe the probable advantage to the ultimate carnivore of the food web over a single food chain as a feeding strategy.

REFERENCES

Ducklow, H. W. 1983. Production and fate of bacteria in the oceans. *BioScience* 33:8, 494–501.

George, D., and George, J. 1979. *Marine life: An illustrated encyclopedia of invertebrates in the sea.* New York: Wiley-Interscience.

Grassle, J. F., et al. 1979. Galapagos '79: Initial findings of a deep-sea biological quest. *Oceanus* 22:2, 2–10.

Jenkins, W. J. 1982. Oxygen utilization rates in North Atlantic subtropical gyre and primary production in oligotrophic systems. *Nature* 300, 246–48.

Littler, M. M., Littler, D. S., Blair, S. M., and Norris, J. N. 1985. Deepest known plant life discovered on an uncharted seamount. *Science* 227:4683, 57–59.

Martinez, L., Silver, M., King, J., and Alldredge, A. 1983. Nitrogen fixation by floating diatom mats: A source of new nitrogen to oligotrophic ocean waters. *Science* 221:4606, 152–54.

Paerl, H. W., and Bebout, B. M. 1988. Direct measurement of O_2-depleted microzones in marine *Oscillatoria:* Relation to N_2 fixation. *Science* 241:4864, 442–45.

Parsons, T. R., Takahashi, M., and Hargrave, B. 1984. *Biological oceanographic processes,* 3rd ed. New York: Pergamon Press.

Platt, T., and Sathyendranath, S. 1988. Oceanic primary production: Estimation by remote sensing at local and regional scales. *Science* 241:4873, 1613–19.

Platt, T., Subba Rao, D. V., and Irwin, B. 1983. Photosynthesis of picoplankton in the oligotrophic ocean. *Nature* 301:5902, 702–4.

Russell-Hunter, W. D. 1970. *Aquatic productivity.* New York: Macmillan.

Sherr, B. F., Sherr, E. B., and Hopkinson, C. S. 1988. Trophic interactions within pelagic microbial communities: Indications of feedback regulation of carbon flow. *Hydrobiologia* 159:1, 19–26.

Shulenberger, E., and Reid, J. L. 1981. The Pacific shallow oxygen maximum, deep chlorophyll maximum, and primary productivity reconsidered. *Deep Sea Research* 28A:9, 901–19.

SUGGESTED READING

Sea Frontiers

Arehart, J. L. 1972. Diatoms and silicon. 18:2, 89–94.
A very readable description of the important role of silicon and other elements in the ecology of diatoms, including microphotographs showing the varied forms of diatoms.

Coleman, B. A., Doetsch, R. N., and Sjblad, R. D. 1986. Red tide: A recurrent marine phenomenon. 32:3, 184–191.
The problem of periodic red tides along the Florida, New England, and California coasts is discussed.

Ebert, C. H. .V. 1978. El Niño: An unwanted visitor. 24:6, 347–51.
A description of the effects of El Niño and the attempts that are being made to understand its causes.

Idyll, C. P. 1971. The harvest of plankton. 17:5, 258–67.
An interesting discussion of the potential of zooplankton as a major fishery.

Jensen, A. C. 1973. WARNING—red tide. 19:3, 164–75.
An informative discussion of what is known of the cause, nature, and effect of red tides.

Johnson, S. 1981. Crustacean symbiosis. 27:6, 351–60.
A description of various symbiotic relationships entered into by tropical shrimps and crabs.

McFadden, G. l987. Not-so-naked ancestors. 33:1, 46–51.
The nature of the coverings of marine phytoplankton cells is revealed by the electron microscope.

Oremland, R. S. 1976. Microorganisms and marine ecology. 22:5, 305–10.
The role of such microorganisms as phytoplankton and bacteria in cycling matter in the oceans is discussed.

Philips, E. 1982. Biological sources of energy from the sea. 28:1, 36–46.
The potential for converting marine biomass to energy sources useful to society is discussed.

Scientific American

Benson, A. A. 1975. Role of wax in oceanic food chains. 232:3, 76–89.
A report on the findings from observations made of the content of wax in the bodies of many marine animals from copepods to small deep water fishes and their implications.

Childress, J. J., Feldback, H., and Somero, G. N. 1987. Symbiosis in the deep sea. 256:5, 114–21.
Deep-sea hydrothermal vent animals have a symbiotic relationship with surfur-oxidizing bacteria that allows them to live in the darkness of the deep ocean.

Levine, R. P. 1969. The mechanism of photosynthesis. 221:6, 58–71.
Reveals what is known of the process by which energy is captured by plants and converted to useful forms of chemical energy while freeing oxygen to the atmosphere.

Pettitt, J., Ducker, S., and Knox, B. 1981. Submarine pollination. 244:3, 134–44.
Discusses the pollination of sea grasses by wave action.

15

ANIMALS OF THE PELAGIC ENVIRONMENT

In the sunlit waters of the pelagic environment live the phytoplankton that provide the basis for over 99 percent of the marine biomass through their photosynthetic activities. This important marine population was discussed in chapter 14, so our emphasis here will be on the animal populations. The discussion will deal primarily with the adaptations that allow them to live successfully in the pelagic environment.

STAYING ABOVE THE OCEAN FLOOR

Since the tissues of the zooplankton and nekton (muscle, cartilage, scales, bone, and shell) are more dense than ocean water, pelagic animals can remain off the bottom only through the application of buoyancy or energy. The role of decreased size aiding in this process by increasing the frictional resistance to sinking was discussed in chapter 13 as helpful to microplankton. However, the animals we will discuss in this chapter are large enough that additional means must be applied if they are to remain off the ocean floor.

Gas Containers

Since air is approximately 0.001 the density of water at sea level, a small amount of it stored in the body of an organism could provide a significant reduction in the average density of the organism and make it easier for the organism to remain in the water column. Some cephalopods have rigid gas containers (figure 15–1). The genus *Nautilus* has an external shell; the cuttlefish *Sepia* and deep-water squid *Spirula* have an internal chambered structure. These animals become neutrally buoyant and can maintain their position in the water column easily. Since the air pressure in their air chambers is always 1 atmosphere, they are limited in the depth to which they may venture. The *Nautilus* must stay above a depth of approximately 500 m (1640 ft), or its chambered shell will collapse as the external pressure approaches 50 atmospheres. The *Nautilus* is rarely observed below a depth of 250 m (820 ft).

Clown fish among sea anemone tentacles. Photo by Christopher Newbert.

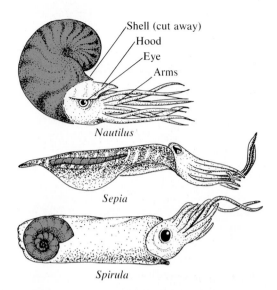

FIGURE 15–1

Gas Containers of the Cephalopods *Nautilus, Sepia,* and *Spirula*. The *Nautilus* has an external chambered shell, and the *Sepia* and *Spirula* have internal chambered structures that can be filled with gas to provide buoyancy.

Neutral buoyancy is achieved by some bony fish through filling a gas bladder, or *swim bladder,* with gases. The swim bladder is normally not present in very active swimmers, such as the tuna, or in fish that live on the bottom. Some fish have a pneumatic duct that connects the swim bladder to the esophagus (figure 15–2). These fish can add or remove air through this duct. In

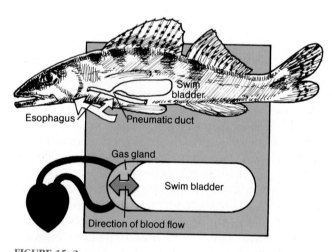

FIGURE 15–2

Swim Bladder of Some Bony Fishes. *A,* example of a swim bladder connected to the esophagus by the pneumatic duct, allowing air to be added or removed rapidly. *B,* in fish with no pneumatic duct, all gas must be added or removed through the blood. This requires more time and is achieved by a network of capillaries associated with the gas gland.

other fish, the gases of the swim bladder must be added or removed more slowly by an interchange with the blood. Since change in depth will cause the gas in the swim bladder to expand or contract, gas must be removed from or added to the swim bladder to maintain a constant volume. Those fish without the pneumatic duct are limited in the rate at which they can make these adjustments and, therefore, cannot withstand rapid changes in depth.

The composition of gases in the swim bladder of shallow-water fishes is similar to that of the atmosphere. With increasing depth (fish with swim bladders have been taken from a depth of 7000 m or 23,000 ft), the concentration of oxygen in the swim bladder gas increases from the 20 percent common near the surface to more than 90 percent. At the 700-atmosphere pressure that exists at a depth of 7000 m (23,000 ft), gas is compressed to a density of 0.7 g/cc. This is about the density of fat, so many deep-water fish have air bladders filled with fat instead of compressed gas. This provides almost as much buoyancy as compressed gas and avoids the energy requirements of maintaining a constant volume of gases in the swim bladder.

Floating Forms

Floating at the ocean surface are some relatively large members of the plankton, the evolutionarily primitive siphonophore and scyphozoan coelenterates, as well as the pelagic tunicates. Somewhat smaller are the ctenophores and arrowworms. They all have soft gelatinous bodies with little if any hard tissue. A major strategy of this body type is the replacement in body fluids of heavy sulfate ions with chloride ions to maintain osmotic equilibrium.

The siphonophore coelenterates are represented in all oceans by the genera *Physalia* (Portuguese man-of-war) and *Velella* (by-the-wind sailor). Their gas floats, *pneumatophores,* serve as floats and sails that allow the wind to push these colonial forms across the ocean surface (figure 15–3). A colony of tiny sea-anemone-like *polyps* and jellyfish-like *medusae* are suspended beneath the float. The long tentacles of the polyps serve primarily to capture food, and the medusae carry on the sexual reproduction required to create a new colony. The tentacles of the *Physalia* may be many meters long and do possess nematocysts long enough to penetrate the skin of humans; they have been known to inflict a painful and occasionally dangerous neurotoxin poisoning. The colonies grow from the initial polyp through asexual budding.

The scyphozoans, or jellyfish, have a medusoid, bell-shaped body with a fringe of tentacles and a mouth at the end of a clapper-like extension hanging down under

FIGURE 15–3
Planktonic Coelenterates. *A*, Physalia and jellyfish. *B*, medusa. (Photo by Larry Ford.)

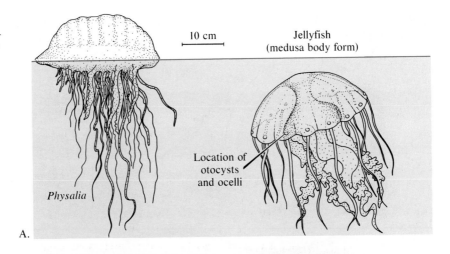

the bell-shaped float. Ranging in size from near microscopic to 2 m (6.5 ft) in diameter with 60 m (200 ft) tentacles, most jellyfish have bells with a diameter of less than 0.5 m (1.6 ft). Jellyfish are capable of movement through muscular contractions. Water enters the cavity under the bell and is forced out by contractions of muscles that circle the bell, jetting the animal ahead in short spurts. To ensure the animal swims generally in an upward direction, there are sensory organs spaced around the outer edge of the bell. These may be light-sensitive *ocelli* or gravity-sensitive *otocysts*. This orientation capacity is important because the jellyfish feed by swimming to the surface and sinking slowly through the life-rich surface waters.

Pelagic *tunicates* are generally transparent and more or less barrel-shaped with an anterior incurrent and posterior excurrent opening (figure 15–4). They move by a feeble form of jet propulsion effected by the contraction of bands of muscles that force water out the

excurrent opening. Reproduction of solitary forms is complicated, involving alternation of sexual and asexual reproduction. The genus *Salpa* includes solitary forms reaching a length of 20 cm (8 in.) and smaller aggregate forms that produce new individuals by budding. Individual members of an aggregate chain may be 7 cm (3 in.) long, and the chain of newly budded members may reach great lengths. The genus *Pyrosoma* is luminescent and colonial. Individual members of the colony have their incurrent openings facing the outside surface of a tube-shaped colony that may be a few meters long. One end of the tube is closed, and the excurrent openings of the thousands of individuals all empty into the tube. Muscular contraction forces water out the open end to provide a means of propulsion.

Ctenophores are comb-bearing animals closely related to the coelenterates in that a radially symmetrical body develops around a mouth. Some forms, however, are apparently more advanced; they possess an anal

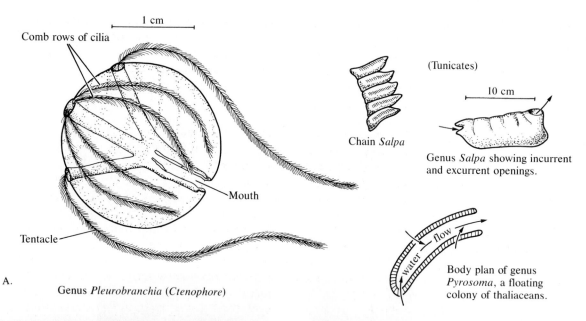

Comb rows of cilia

1 cm

Mouth

Tentacle

A.

Genus *Pleurobranchia* (*Ctenophore*)

(Tunicates)

10 cm

Chain *Salpa*

Genus *Salpa* showing incurrent and excurrent openings.

water — flow

Body plan of genus *Pyrosoma*, a floating colony of thaliaceans.

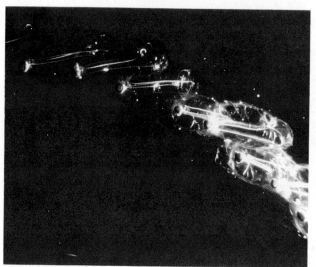

B.

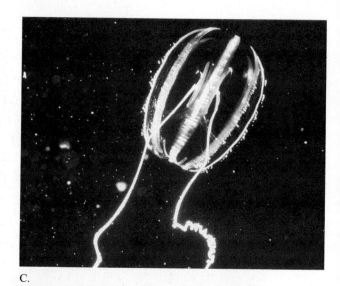

C.

FIGURE 15–4
Pelagic Tunicates and Ctenophores. *A*, body structure. *B*, chain of salps. *C*, ctenophore, *Pleurobranchia*. *D*, ctenophore, *Beroe*. (Photos in *B*, *C*, and *D* by James M. King.)

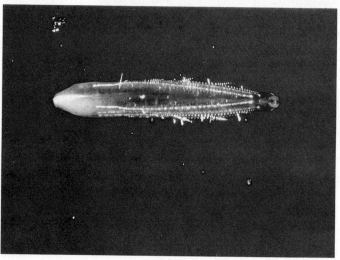

D.

opening from the digestive cavity. Often referred to as *comb jellies*, this group of animals is entirely pelagic and confined to the marine environment. The body form of most members of the phylum Ctenophora is basically spherical, with eight rows of cilia spaced evenly around the sphere. It is from these structures that the name *comb-bearing* is derived; when magnified, they look like miniature combs. If a ctenophore possesses tentacles, there will be two that contain adhesive organs instead of stinging cells to capture prey. The genera *Pleurobranchia* and the pink, elongated *Beroe* range in size from the gooseberry dimension of the former to over 15 cm (6 in.) in the latter.

The *arrowworms* (chaetognaths) are transparent and difficult to see, although they may grow to more than 2.5 cm (1 in.) in length (figure 15–5). The name *chaetognath* refers to the hairlike attachments around the mouth that are used to grasp prey while it is being devoured. They are voracious feeders, feeding primarily on the small zooplankton and in turn, are fed upon by fishes and larger planktonic animals such as jellyfish. These exclusively marine hermaphroditic animals are usually more abundant in the surface waters some distance from shore.

Swimming Forms

In this section we discuss larger pelagic animals, referred to as the *nekton*, with substantial powers of locomotion. They include rapid swimming invertebrate squids, as well as fish and marine mammals. Rapid-swimming animals must have streamlined bodies and an effective mechanism for locomotion.

Swimming squid include the genera *Loligo* (common squid), *Ommastrephes* (flying squid), and the *Architeuthis* (giant squid). Active predators on small fish, the smaller varieties of squid have long slender bodies with paired fins (figure 15–6). They have no hollow chambers in their bodies, as were described for *Sepia* and *Spirula*, and therefore require more energy to remain

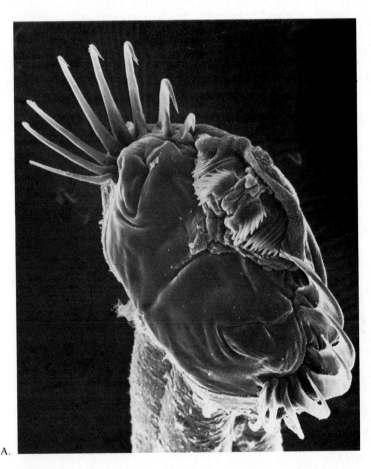

A.

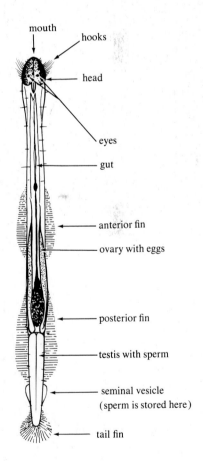

B. General Anatomy of an Arrowworm

FIGURE 15–5

Arrowworm. *A*, head of chaetognath *Sagitta tenuis* from Gulf of Mexico, magnified 161 times. (Photo by Howard J. Spero, University of California at Santa Barbara.) *B*, adult chaetognaths reach lengths ranging from 1 to 5 cm.

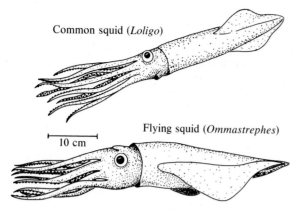

Common squid (*Loligo*)

Flying squid (*Ommastrephes*)

10 cm

FIGURE 15–6
Squids.

in the upper water of the oceans. Trapping water in their mantle cavity and forcing it out through a siphon, these invertebrates can swim about as fast as any fish their size. Two long arms with pads containing suction cups at the end are used to capture prey, and eight shorter arms containing suckers take the prey to the mouth, where it is crushed by a beaklike mouthpiece.

The consideration of locomotion in fish is more complex than that of the squid because of the body motion and the role of fins. The basic movement of swimming fish is the passage of a wave of lateral body curvature from the front to the back of the fish. This is achieved

by the alternate contraction and relaxation of muscle segments, *myomeres,* along the sides of the body. The backward pressure of the fish's body and fins produced by the movement of this wave provides the forward thrust (figure 15–7).

Most active swimming fish have two sets of paired fins, *pelvic* and *pectoral,* used in maneuvers such as turning, braking, or balancing. When they are not in use, they can be folded against the body. Vertically oriented fins, *dorsal* and *anal,* serve primarily as stabilizers. The fin most important in propelling the high-speed fish is the tail, or *caudal fin.* Caudal fins flare dorsally and ventrally to increase the surface area available to develop thrust. Increased surface area also increases frictional drag. The efficiency of the design of a caudal fin depends on its shape and can be expressed mathematically as the *aspect ratio,* calculated as follows:

$$\frac{(\text{Fin height})^2}{\text{Fin area}}$$

There are five basic shapes of caudal fins: rounded, truncate, forked, lunate, and heterocercal; they are found, respectively, on sculpin, bass, yellow-tail, tuna, and shark. The rounded fin, with an aspect ratio of approximately 1, is flexible (figure 15–8) and useful in accelerating and maneuvering at slow speeds. The somewhat flexible truncate (aspect ratio 3) and forked (aspect ratio 5) tails will be found on faster fish and may still be used for maneuvering. The lunate caudal fin, with an

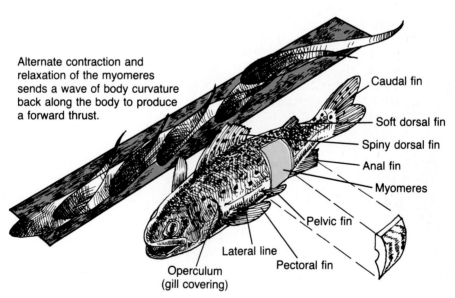

Alternate contraction and relaxation of the myomeres sends a wave of body curvature back along the body to produce a forward thrust.

Caudal fin

Soft dorsal fin

Spiny dorsal fin

Anal fin

Myomeres

Pelvic fin

Lateral line

Pectoral fin

Operculum (gill covering)

FIGURE 15–7
Fish—Swimming Motions and Fins.

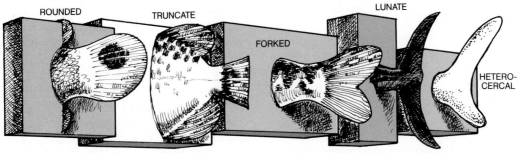

Rounded—Aspect ratio 1, typical of bottom-dwelling fish like sculpin and flounder

Truncate—Aspect ratio 3, typical of salmon and bass

Forked—Aspect ratio 5, typical of herring and yellowtail

Lunate—Aspect ratio 7–10, typical of most tuna

Heterocercal—Aspect ratio varies, typical of sharks that need an upward thrust from their swimming motion

FIGURE 15–8
Caudal Fin Shapes and Aspect Ratios.

aspect ratio of up to 10, is found on the fast cruising fishes such as tuna, marlin, and swordfish; it is very rigid and useless in maneuverability but very efficient in propelling. The heterocercal caudal fin is asymmetrical, with most of its mass and surface area in the ventral lobe.

The heterocercal fin produces a significant lift to sharks as it is moved from side to side. This is important because sharks have no swim bladder and tend to sink when they stop moving. To aid this lifting, the pectoral fins are large and flat. Located on the anterior of the shark's body like the wings of an airplane, they serve as hydrofoils that lift the front of the shark's body to balance the posterior lift applied by the caudal fin. The shark loses a lot in maneuverability as a result of this modification of the pectoral fins.

Pectoral fins are modified in many fish for functions other than maneuvering (figure 15–9). Wrass and sculpins use them to "row" themselves through the water in jerky motions while the rest of the body remains motionless. Skates and rays swim by sending undulating

motions across the greatly modified pectoral fins, and manta rays flap them like wings. The flying fish, *Exocoetus*, uses greatly enlarged pectoral fins to glide for up to 400 m (1300 ft) across the ocean surface after propelling itself with the enlarged ventral lobe of its caudal fin into the air to escape dolphins and other predators. Other fish have pectoral fins modified into fingerlike structures used to walk on the ocean floor.

Many fish in no great hurry propel themselves by undulation of their dorsal fin. Examples are triggerfish, seahorses, ocean sunfish, and trunkfish. The sunfish also uses its anal fin and stubby caudal fin to aid in the process, still without impressive results.

Modifications Associated with Swimming Behavior

Some fish spend most of their time waiting patiently for prey and exert themselves only in short bursts as they lunge at the prey; others cruise relentlessly through the

Flying fish—
Pectoral fins modified for gliding

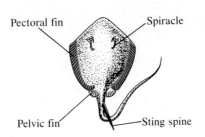

Stingray—
Pectoral fins modified for swimming

Gurnard—
Pectoral fins modified for walking

FIGURE 15–9
Pectoral Fin Modifications.

water seeking out prey. There is a marked difference in the nature of the musculature of fishes with these different styles of seeking out food. A grouper, representative of the *lungers*, has a truncate caudal fin (aspect ratio 3), and almost all its muscle tissue is white. Tuna, on the other hand, are *cruisers*, and less than half their muscle tissue is white; it is mostly of a variety referred to as *red* (figure 15–10).

Red muscle fibers are much smaller in diameter (25–50μm 0.001–0.002 in.) than white muscle fibers (135μm) and contain higher concentrations of *myoglobin*, a red pigment with an affinity for oxygen (O_2). These factors allow the red fibers to obtain a much greater oxygen supply than white fibers and therefore achieve a metabolic rate six times that of white fibers. The red muscle tissue is abundant in cruisers that constantly swim. The lungers can get along quite well with little red tissue because they do not require the ability to be constantly moving. White tissue, which fatigues much more rapidly than red tissue, is used by the tuna for short periods of acceleration while on the attack. It is also quite adequate for propelling the grouper and other lungers during their quick passes at prey.

The speed of fish is thought to be closely related to body length. For tuna, fish that are well-adapted for sustained cruising and short bursts of high-speed swimming, cruising speed averages about 3 body lengths/second, and maximum speed of about 10 body lengths/second can be maintained for only 1 s. A yellowfin tuna, *Thunnus albacares,* has been clocked at 74.6 km/h (46.3 mi/h) or more than 20 body lengths/second over a period of 0.19 s. Theoretically, a 4-m (13-ft) bluefin tuna, *Thunnus thynnus,* could reach speeds of up to 143.9

km/h (86.8 mi/h). Many of the toothed whales are known to be capable of high rates of speed. The dolphin, *Stenella,* has been clocked at 40 km/h (25 mi/h), and it is believed that the top speed of killer whales may be in excess of 55 km/h (34 mi/h).

As is true with many chemical reactions, metabolic processes can proceed at greater rates at higher temperatures. Although fish are generally considered to be *poikilothermic* (cold blooded), having body temperatures that conform to the temperature of their environment, there are some fast swimmers with body temperatures above that of the surrounding water. Some are only slightly above the temperature of the water: the mackerel (*Scomber*), yellowtail (*Seriola*), and bonito (*Sarda*) have body temperature elevations of 1.3, 1.4, and 1.8°C (2.3, 2.5 and 3.2°F), respectively. Other fast swimmers that have much higher elevations are members of the mackerel shark genera, *Lamna* and *Isurus,* as well as the tuna, *Thunnus.* The bluefin tuna has been observed to maintain a body temperature of 30–32°C (86–90°F) (*homeothermic*) regardless of the temperature of the water in which it is swimming. Although it is more commonly found in warmer water where the temperature difference is no more than 5°C (41°F), body temperatures of 30°C (86°F) have been measured in bluefin tuna swimming in 7°C (45°F) water.

Why do these fish exert so much energy to maintain their body temperatures at high levels when other fishes do quite well with ambient body temperatures? Their mode of behavior is that of a cruiser, and any adaptation that can increase the power output of their muscle tissue aids them in their efforts of searching out and capturing prey.

A.

B.

FIGURE 15–10
Feeding Styles—Lunger and Cruiser. *A,* lungers like the grouper sit patiently on the bottom and capture prey with quick, short lunges. Their muscle tissue is predominantly white. *B,* cruisers like the tuna swim constantly in search of prey and capture it with short periods of high-speed swimming. Their muscle tissue is predominantly red. (Photographs by Marty Snyderman.)

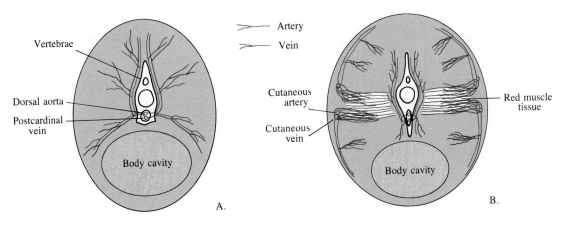

FIGURE 15–11
Circulatory Modifications in Warm-Blooded Fishes. *A,* most cold-blooded (poikilothermic) fishes have major blood vessels arranged in the pattern represented on the left. All blood flows to muscle tissue from the dorsal aorta and returns to the postcardinal vein beneath the vertebral column. *B,* warm-blooded (homeothermic) fishes like the tuna have cutaneous arteries and veins that help maintain high blood temperature by using heat energy generated by contracting muscle tissue. Their blood vessel pattern is represented on the right.

The mackerel sharks and tuna are aided in conserving the heat energy needed to maintain their high body heats by a modification of the circulatory system (figure 15–11). While most fish have a dorsal aorta just beneath the vertebral column that provides blood to the swimming muscles, the mackerel sharks and tuna have a *cutaneous artery* just beneath the skin on either side of the body. As the cool blood flows into the red muscle tissue, its temperature is increased by heat generated by muscle metabolism (muscle contractions). A fine network of tiny blood vessels within the muscle tissue is designed to minimize heat loss. The vessels which return the blood to the cutaneous vein, parallel to the cutaneous artery along the side of the fish, are all paired with small vessels carrying blood into the muscle tissue. In this way, the warm blood leaving the tissue helps to heat the cooler blood entering from the cutaneous artery.

MARINE MAMMALS

Mammals are represented in the ocean by the coastal-dwelling herbivores, the sirenans (sea cows), and a variety of carnivorous forms (figure 15–12). Spending all or most of their lives in coastal waters and coming ashore to breed are the sea otter and pinnipeds (sea lions, seals, and walruses). The truly oceanic mammals are the cetaceans—whales, porpoises, and dolphins. Since mammals evolved from reptiles on land some 200 million years ago, and no marine forms are known earlier than 50 million years ago, it is believed that all marine forms evolved from some ancient land dwellers.

Cetaceans

Well known as the mammals best adapted to life in the oceans, whales are of two basic types. The toothed whales, the *odonticeti,* include 74 species grouped under three common categories: sperm whales, porpoises, and dolphins. The baleen, the *mysticeti,* probably evolved from the toothed whales some 30 million years ago. In place of teeth, they have baleen, plates of horny material, that hang down from the upper jaw to serve as a sieve. The toothed whales are active predators that feed mostly on smaller fish and squid, although the killer whale is known to feed on a variety of larger animals, including other whales. The baleen whales feed primarily by filtering crustaceans at depths ranging from the surface down to and including the sediment of shallow ocean basins. (See plates 27B, 28, and 29.)

The body of whales is more or less cigar-shaped, nearly hairless, and insulated with a thick layer of blubber. The forelimbs are modified into flippers that may be moved only at the "shoulder" joint. The hind limbs are vestigial, not attached to the rest of the skeleton and not externally visible. The skull is highly modified with the nasal opening or openings near the top, and whales propel themselves by vertical movements of a horizontal *fluke* (tail fin).

Modifications to Increase Swimming Speed Cetaceans' muscles are not vastly more powerful than the muscles of other mammals, so it is believed that their ability to swim with high speed must result from modifications that reduce frictional drag. To illustrate the importance of streamlining in reducing the energy re-

A. Sea cow

B. Elephant seals

C. Baby fur seal

D. Sea otter

FIGURE 15–12

Marine Mammals—Sea Cow, Pinnipeds, and Sea Otter. (*A*, photo by Marty Snyderman. *B* and *C*, photos courtesy of Wayne Perryman, National Marine Fisheries Service. *D*, photo from Scripps Institute of Oceanography, University of California, San Diego.)

quirements of swimmers, a small dolphin would require muscles five times as powerful to swim at 40 km/h (25 mi/h) in turbulent flow than in streamlined laminar flow. In addition to the streamlined body, it is believed that cetaceans achieve near-laminar flow of water across their bodies with the aid of a specialized skin structure. The skin is composed of two layers: a soft outer layer that is 80 percent water and has narrow canals filled with spongy material and a stiffer inner layer composed mostly of tough connective tissue. The soft layer tends to reduce the pressure differences at the skin-water interface by compressing under regions of higher pressure and expanding into regions of low pressure.

Modifications to Allow Deep Diving Whereas humans can free dive to a maximum recorded depth of about 60 m (200 ft) and hold their breath in rare instances for 6 min, the sperm whale, *Physeter*, is known

to dive deeper than 2200 m (7200 ft), and the North Atlantic bottle-nosed dolphin, *Hyperoodon ampullatus*, can stay submerged for up to 2 h.

One of the characteristics associated with the ability of whales to dive deep and stay submerged for prolonged periods is the ability to alternate between periods of *eupnea* (normal breathing) and *apnea* (cessation of breathing). The periods of apnea occur, of course, while the animal is submerged. To understand how some cetaceans are able to go for long periods of time without breathing requires a general knowledge of the lungs and their associated structures. Air that is inhaled passes through the *trachea*, into the *bronchi*, and through a series of smaller *branchioles* and *alveolar ducts* to tiny terminal chambers, the *alveoli* (figure 15–13). The alveoli are lined with a thin *alveolar membrane* that is in contact with a dense bed of capillaries; the exchange of gases between the inspired air and the blood occurs across the alveolar membrane. Some cetaceans have an exceptionally large concentration of capillaries surrounding the alveoli, which may have muscles that move air against the membrane by repeatedly contracting and expanding. Cetaceans take from 1 to 3 breaths per minute while resting, compared with about 15 in humans. Because of the longer time inhaled breath is held, the large capillary mass in contact with the alveolar membrane, and the circulation of the air by muscular action, many cetaceans can extract almost 90 percent of the oxygen in each breath, compared with the 4 to 20 percent extracted by terrestrial mammals.

To use this large amount of oxygen that can be taken into the blood efficiently during long periods of submergence, cetaceans apply two strategies. They (1) store large amounts of oxygen and (2) reduce oxygen use. The storage of so much oxygen is made possible by the fact that prolonged divers have a much greater blood volume per unit of body mass than those that dive for only short periods. Compared with terrestrial animals, some cetaceans have twice as many red blood cells per unit of blood volume and up to nine times as much myoglobin in the muscle tissue. Thus, large supplies of oxygen can be chemically stored in the hemoglobin of the red blood cells and the myoglobin.

Additionally, the muscles that start a dive with a significant oxygen supply can continue to function through anaerobic glycolysis when the oxygen is used up. The muscle tissue is relatively insensitive to high levels of carbon dioxide, a product of aerobic respiration, and lactic acid, a product of anaerobic respiration.

Because the swimming muscles can function without oxygen during a dive, they and other organs, such as the digestive tract and kidneys, may be sealed off from the circulatory system by constriction of key arteries. The circulatory system services primarily the heart and brain. Because of the decreased circulatory requirements, the heart rate can be reduced to 20 to 50 percent of its normal rate. However, recent research has shown that no such reduction in heart rate occurs when the common dolphin (*Delphinus delphis*), white whale, (*Delphinapterus leucas*), and the bottle-nosed dolphin (*Tursiops truncatus*) dive.

Another problem normally associated with deep and prolonged dives is the absorption of compressed gases into the blood. When humans make dives using compressed air, the prolonged breathing of compressed air, particularly nitrogen, can result in *nitrogen narcosis* or *decompression sickness* (the bends). Nitrogen narcosis is an effect similar to drunkenness, and it can occur when a diver either goes too deep and/or stays too long at depths greater than 30 m (100 ft). If a diver surfaces too rapidly, the lungs cannot remove the excess gases rapidly enough, and the reduced pressure may cause small bubbles to form in the blood and tissue. The bubbles interfere with blood circulation, and the resulting decompression sickness can cause excruciating pain, severe physical debilitation, or even death.

Cetaceans and other marine mammals do not suffer from these difficulties. Their main defense against absorbing too much nitrogen seems to be associated with a more flexible rib cage. By the time a cetacean has reached a depth of 100 m (328 ft), the rib cage has collapsed under the 11 atmospheres of pressure. The lungs within the rib cage have also collapsed, removing all air from the alveoli. Since most absorption of gases by the blood occurs across the alveolar membrane, the blood cannot absorb the compressed gases, and the problems of nitrogen narcosis and decompression sickness are avoided.

Echolocation   Based on present evidence, fully 20 percent of the mammalian species are thought to echolocate. Of the terrestrial forms, bats are the best known,

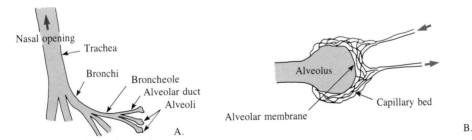

FIGURE 15–13
Cetacean Modifications to Allow Prolonged Submergence. *A*, on the left is a diagram of basic lung design. Air enters the lung through the trachea, and oxygen is absorbed into the blood through the walls of the alveoli. *B*, the figure on the right represents oxygen exchange in the alveolus. A dense mat of capillaries receives oxygen through the alveolar membrane. Because air is normally held in the lungs of cetaceans for up to 1 min before being exhaled, as much as 90 percent of the oxygen can be extracted. Whales have up to twice as many red blood cells and nine times as much myoglobin in their muscle tissue in which to store this oxygen.

but shrews, flying lemurs, the fat dormouse, and golden hamster are also known to use echolocation. Marine mammals that are known or strongly suspected to use echolocation are all the toothed whales, the Weddell seal, California sea lion, the walrus, and the gray, blue, and minke whales. These animals have within their brain the ability to involuntarily calculate the distance of an object by multiplying the velocity with which a sound signal travels to and returns from the object and dividing this product by 2.

$$D = \frac{V \times T}{2}$$

Although all marine mammals have good vision, conditions within the marine environment often limit its effectiveness. In coastal waters, where suspended sediment and dense plankton blooms make the water turbid, and in the deeper waters, where light is limited or absent, echolocation would surely be beneficial in pursuit of prey or the location of objects in the water.

Cetaceans emit a wide variety of sounds including what might be called wails, creaks, squeals, moans, clicks, and even songs in the case of the humpback whale, *Megaptera*. Although the other sounds are thought to be of value in communication, only the clicks are thought to be used for echolocation.

The common dolphin, *Tursiops truncatus*, has been closely studied by echolocation researchers. For a general scanning of its environment, *Tursiops* emits low-frequency clicks to locate objects at a distance; upon close approach, higher frequency clicks are produced to determine fine details. Using only its echolocation abilities, the common dolphin has been able to discriminate between such similar objects as fish of similar size and shape, identically shaped plates of different metallic composition, and plates of the same metallic composition with slightly different thickness. The clicks of the small common dolphin are of a frequency range that is partly audible to humans; however, some are of ultrasonic frequencies that can be repeated up to 800 times per second. Large toothed whales emit lower frequency clicks at lower repetition rates. The killer whale, *Orcinus orca*, produces from 6 to 18 clicks per second, and the sperm whale, *Physeter catodon*, may produce from less than one to more than 40 clicks per second. Each click of the sperm whale contains 9 pulses or clickettes. It has been estimated that the sperm whales can detect their main prey, squid, from a distance of up to 400 m (1300 ft) by use of their low-frequency scanning clicks.

How the pulses of clicks are produced and how the returning sound is received are not fully understood. However, there are some widely accepted theories that attempt to answer both questions for certain species of toothed whales. Measurements of the intensity of clicks emitted by smaller-toothed whales indicate that the sound originates in the forehead region. The source of sound in this area could be the complex of air sacs associated with the nasal passage functioning in conjunction with the muscular nasal plugs that close off the blowhole (figure 15–14A). Sounds produced in the blowhole could be reflected off the concave surface at the front of the skull of toothed whales and focused by the lens-shaped *melon*. The melon is a dome-shaped fatty structure located in the center of the forehead of many toothed whales. Studies using hydrophones have indicated that the clicks of dolphins are focused into a forward-directed beam. Although most researchers think it is a less likely source of echolocation clicks, the larynx of toothed whales is well muscled and complex; even without vocal chords, it could be capable of producing sound.

In the sperm whale, the *spermaceti organ*, once prized by whalers for the fine quality of its oil, may play a major role in focusing the clicks (figure 15–14B). The spermaceti organ is encased in ligaments and rests in the *rostrum*, a forward extension from the base of the frontal skull. Norris and Harvey (1972) suggested that a special complex arrangement, part of the respiratory system, produces the click. The blowhole high on the front of the sperm whale's bulbous head opens into the *distal air sac* located in front of the spermaceti organ and the *left nasal passage* that leads to the trachea. Smaller than the left nasal passage, the *right nasal passage* extends from the *museau du singe* at the base of the distal air sac, along the base of the spermaceti organ, to the *frontal air sac* at the posterior end of the spermaceti organ, then on to the trachea. The sound-making structure in this system is the museau du singe or monkey's muzzle. It is composed of a pair of hard, tightly closed lips.

A means by which this complex system could generate sound involves the forcing of air from the trachea through the right nasal passage. As air is forced past the lips of the monkey's muzzle, they separate and snap back together, producing a click. This sound is the first pulse of the nine-pulse click characteristic of sperm whales. The other eight pulses, of progressively lower intensity, may result from the reverberation of the initial sound back and forth between the distal air sac and the frontal air sac across the spermaceti organ. The spermaceti organ may serve as the sound channel that guides the reflected sound signals. After the air passes through the monkey's muzzle, it is not lost through the blowhole. The blowhole is sealed off, and the air is returned to the trachea through the left nasal passage.

Whereas the human ear is sensitive to frequencies from 16 Hz to 20 kHz, some toothed whales respond to frequencies as high as 150 kHz. In most mammals, the

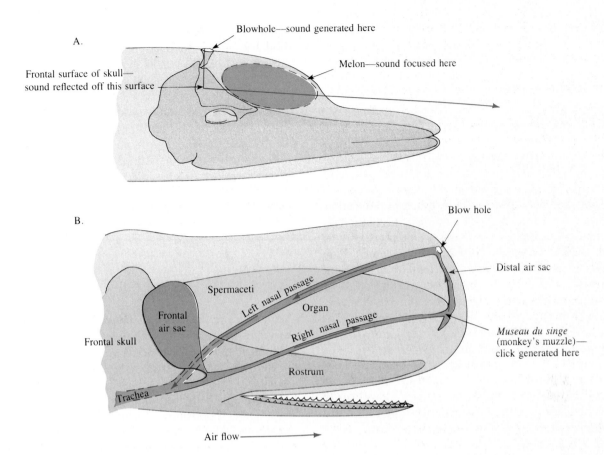

A.

Blowhole—sound generated here

Melon—sound focused here

Frontal surface of skull—
sound reflected off this surface

B.

Blow hole

Distal air sac

Spermaceti

Left nasal passage

Organ

Right nasal passage

Frontal
air sac

Frontal skull

Museau du singe
(monkey's muzzle)—
click generated here

Rostrum

Trachea

Air flow

FIGURE 15–14

Generation of Echolocation Clicks in Small-Toothed and Sperm Whales. *A*, in small-toothed whales, the clicks may be generated within the blowhole mechanism, reflected off the frontal surface of the skull, and focused by the fatty melon into a forward-directed beam. *B*, this drawing represents structures that may be related to the generation of clicks by sperm whales.

bony housing of the inner ear structure is fused to the skull, so that, when the mammals are submerged, sounds transmitted through the water are first picked up by the skull and travel to the hearing structure from many directions. This makes it impossible for such mammals to locate accurately the source of the sounds. Obviously such an arrangement for the hearing structure would not be suitable for an animal that depended on echolocation to accurately determine the position of objects in the water. All cetaceans have evolved structural features that insulate the inner ear housing, the *tympanic bulla,* from the rest of the skull. In the toothed whales, the tympanic bullae are separated from the rest of the skull by an appreciable distance and surrounded by an extensive system of air sinuses. The sinuses are filled with an insulating emulsion of oil, mucus, and air and are surrounded by fibrous connective tissue and venous networks.

There is some disagreement on how sound reaches the inner ear to be processed. Some investigators be-

lieve it passes through the external auditory canal as it does in other mammals, yet this canal is usually blocked by a plug of ear wax in baleen whales. In toothed whales it may be nearly or completely covered by skin. A thin flaring of the lower jaw, less than 0.1 mm (0.004 in.) thick, is directed toward the tympanic bulla on either side of the head and connected with it by a fat- or oil-containing body. It is believed by most investigators that sound is picked up by the thin, flaring jawbone and passed to the inner ear via the connecting fat body. Tests of acoustical sensitivity have indicated the lower jaw is up to six times as sensitive as the external auditory canal.

GROUP BEHAVIOR

Depending on the size, feeding behavior, nature of the reproductive process, and other requirements of maintaining the species, many animals inhabiting the open ocean have developed patterns of group behavior that allow them most efficiently to exploit their environment. Two of these patterns are *schooling* and *migration*.

Schooling

Obtaining food is an activity that occupies most of the time of many inhabitants of the open ocean. This is achieved by the fast and agile animals through active predation; other animals move more leisurely while filtering small food particles from the water. Examples of predators and filter-feeders can be found in populations of pelagic animals from the tiny zooplankton to the massive whales. Examples of dense concentrations of predators and filter-feeders cover the same wide range. Patches of zooplankton may occur simply because the nutrients that support the phytoplankton they feed on may be in high concentrations in certain coastal waters. We usually do not refer to these patches as schools. The term *school* is usually reserved for well-defined social organizations of fish, squid, and crustaceans.

The number of individuals in a school can vary from a few larger predaceous fish to hundreds of thousands of small filter-feeders. Within the school, individuals of the same size move in the same direction with equal spacing between each. This spacing is probably maintained through visual contact, and in the case of fish, by use of the lateral line system that detects vibrations of swimming neighbors. The school can turn abruptly or reverse direction as individuals at the flank or the rear of the school assume leadership positions.

The advantage of schooling seems obvious from the reproductive point of view. During spawning, it assures that there will be males to release sperm to fertilize the eggs shed into the water or deposited on the bottom by females. However, most investigators believe the most important function of schooling in small fish is protection from predators. On first consideration, it seems illogical that the formation of schools would serve such a purpose. Aren't the smaller fish making it easier for the predator by forming a large mass? This is a difficult question; in fact, no experimental evidence provides the basis for an unequivocal answer. However, the consensus, based primarily on conjecture, is no.

The belief that small fish are safer from predators if they form schools is based in part on the following reasoning: Over 4000 species of fish are known to form schools. This fact alone indicates that this behavior has evolved and is so pervasive today among fish populations because, for fish with no other means of defense, it somehow provided a better chance of survival than swimming alone. How schooling is protective may grow out of the following considerations: (1) If members of a species form schools, they reduce the percentage of the volume of the ocean in which a cruising predator might find one of their kind. (2) Should a predator encounter a large school, it is less likely to consume the entire unit than if it encounters a small school or an individual. (3) The school may appear as a single large and dangerous opponent to the potential predator; this perception could prevent some attacks. (4) Predators may find the continually changing position and direction of movement of fish within the school confusing, making attack particularly difficult for predators that can attack only one fish at a time. It is surely possible that there are other more subtle reasons for schooling. Whatever they might be, this gregarious behavior must enhance species survival because it is widely observed among pelagic animals.

Migration

Many oceanic animals undertake migrations of varying magnitudes. This behavior is observed among sea turtles, fish, and mammals. Some animals, for instance the sea turtles, have simply not evolved far enough from their land-based ancestors to carry out reproduction at sea and must return to dry beach deposits to lay their eggs. Other populations, including many of the baleen whales, migrate because the physical environment and nature of available food suitable for adults do not meet the needs of young members of their species.

Migratory routes of commercially important baleen whales have been well known since the mid-19th century. The paths of these and other air-breathing mammals are easy to observe. However, the migratory paths of many less visible fish, even though they may also have high commercial value, have been more difficult to identify. Tagging studies have helped us understand the patterns of movement of these fish. Sampling the distribution patterns of eggs, larvae, young, and adult populations and radio-tracking individuals have also helped to identify migratory routes.

One aspect of migratory behavior that is of interest to investigators is orientation: How do migratory species orient themselves in time and space? To put the problem another way, how do they know where they are in relationship to where they want to go? How do they know when to leave their present location to get where they are going at the proper time?

Although these answers are still being sought, it is believed that all the migratory species have an innate sense of time referred to as a *biological clock*. External manifestations of the biological clock of a species, called *circadian rhythms,* are physiological variations that occur independent of changes in the external environment. Circadian rhythms include cyclic changes in respiration rate and body temperature. However, some changes in the external environment can alter the circadian rhythms. It is possible that changes in food availability, water temperature, and length of daylight periods may serve to trigger seasonal migrations.

It seems that orientation in space must be a more complex problem than orientation to time. Animals such as mammals and turtles that migrate at the surface could well use their sight as a means of orientation. Gray whales that migrate close to land over a large part of their migratory route could possibly identify landmarks along the shore. When out of sight of land, mammals and turtles could use the relative positions of the sun, moon, and stars to guide them on their way. Fish that migrate beneath the surface are thought to use smell, relative movement within ocean currents, and small induced currents (generated as the ions dissolved in ocean water are carried through the earth's magnetic field by ocean currents) to aid them in orienting themselves to ocean space.

Many of the important food fish of the high northern latitudes deposit pelagic egg masses that are transported by currents. Such fish will actively migrate up-current during spring or summer and spawn in a location that will assure the current will carry the eggs back to a nurseryground where the food supply will be appropriate for newly hatched fry. Migrations of this type are usually over distances of a few tens or hundreds of kilometers. This behavior is observed in the population of Atlantic cod, *Gadus morhua*, that spawns along the south shore of Iceland. Adults that occupy feeding grounds along the north and east coasts of Iceland and around the southern tip of Greenland migrate to the spawning grounds in the late winter and early spring. This migration involves swimming against the East Greenland and Irminger currents and includes only mature adults at least 8 years of age. These spawning migrations are undertaken annually until the adults die at an age of 18 to 20 years (figure 15–15A, B).

The spawning grounds are bathed in the relatively warm water of the Irminger Current, a northward-flowing branch of the Gulf Stream. Each female releases up to 15 million eggs which float in the surface current. The drifting eggs, carried by the East Greenland and Irminger currents, hatch in about 2 weeks. The larvae feed at midwater depths while the currents are carrying them to the adult feeding grounds. They will remain there, undertaking only short onshore-offshore migrations, until they mature and make their first spawning migration.

Longer migrations are undertaken by two Atlantic bluefin tuna populations that spawn in tropical waters near the Azores in the eastern Atlantic and the Bahamas

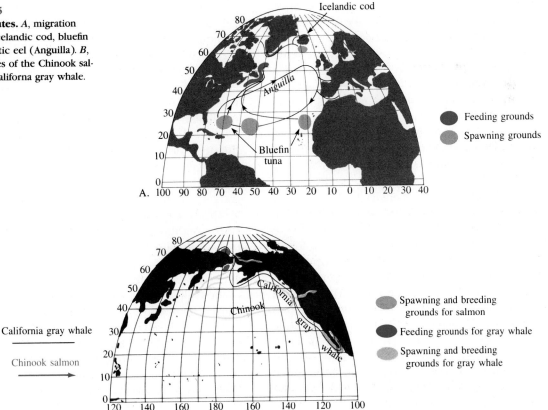

FIGURE 15–15

Migration Routes. *A,* migration routes of the Icelandic cod, bluefin tuna, and Atlantic eel (Anguilla). *B,* migration routes of the Chinook salmon and the Californa gray whale.

in the western Atlantic. Little is known about where the newly hatched tuna feed, but the adults are known to move north along the seaward edge of the Gulf Stream in May and June and feed on the herring and mackerel populations off the coast of Newfoundland and Nova Scotia. They may move even farther north, but the route used by them to return to the spawning grounds is unknown.

Anguilla, the Atlantic eel, undertakes what appears to be a single round-trip migration during its lifetime. Its behavior is *catadromous:* After being spawned in the ocean, it enters freshwater streams to spend its adult life and returns near the end of its life to the ocean spawning site.

Spawning is thought to occur in the Sargasso Sea southeast of Bermuda. The spawning has not been observed, and the earliest stage of *Anguilla* that has been observed is the leaf-shaped, transparent leptocephalus larva. The 5-mm (0.2-in.) long larva are carried into the Gulf Stream and north along the North American coast. After 1 year in the ocean, some of the *leptocephalus* metamorphose into young eels, *elvers,* and move into the streams of North America. Others spend another year migrating to the European coast before metamorphosing and moving into European streams. A third population spends another year moving into the Mediterranean Sea before undergoing metamorphosis. The American and European populations are considered by some to be different species—*A. rostrata* and *A. anguilla,* respectively—based primarily on the fact the American eel has 8 to 10 fewer vertebra than the European eel.

After spending up to 10 years in the freshwater environment, the mature eels undergo yet another change. They develop a silvery color pattern and enlarged eyes that are typical of fish inhabiting the mesopelagic zone. They swim downriver to the ocean and presumably back to the spawning ground, where they are thought to die.

An opposite behavior pattern is characteristic of salmon of the north Atlantic and Pacific oceans. This *anadromous* behavior includes spawning in freshwater streams and spending most of their adult life in the ocean. Pacific species die after spawning; Atlantic salmon return to the ocean.

Six species of Pacific salmon, *Oncorhynchus,* follow similar paths in their migrations. Spawning occurs during the late summer and early fall. Eggs are deposited and fertilized in gravel beds far upstream. The Chinook salmon spawns from as far south as central California to Alaska. The young Chinook that hatches in the spring may head downstream as a silvery, filter-feeding smolt during its first or second year. By the time it reaches the ocean, the young Chinook has become a predator, feeding on smaller fish. After spending about 4 years in the North Pacific, mature salmon of more than 10 kg (22 lb)

in weight and 1 m (3.3 ft) in length return to their home streams to spawn (figure 15–15). Although it is not known how salmon achieve this homing, most investigators think the most likely sources of information processed by the salmon during their return trip to spawning grounds are odors and currents.

Some of the longest migrations known to occur in the open ocean are the seasonal migrations of baleen whales. The benefit of these migrations to the maintenance of the species seems clear. Essentially all baleen species feed in the colder high-latitude waters and breed and calve in warm tropical waters. Feeding occurs during summer while the long hours of sunlight illuminating the nutrient-rich waters produce a vast population of crustaceans to be fed upon. Only the vast supplies of food available in these waters make it possible for the whales to maintain their great bulk. The relatively small size and thin layers of blubber characteristic of newborn whales necessitate calving in warmer waters. Although most newborn whales weigh over 2 tn they are still small enough and their blubber layer thin enough that they would lose body heat at too great a rate in the cold high-latitude waters. They can survive only if they are born into warm tropical water. Because of the energy demands faced by female baleen whales in producing large offspring (the gestation period is up to 11 months) and providing them with fat-rich milk for several months, it is not uncommon for them to mate only once every 2 or 3 years.

The migration route of the California gray whale demonstrates how the conditions discussed above are met by whale migration patterns. Gray whales are moderately large whales, reaching lengths of 15 m (50 ft) and weighing over 30 metric tons. Populations of gray whales feed during the summer months in the extreme north Pacific Ocean, Sea of Okhotsk, Bering Sea, and Chukchi Sea. They are unique among baleen whales in that they do not feed by straining pelagic crustaceans and small fish from the water; instead they stir up bottom sediment with their snouts and feed on bottom-dwelling amphipods.

The western populations winter along the coast of Korea, but the population known as the *California gray whale* migrates from the Chukchi and Bering seas to winter in lagoons of the Pacific coast of Baja California and the Mexican mainland coast near the southern end of the Gulf of California (figure 15–15B). The migration usually begins in September when pack ice begins to form over the continental shelf areas that are their feeding grounds. This migration, which is the longest known migration undertaken by any mammal, may involve a round trip distance of 22,000 km (13,660 mi). First to leave are the pregnant females. They are followed by a procession of mature females that are not pregnant, immature females, mature males, and immature males. After cutting through the Aleutian Islands by way of Unimak Pass, they follow the coast throughout their southern journey. Traveling at an average rate of about 200 km/d (125 mi/d), most of the whales reach the lagoons of Baja California by the end of January.

In these warm water lagoons, the pregnant females give birth to 2-tn calves. The calves nurse and put on weight quickly during the next 2 months. While the calves are nursing, the mature males breed with the mature females that did not bear calves. Late in March, the return to the feeding grounds begins, with the procession order reversed. Most of the whales are back in the feeding grounds by the end of June, and they feed on prodigious amounts of amphipods to replenish their depleted store of fat and blubber before the next trip south.

In the case of the gray whales and other mammals that have been studied, the reason for migration is clearly to find an optimum environment for feeding and reproduction. It is probably the same for other migratory species, although it is sometimes more difficult to identify the reason for the choice of environments.

Reproduction As in the examples just described, reproduction of pelagic animals involves bringing the males and females of the species together for this purpose on some periodic basis. These gatherings usually occur during the warmer spring or summer months when water temperatures are higher and primary productivity at its peak.

Most of the invertebrates and fish that inhabit the open ocean are *oviparous*: They lay eggs that hatch in the open water or on the bottom into larval forms that become meroplankters. Animals that reproduce in this way usually produce enormous numbers of eggs—15 million eggs per season in the migrating Icelandic cod—because most will be lost to predators before they hatch, or the larvae will be eaten by other members of the plankton community. However, some oviparous fish may produce only one or a few well-protected eggs per season; the cartilaginous sharks, skates, and rays include many species that are examples of this behavior (figure 15–16). The horn shark, *Heterodontus,* produces a horny egg capsule commonly called a *mermaid's purse* to house its eggs.

The females of other fish species keep their fertilized eggs in their reproductive tract until they hatch. Although the young come into the world live, the process by which the embryos develop is the same as for those that develop from eggs that are laid. This behavior is called *ovoviviparous*, a method of reproduction in which the eggs are incubated internally. The sea horse and pipefish exemplify a special twist to internal incu-

FIGURE 15–16
Skate Egg Case. This horny mermaid's purse protects the egg that develops within it until it is ready to hatch. (Photo courtesy of James King, Graphic Impressions, Santa Barbara, California.)

bation where the female deposits the eggs in a ventral pouch of the male (figure 15–17). The male carries the eggs until they hatch, and the young fish enter the ocean directly from this pouch. The additional protection against loss of eggs through pelagic predation results in the reduction of the number of eggs that must be produced; for example, the dogfish shark, *Squalus,* usually produces fewer than a dozen eggs at a time.

FIGURE 15–17
Ovoviviparous Behavior in Seahorses. The female deposits the eggs in the ventral pouch of the male, where they are protected during incubation. (Photo by Larry Ford.)

Viviparous animals give birth to live young, but the behavior is considered by most to require also that the young be given more than space in the mother's reproductive tract to incubate. Part of the behavior includes providing embryonic nutrient in addition to the yolk of the egg. Some sharks and rays have projections from the wall of the uterus called *villi* that secrete a rich milk. With this additional nutrition, the stingray, *Pteroplatea,* produces young that enter the ocean with a mass up to 50 times that of the mass of the yolk of the egg from which they came. In the case of the white-tip shark, *Carcharhinus,* this additional nutrition passes to the embryo through a yolk sac attached to the wall of the uterus.

Mammals exhibit the highest level of viviparous behavior, in that the embryo is encased in the *placental sac.* The mother's blood flows through the *placenta,* and the embryo receives nutrition from the mother through an *umbilical cord* by which it is attached to the placenta; essentially all nutrition is provided by the mother. This behavior provides a high degree of protection to the developing young but puts a high energy demand on the mother. Thus, mammalian births typically involve only one or a few young who will remain dependent on their mothers for protection and food for some time after their birth.

SUMMARY

Frictional resistance to sinking—a major factor in helping the tiny plankton stay near the surface—is not a major factor in keeping the larger nekton from sinking; they depend primarily on **buoyancy** or **swimming**. **Rigid gas containers** found in some cephalopods and **expandable swim bladders** of many bony fishes are adaptations that help increase buoyancy. Many invertebrates such as the jellyfish, tunicates, and arrowworms have soft, gelatinous bodies that reduce their density. The Portuguese man-of-war also has a **gas-filled float** that supports this pelagic colony. Many of these invertebrate forms possess weak swimming ability and depend primarily on buoyancy to maintain their position near the surface.

Strong swimmers, the nekton—squid, fish, and mammals—depend on this expenditure of energy to maintain their position in the water column. Squid swim by trapping water in their mantle cavity and forcing it out through a siphon. Most fish swim by creating a wave of body curvature that passes from the front to the back of the body and provides a forward thrust. The **caudal fin** is the most important in providing thrust, while the paired **pelvic and pectoral fins** are used for maneuvering. The vertically oriented **dorsal and anal fins** serve primarily as stabilizers. The efficiency with which a caudal fin produces thrust is indicated by its **aspect ratio**, equal to the square of the fin height divided by the fin area. The **rounded caudal** found on the sculpin is flexible, has a low aspect ratio of about 1, and can be used for maneuvering at slow speeds. The **lunate fin** is rigid, has a high aspect ratio of about 10, and is of little use in maneuvering. It is very efficient in producing thrust for fast swimmers such as the tuna.

Fish such as groupers are **lungers** that sit motionless and make short, quick passes at passing prey. They have mostly white muscle tissue. Tuna and other **cruisers,** constantly swimming in search of prey, possess mostly red muscle tissue. Red fibers have a greater affinity for oxygen and tire less rapidly than white fibers, allowing tuna to maintain a cruising speed of about 3 body lengths per second. Using their white muscles to help in short periods of rapid swimming, they can reach speeds of up to 10 body lengths per second. Although most fish are poikilothermic (cold blooded), the tuna, **Thunnus,** is homeothermic; that is, it maintains a body temperature well above the temperature of the water.

The best adapted mammals for life in the open ocean are the whales (Cetacea). The **toothed whales** (odonticeti) and **baleen whales** (mysticeti) propel their highly streamlined bodies through the water with vertical movement of their horizontal flukes.

Whales are able to dive deep and stay submerged for unusually long periods of time because of modifications that allow them to absorb 90 percent of the oxygen in the air they inhale, store large quantities of oxygen, reduce oxygen use, and collapse their lungs below depths of 100 m.

Most whales and many other marine mammals are thought to use **echolocation** in finding their way through the ocean and locating prey. The clicking sounds emitted by whales are bounced off objects, and the animal can determine the size, shape, and distance of the objects by the nature of the returning signals and the time elapsed.

Schooling of active swimmers such as fish, squid, and crustaceans is not fully understood, but it has obvious advantages as a reproductive strategy. Schooling more likely serves a protective function, although the ways in which it may meet this end are speculative.

Migrations are observed among sea turtles, fish, and mammals and are related to the needs of reproduction and finding food. It is believed that orientation is maintained during migrations through use of visible landmarks, smell, and the earth's magnetic field. Baleen whales may migrate from their cold water summer feeding grounds to warm low-latitude lagoons in winter so their young can be born into warm water; they might not be able to survive the cold water of higher latitudes at birth. Fishes such as the Atlantic cod swim up-current to deposit their eggs so that, when they hatch, the cod fry will be in water where suitable food is available.

The North Atlantic eel, *Anguilla,* is **catadromous**: It spawns in the deep waters of the Sargasso Sea and spends its adult life in freshwater streams of North America and Europe. The Atlantic and Pacific salmon are **anadromous**: They spawn in the freshwater streams of North America, Europe, and Asia and spend their adult life in the open ocean.

Most fish are **oviparous**, depositing their eggs in the ocean. Some sharks and rays maintain their eggs in a body cavity until they hatch; this ovoviviparous behavior provides a greater protection for the eggs. Whereas oviparous fish may produce millions of eggs, **ovoviviparous** fish may produce fewer than a dozen. The stingray, white-tip shark, and mammals are **viviparous**, not only providing space in their bodies for the eggs to develop but also providing nutrition in addition to the egg yolk. In mammals, the young will still be dependent on their mothers for nutrition for some time after their birth.

QUESTIONS AND EXERCISES

1. Discuss how the rigid gas chambers in cephalopods might be more effective in limiting the depth to which they can descend than the flexible swim bladders of bony fish.

2. Describe the body form and lifestyle of the following plankters: *Physilia,* jellyfish, tunicates, ctenophorans, and arrowworms.

3. What are the major structural and physiological differences between the fast-swimming cruisers and the lungers that patiently lie in wait for their prey?

4. List the modifications that are thought to allow some cetaceans to (1) dive to great depths without suffering the bends and (2) stay submerged for unusually long periods of time.

5. Describe the process by which the sperm whale is thought to produce echolocation clicks.

6. Although there is disagreement about how it is achieved, discuss what most investigators believe to be the method by which sound reaches the inner ear of toothed whales.

7. Summarize the reasons some investigators believe schooling increases the safety of fishes from predators.

8. What are the methods believed to be used by migrating animals to maintain their orientation?

9. Why do fishes such as Icelandic cod swim up-current to spawn?

10. How are the migrations of the North Atlantic eels and Pacific salmon fundamentally different?

11. Why don't the California gray whales remain in the cold water feeding grounds during the winter season?

12. Compare the reproductive behavior of oviparous, ovoviviparous, and viviparous animals.

REFERENCES

Carey, F. G. 1973. Fishes with warm bodies. *Scientific American* 228:2, 36–44.

Coker, R. E. 1962. *This great and wide sea: An introduction to oceanography and marine biology.* New York: Harper and Row.

Denton, E. J., and Shaw, T. I. 1962. The buoyancy of gelatinous marine animals. *Journal of Physiology* 161:14P–15P.

George, D., and George, J. 1979. *Marine life: an illustrated encyclopedia of invertebrates in the sea.* New York: Wiley-Interscience.

Herald, E. S. 1961. *Living fishes of the world.* Garden City, N.Y.: Doubleday.

Kanwisher, J. W., and Ridgway, S. H. 1983. The physiological ecology of whales and porpoises. *Scientific American* 248:6, 110- 21.

MacGinitie, G. E., and MacGinitie, N. 1968. *Natural history of marine animals,* 2nd ed. New York: McGraw-Hill.

Norris, K. S., and Harvey, G. W. 1972. A theory for the function of the spermaceti organ of the sperm whale (*Physeter catodon* L.). *Animal Orientation and Navigation,* pp. 397–417. Washington, D.C.: National Aeronautics and Space Administration.

Pike, G. C. 1962. Migration and feeding of the gray whale (*Eschrichtius gibbosus*). *Journal of the Fisheries Research Board of Canada* 19:815–38.

Royce, W., Smith, L. S., and Hartt, A. C. 1968. Models of oceanic migrations of Pacific salmon and comments on guidance mechanisms. *Fishery Bulletin* 66:441–62.

Thorson, G. 1971. *Life in the sea.* New York: McGraw-Hill.

Vaughan, T. A. 1972. *Mammalogy.* Philadelphia: W. B. Saunders.

SUGGESTED READING

Sea Frontiers

Bachand, R. G. 1985. Vision in marine animals. 31:2, 68–74.
An overview of the types of eyes possessed by marine animals.

Bleecker, S. E. 1975. Fishes with electric know-how. 21:3, 142–48.
A survey of fishes that use electrical fields to navigate, to capture prey, and to defend themselves.

Klimley, A. P. 1976. The white shark—a matter of size. 22:1, 2–8.
Describes the procedure used by Dr. John E. Randall for determining the size of sharks from the perimeter of the upper jaw and height of teeth.

Lineaweaver, T. H. III. 1971. The hotbloods. 17:2, 66–71.
Discusses the physiology of the bluefin tuna and sharks that have high body temperatures.

Maranto, G. 1988. The Pacific walrus. 34:3, 152–59.
A summary of the natural history of walruses of the Bering Sea and the role of humans as their predators.

Netboy, A. 1976. The mysterious eels. 22:3, 172–82.
An informative discussion describing what is known of the migrations of the catadromous eels and their importance as a fishery.

O'Feldman, R. 1980. The dolphin project. 26:2, 114–18.
A description of the response of Atlantic spotted dolphin to musical sounds.

Reeve, M. 1971. The deadly arrow worm. 17:3, 175–83.
This important group of plankton is described in terms of body structure and life style.

Winkler, M., Sitko, S. E., and Sund, P. N. 1983. Tunas—Nomads of the sea. 29:1, 51–56.
The varieties of tunas, their physiology, and ocean ranges are discussed.

Scientific American

Denton, E. 1960. The buoyancy of marine animals. 303:11, 118–28.
The methods used by various marine animals to use buoyancy to maintain their position in the water column.

Donaldson, Lauren R., and Joyner, Timothy. 1983. The salmonid fishes as a natural livestock. 249:1, 50–69.
The genetic adaptability of the salmonid fishes may help them adapt to "ranching" operations.

Gosline, John M., and DeMont, M. Edwin. 1985. Jet propelled swimming in squids. 252:1, 96–103.
By using jet propulsion resulting from expelling water through their siphons, squids can move as fast as the speediest fishes.

Gray, J. 1957. How fishes swim. 197:2, 48–54.
The roles of musculature and fins in the swimming of fishes.

Kanwisher, John W., and Ridgeway, Sam H. 1983. The physiological ecology of whales and porpoises. 248:6, 110–21.
Some whales can stay submerged for up to 1 h and surface rapidly from long, deep dives.

Kooyman, G. L. 1969. The Weddell seal. 221:2, 100–107.
The life style and problems related to this mammal's living in water permanently covered with ice.

Leggett, W. C. 1973. Migration of shad. 228:3, 92–100.
The routes of the anadromous shad in the rivers and Atlantic waters are described, along with factors that may control the migrations.

Rudd, J. T. 1956. The blue whale. 195:6, 46–65.
A description of the ecology of the largest animal that has ever lived.

Shaw, E. 1962. The schooling of fishes. 206:6, 128–36.
A consideration of theories seeking to explain the schooling behavior of fishes.

Whitehead, Hal. 1985. Why whales leap. 252:3, 84–93.
Whales seem to communicate with their spectacular lunges above the ocean's surface.

Würsig, B. 1979. Dolphins. 240:3, 136–48.
A summary of the facts concerning dolphin intelligence.

Zapol, W. M. 1987. Diving adaptations of the Weddell seal. 256:6, 100–107.
The physiological adaptations that enable the Weddell seal to make deep, long dives are discussed.

16
ANIMALS OF THE BENTHIC ENVIRONMENT

More than 98 percent of the 200,000 species of animals found in the ocean live in the two-dimensional world of the ocean floor. Ranging from the rocky, sandy, and muddy environments of the intertidal zone to the muddy deposits of deep ocean trenches more than 11 km (6.8 mi) deep, the ocean floor provides a varied environment that is home for a diverse benthic community.

Living at or near the interface of the ocean floor and the ocean water, the success of an organism is closely tied to its ability to cope with the physical conditions of the water, the ocean floor, and the other members of the biological community. The vast majority of benthic species live on the shallow bottom of the continental shelf, and approximately 400 of the known 196,000 benthic species are found in the hadal zone of the deep ocean trenches.

One of the most prominent variables affecting species diversification is temperature. We have previously discussed its effect on species diversity on a latitudinal basis; however, even at the same latitude, a significant difference in the number of benthic species is found on opposite sides of an ocean basin because of the effect of ocean currents on the coastal water temperature. Along the European coast where the Gulf Stream warms the water from the northern tip of Norway to the Spanish coast, over three times the number of species of benthos exist than are found along the similar latitudinal range of the Atlantic coast of North America, where the Labrador Current cools the water as far south as Cape Cod.

ROCKY SHORES

Most of the rocky shore is confined to the supralittoral and littoral zones. The supralittoral corresponds with the backshore, or *spray zone,* above the spring high tide line and is covered by water only during storms. The littoral zone, also known as the *foreshore* or the *intertidal zone,* lies between the high and low tidal extremes. Along most shores, the intertidal zone can be divided into the high tide zone, mostly dry—covered

Potamilla reniformis. Photograph courtesy of Woods Hole Oceanographic Institution

by the highest high tide but not by the lowest high tides; the middle tide zone, exposed and covered equally—covered by all high tides and exposed during all low tides; and the low tide zone, mostly wet—covered during the highest low tides and exposed during the lowest low tides (figure 16–1).

Along rocky shores these divisions of the intertidal zone are often obvious due to the sharp boundaries between populations of attached organisms. Because each centimeter of the littoral rocky shore has a significantly different character from the centimeter above and below it, evolution has been able to produce organisms with the abilities to withstand very specific degrees of exposure to the atmosphere. This condition results in the most finely defined biozones known in the marine environment.

Supralittoral (Spray) Zone

Throughout the world the most obvious inhabitant of the rocky supralittoral zone is the periwinkle snail. The supralittoral zone can easily be identified if one considers the middle of the periwinkle belt to be the boundary between the littoral and the supralittoral. The genus *Littorina* includes species able to breathe air, like land snails. They need the spray of breaking waves as their only tie to the sea, since they are viviparous (give birth to live young) rather than depositing their eggs into the water as do more marine periwinkles and other snails (figure 16–2).

The periwinkles feed by rasping algae and lichens off the rocks of the spray zone with their filelike *radula*, a calcium carbonate mouthpiece common to many snails. These little snails can be as small as a grain of sand or more than a centimeter long and are a significant factor in the erosion of upper intertidal and spray zone rocks.

Hiding among the cobbles and boulders covering the floors of sea caves well above the high tide line are isopods belonging to the genus *Ligia*; their common names are *rock lice* and *sea roaches*. Neither name is particularly flattering to these little scavengers that reach lengths of 3 cm (1.2 in.) and scurry about at night feeding on organic debris.

Another indicator of the spray zone is a distant relative of the periwinkle snails, a limpet belonging to the genus *Acmaea*. Having a flattened conical shell, the limpets feed in a manner similar to that of the periwinkles. During daylight hours, limpets will remain motionless with their shells pulled down tight against the rock on which they make their home. Most limpets are known to leave their home spot on the rock at night to feed. They will then return to their home, marked by a

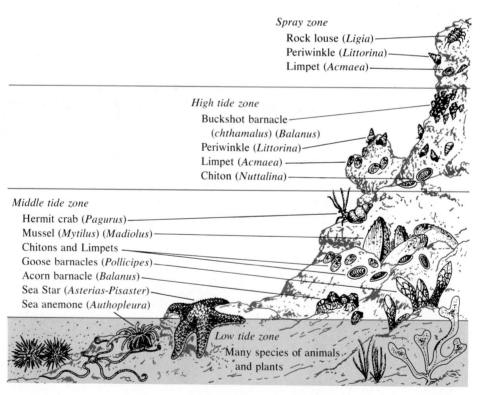

FIGURE 16–1
The Rocky Shore. A typical rocky intertidal zone and some organisms typically found in its subzones.

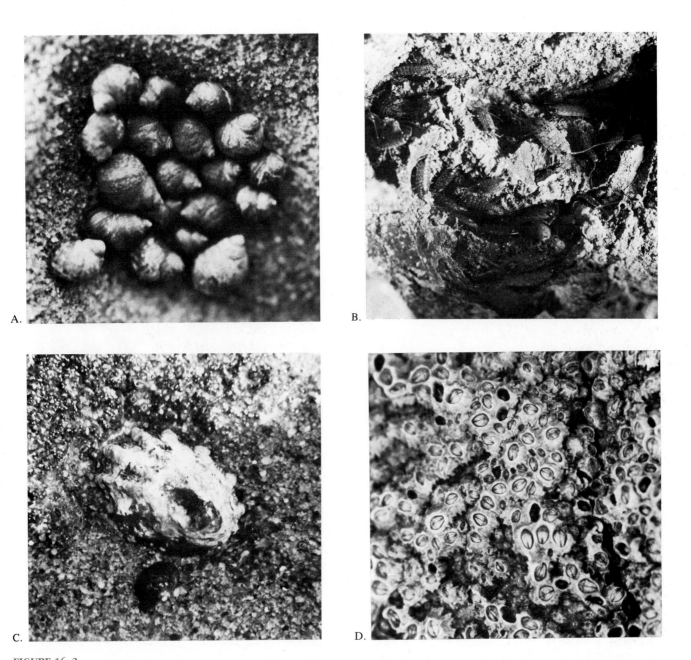

FIGURE 16–2

Animals of the Supralittoral and High Tide Zones. *A*, periwinkles *(Littorina)* nestled in a depression near the upper limit of the high-tide zone. This behavior helps reduce exposure to direct sunlight. *B*, rock lice *(Ligia)* on the roof of a sea cave at Laguna Beach, California. *C*, limpet *(Acmaea)*. *D*, buckshot barnacles *(Chthamalus)*. Photos *B* by John D. Roche, Jr., *C* by Greg West.

depression or discoloration on the rock, to wait out the drying period of the sunlit day.

High Tide Zone Unlike the periwinkles, some of which give birth to live young, the barnacles cannot abandon the sea for dry land. Their limit is the high tide shoreline. These hermaphroditic animals fertilize their neighbor's eggs by inserting a penis into a neighboring barnacle. The fertilized eggs hatch within the housing of

the barnacle and are released into the water as *nauplii* larvae; these eventually transform into *cypris* larvae and abandon the planktonic existence to attach to the rocky shore. In addition to needing water for the planktonic existence of their larval forms, barnacles also are confined to their marine existence through their method of feeding.

At the highest intertidal level exists the most prominent inhabitant of this zone, the buckshot barnacle (fig-

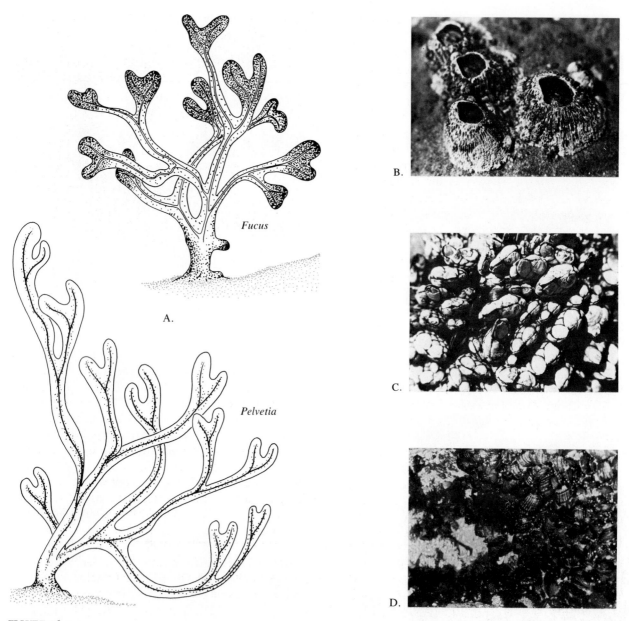

FIGURE 16–3
Life of the High- and Middle-Tide Zones. *A,* rock weeds *Pelvetia* and *Fucus.* Sharing the middle-tide zone with these rock weeds are *B,* acorn barnacles (*Balanus*); *C,* goose barnacles (*Pollicipes*); and *D,* a mussel bed. Photos *B* and *C* by Greg West.

ure 16–2), so called because of its small size, which seldom exceeds 0.5 cm (0.2 in.). This unusual arthropod attaches itself to the rocks by the back of its neck, secretes a protective calcium carbonate housing that looks like a tiny volcano, and extends its "legs," modified into featherlike cirri, to comb microscopic plankton from the water.

The most conspicuous plants in the high tide zone are members of the genus *Fucus* in colder latitudes and *Pelvetia* in warmer latitudes (figure 16–3). Both have thick cell walls to reduce water loss during periods of low tide. These plants continue to flourish in the middle tide zone, where the variety of life forms is much greater than in the upper tide zone. Not only does the variety increase, but the total biomass is also much greater; there is, therefore, a greater competition for rock space by sessile forms.

Middle Tide Zone On a clean rocky shore, the *Fucus* or *Pelvetia* establishes itself before the sessile animal forms; however, it seems doomed once barnacles or mussels move in. Although the smaller acorn barnacles usually found in the upper tide zone may occasionally be found in the upper reaches of the middle tide zone, larger species of the genus *Balanus* (figure 16–3) are the more common acorn barnacles of the middle tide zone. The barnacle most characteristic of this zone is the goose barnacle, genus *Pollicipes* (figure 16–3), with two valves composed of numerous plates that house the animal, which attaches itself to the rock surface by a long muscular neck. The total length of this structure may exceed 10 cm (4 in.). Even more successful than barnacles in the competition for space in the middle tide zone are the various species of mussels belonging to the genera *Mytilus* and *Modiolus;* they will attach to bare rock, algae, or barnacles. Settling on these surfaces as larval forms, they attach themselves by tough proteinous *byssal* threads. Mussels extend their foot and press it against the hard surface. They then secrete a fluid that flows down a groove on the surface of the foot to the attachment surface. An attachment thread is soon formed when the ocean water causes the fluid to harden. This process is repeated until many threads hold the mussel firmly in place. Given time, mussels would eventually overgrow all other sessile forms.

Life in the intertidal zone is not simple. Mussels are fed upon by predators such as sea stars and carnivorous snails. Two common genera of sea stars are the *Pisaster* and *Asterias.*

Sea stars feed by attaching tube feet that radiate from the mouth at the center of the bottom side of the animal. There are five double rows located in an oral groove beneath each ray or arm. To get to the mussel tissue protected by calcium carbonate bivalve covering, sea stars exert a continuous pull on the valves by alternating the use of the tube feet such that some are always pulling while others are resting. The mussel is eventually fatigued and can no longer hold the valves closed. The valves open ever so slightly, and the sea star everts its stomach, slips it through the crack, and digests the mussel without having to take it into its mouth (figure 16–4).

Boring carnivorous snails feed on both barnacles and mussels. Their preferred food, barnacles, can be consumed after the snail envelops it with its fleshy foot and forces the valves open. A narcotic secretion called *purpurin* may also be secreted to make the opening of the valves easier. To feed on the mussels, the snail must use its radula to drill a hole through the shell, through which it can rasp away the flesh. The more widely distributed carnivorous snails are the dog-whelk, *Nucella,* and the similar Pacific coast species *Thais lamellosa.*

Even though mussels have survival problems in the intertidal zone, the dominant feature of the middle tidal zone along most rocky coasts is a mussel bed that thickens toward the bottom until it reaches an abrupt bottom

FIGURE 16–4
Sea Star Feeding on a Clam.

limit. This may be so pronounced that it appears as though an invisible horizontal plane has prevented the mussels from growing below this depth. Protruding from the mussel bed will be numerous goose barnacles, and concentrated in the lower levels of the bed will be the sea stars browsing on the mussels. Less conspicuous forms common to the mussel beds are varieties of algae, hydroids, worms, clams, and crustaceans.

Where the rock surface flattens out within the middle tidal zone, tide pools trapping water as the tide ebbs serve as interesting microecosystems containing a wide variety of organisms. The largest member of this community will often be the sedentary relative of the jellyfish, the sea anemone. Shaped like a sack, anemones have a flat pedal disc that provides a suction attachment to the rock surface. Directed upward, the open end of the sack has the only opening to the gut cavity, the mouth, surrounded by rows of tentacles. The tentacles are covered with *cnidoblast* cells which contain a stinging threadlike *nematocyst* that automatically is released to penetrate any organism that brushes against the tentacles. Unlike their pelagic relatives, the tide pool coelenterates do not have nematocysts long enough to penetrate human skin and therefore are not dangerous (figure 16–5).

Swimming in the tide pools are a variety of small fish. The opaleye reaches length of 5 cm (2 in.) in the pools and grows larger in deeper water offshore. It is easily identified by the tiny white spots on its back on either side of the dorsal fin. The woolly sculpin is a permanent resident of tide pools and grows longer than 15 cm (6 in.). It is covered with hairlike cirri and usually is resting on the bottom or walking on its large pectoral fins.

FIGURE 16–5
Sea Anemone. *A,* sea anemone. *B,*
this diagram shows the basic body
plan of an anemone and a detail of
the stinging mechanism.

A.

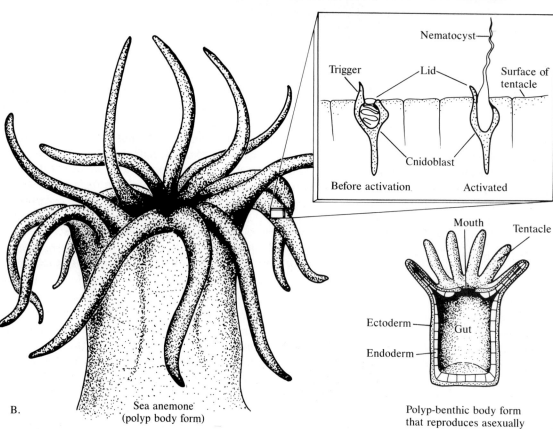

B.   Sea anemone
(polyp body form)

Polyp-benthic body form
that reproduces asexually

Readily identified by its blunt head and continuous dorsal fin is the 15-cm (6-in.) rockpool blenny (figure 16–6).

The most interesting inhabitant of the tide pools is the hermit crab, *Pagurus* sp. With a well-armored pair of claws and cephalothorax, hermit crabs have a soft abdomen that is protected by some hard protective container, usually a snail shell. The abdomen has even developed a curl to the right to make it fit properly into snail shells. Once in the snail shell, the crab can close

off the opening with its large claws. However, one primary enemy is the octopus, which is not bothered by the shell that the hermit crab has adopted. The octopus can cover the crab with the web surrounding its mouth and release a poisonous secretion that paralyzes the crab so that it can be pulled from the shell. Male hermit crabs may be seen fighting over females or carrying them around by grasping the edge of the shell in which the female lives. The male is waiting for the female to molt so he may deposit his sperm on her abdomen. She will later use the sperm to fertilize her eggs (figure 16–7).

In tide pools near the lower limit of the middle tide zone, sea urchins (figure 16–7), with their five-toothed structure centered on the bottom side of a hard spherical covering that supports many spines, may be found feeding on algae. The hard protective covering is called a *test* and is composed of fused calcium carbonate plates that are perforated to allow tube feet and dermal gills to pass through.

Low Tide Zone Unlike the upper and middle tide zones, the low tide zone is dominated by plants rather than animals. A diverse community of animals exist, but they are less obvious because they are hidden by the great variety of seaweeds and surf grass, *Phylospadix.* The encrusting red algae, *Lithothamnion,* which is also seen in middle zone tide pools, becomes very abundant in the lower tide pools. In temperate latitudes, moderate-sized red and brown algae provide a drooping canopy beneath which much of the animal life is found (figure 16–8).

In open tide pools of the low tide zone, the larger sea anemones will also be found, reaching diameters greater than 25 cm (10 in.). Nudibranchs, snails with no shells and exposed gills on their backs, are common in quiet tide pools. They feed on sea anemones, sponges, and hydroids, delicate relatives of the sea anemones. Actually, almost all the phyla of animals are represented in the low tide zone, where life is less rigorous than in the zones previously discussed. There is no great distinction between the fauna and flora immediately above and below the spring low tide shoreline.

Scampering from crevice to crevice and in and out of tide pools across the full range of the intertidal zone are various species of shore crabs (figure 16–9). Scavengers that help keep the shore clean, these creatures can spend long periods of time out of the sea. Shore crabs spend most of the daylight hours hiding in cracks or beneath overhangs. At night they engage in most of their eating activity, shoveling in algae as rapidly as they can tear it from the rock surface with their *chelae,* large front claws. Shore crabs need to return to ocean water only periodically to wet their gills, and females also must deposit the eggs they carry under their abdomens when it is time for the eggs to hatch. A characteristic they share with many crustaceans is that of *autotomy,* the ability to shed appendages involuntarily along a pre-

FIGURE 16–7

Hermit Crab and Sea Urchin. *A,* hermit crab (*Plagurus*) with its protective snail shell home. *B,* hermit crab out of its shell; note the soft curved abdomen. *C,* sea urchins burrowed into the bottom of a lower middle-tide zone tide pool. *D,* the bottom side of this dead sea urchin shows the mouth at the center and the hard calcium carbonate test beneath the spiny exterior.

determined breaking plane as necessary for survival. It is therefore not unusual to see them with a claw missing or much smaller than its paired counterpart. With each molt the new limb grows until it ultimately becomes full-sized.

SEDIMENT-COVERED SHORE

The Sediment

Rocky shores, where the wave energy level is relatively high, are usually dominated by erosion, and sediments, if found at all, are composed of large cobbles or boulders. The sediment-covered shore described next includes what are commonly called *beaches, salt marshes,* or *mud flats* that represent lower energy environments. As the energy level decreases, the particle size and sediment slope decrease, and sediment stability increases; the energy level refers to the strength of the longshore current. Another characteristic of sediments closely related to particle size is *permeability,* the ease with which fluid can move through the deposit. Because of the strength of cohesive attraction between flat clay-sized

FIGURE 16–8
Algal Canopy of the Low-Tide Zone Exposed during an Extremely Low Tide.

particles and the greater angularity of small silt- and fine sand-sized particles compared with the more rounded shape of coarser sand-sized particles, the permeability of sediments increases with increased particle size. Therefore, breaking waves can more readily percolate down through coarse sands and replace the oxygen used up by animals buried in the sediment. This readily available oxygen also enhances the process of bacterial decomposition of dead tissue.

Because the finer-grained sediment of mud deposits cannot be flushed of organic matter by percolating ocean water, the top 1 cm (0.4 in.) of the deposit traps a high concentration of organic matter. Below this surface layer organic matter concentration decreases until, at a depth of about 50 cm (20 in.), it reaches negligible levels in most mud deposits. At the surface, the sediment may be gray or light brown because of the availability of oxygen and the existence of aerobic bacteria that decompose the organic matter. Below the depth of oxygen penetration, anaerobic bacteria and fungi con-

tinue the breakdown of organic matter, producing the black color and rotten egg smell of hydrogen sulfide gas. Instead of using oxygen to carry on their respiration, these anaerobes use sulfate ions (SO_4^{2-}), which they reduce to hydrogen sulfide (H_2S) in the process of decomposing the organic matter.

Life in the Sediment

Although sediment-covered shores rarely sustain the relatively large red and brown algae characteristic of the rocky shore, fragments of these plants broken from their rocky substrate by heavy wave action are commonly piled high on beaches during storms. Thus, some of the suspension feeders that filter plankton from the clear water of the rocky shore are replaced by deposit feeders that eat detritus on many sandy beaches and ingest sediment in mud flats.

Life on and in the sediment requires very different adaptations than on the rocky coast. The sandy beach supports fewer species than the rocky shore, and mud flats fewer still; however, the number of individuals may be as high. In the low tide zone of some beaches and on mud flats, as many as 5000–8000 burrowing clams have been counted in $1m^2$ (10.8 ft^2). Burrowing is the most successful adaptation for life in the sediment-covered shore, so life is less visible. By burrowing a few centimeters beneath the surface, organisms can find a stable environment where they are not bothered by fluctuations of temperature and salinity and the threat of desiccation.

Burrowing in the sediment does not prevent animals from suspension feeding or straining plankton from the clear water above the sediment as do many of the animals of the rocky shore. By the use of various techniques, the water above the sediment surface can be stripped of its plankton content by buried animals. Deposit feeders collect debris from the sediment surface and eat their way through the sediment, digesting its

A.

B.

C.

FIGURE 16–9
A, striped shore crab, *Pachygrapsus,* backed into a rock crevasse. *B,* dorsal view of a preserved *Pachygrapsus. C,* preserved female *Pachygrapsus* showing eggs under the broad abdomen.

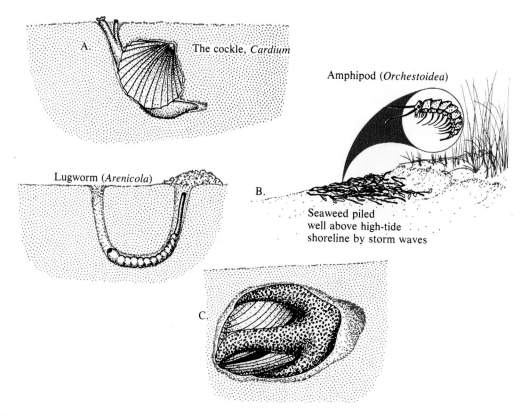

FIGURE 16-10

Modes of Feeding along the Sediment-Covered Shore. *A,* suspension feeding. This method is used by clams that bury themselves in sediment and extend siphons through the surface to pump in overlying water. They feed by filtering plankton and other organic matter that is suspended in the water. *B,* deposit feeding. Some deposit feeders such as the segmented worm *Arenicola* feed by ingesting the sediment and extracting organic matter contained within it. Others, like the amphipod *Orchestoidea,* feed on more concentrated deposits of organic mater, detritus, found on the sediment surface. *C,* carnivorous feeding. The sand star, *Astropecten,* cannot climb rocks like its sea star relatives, but it can burrow rapidly into the sand, where it feeds voraciously on crustaceans, mollusks, worms, and other echinoderms.

organic content and excreting inorganic particles. Burrowing carnivores chemically sense the presence of their prey in the darkness of the infaunal habitat (figure 16–10).

The Sandy Beach

Bivalve Mollusks Of all the animals that live in the sediment-covered shore, the bivalve mollusks or *pelecypods,* are the best adapted. The greatest variety of clams are found buried beneath the low tide region of sandy beaches; their numbers decrease as the sands become muddier. The bivalve mollusks possess a soft body, a portion of which is called the *mantle*; this part secretes the calcium carbonate lateral valves that hinge together on the dorsal region of the body. The foot extends anteriorly to dig into the sediment and pull the valves and posterior siphons down after it. The tip of the foot is then pumped full of blood to make it swell

and anchor the mollusk, while retractor muscles running the length of the foot contract and pull the animal down. How deep the bivalve can bury itself depends on the length of its siphons; they must reach above the sediment surface to pull in water from which plankton will be filtered. Oxygen is also extracted in the gill chamber before the water is expelled through the excurrent siphon. Undigestible particulate matter is forced back out the incurrent siphon periodically by quick muscular contractions (figure 16–11).

Annelid Worms A variety of *annelids,* or segmented worms, are also well adapted for life in the sediment. Most common of the sand worms is *Arenicola* sp., or lug-worm. It lives in a U-shaped burrow, the walls of which are strengthened by mucus. With its finely hooked parapods gripping the sides of the tube, it moves forward and backward. The worm moves forward to feed and extends its proboscis up into the head shaft

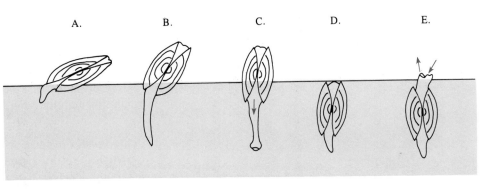

FIGURE 16–11

How a Clam Burrows. Clams exposed at the sediment surface will quickly burrow into the sediment by (*A*) extending their point-shaped foot into the sediment and (*B*) forcing the foot deeper into the sediment and using this increasing leverage to bring the exposed, shell-clad body toward vertical. When the foot has penetrated deep enough, (*C*) a bulbous anchor forms at the tip, and a quick muscular contraction pulls the entire animal into the sediment (*D*). The siphon is then pushed up above the sediment to pump in water from which the clam will extract food and oxygen (*E*).

of the burrow to loosen sand with quick pulsing movements. A cone-shaped depression forms at the surface over the head end of the burrow as sand continually slides into the burrow and is ingested by the worm. As the sand passes through the digestive tract, the organic content is digested, and the processed sand is deposited as castings at the surface surrounding the opening or openings to the tail shaft (figure 16–10).

To help prevent asphyxiation resulting from using up the oxygen in the water that fills the tube when the tide is out, lugworms have red, hemoglobin-rich blood that is able to store oxygen. They rarely are forced to leave their tubes, since each rising of the tide will result in wave action transporting a new supply of sediment to be ingested. Even reproduction is achieved without leaving the burrow. During the month of October, all the worms release their sex cells into the overlying water, where fertilization occurs, and the larval forms produced settle to the bottom to perpetuate this population of highly specialized sand-dwelling worms.

Crustaceans Staying high on the beach and feeding on kelp cast up by storm or high tide waves, numerous amphipod crustaceans called *beach hoppers* are found. They are known to jump distances more than 2 m (6.6 ft). A common genus is *Orchestoidea*, which usually range from 2 to 3 cm (0.8–1.2 in.) in length. Laterally flattened, they usually spend the day buried in the sand or hidden in the kelp on which they feed. They become active at night and may form large clouds above the masses of seaweed on which they are feeding. They are given the name *amphipods* (both legs) because they have one set of appendages specialized for burrowing and another set that serve well for swimming (figure 16–10).

A larger crustacean that may be harder to find on sandy beaches is a member of the variety of sand crabs belonging to the genera *Blepharipoda, Emerita,* and

Lepidopa. Ranging in length from 2.5 to 8 cm (1–3 in.), they move up and down the beach near the shoreline. Burying their bodies into the sand and leaving their long curved V-shaped antennae pointing up the beach slope, these little crabs filter food particles out of the water (figure 16–12).

Echinoderms Echinoderms are represented in the beach deposits by the sand star, *Astropecten,* and heart urchins. Members of the genus *Astropecten* are called sand stars because they are well adapted to prey on invertebrates buried in the low tide region of sandy beaches. With five rays and a smooth dorsal surface, sand stars have sturdy spines along the margins of the rays and tube-feet tapered to a point to aid them in moving through the sediment (figure 16–10*C*).

The most widely distributed heart urchins belong to the genus *Echinocardium.* More flattened and elongated than the sea urchins of the rocky shore, the heart urchins live buried in the sand near the low tide line. Most of the tube-feet emerge through a five-petaled flower-

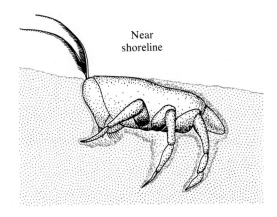

Near shoreline

FIGURE 16–12

Sand Crab, *Emerita.*

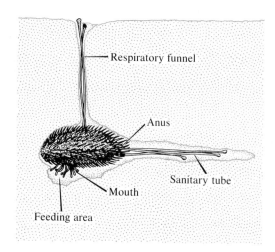

FIGURE 16–13
Heart Urchin, *Echinocardium*.

like arrangement of holes on the dorsal surface of the test. Some tube-feet are very long, forming and maintaining a respiratory tube connecting the chamber in which the urchin lives to the beach surface some 15 cm (6 in.) above the animal. There are a few more tube-feet surrounding the ventral mouth; they gather sand grains into the mouth, where the coating of organic matter is scraped off and ingested. The posterior anus is surrounded by enlarged spines and long tube-feet that create a sanitary tube about 12 cm (5 in.) long. The feces are passed into this tube so that the main chamber is not fouled. The walls of the chamber, respiratory tube, and sanitary tube are supported by mucus applied by the tube-feet (figure 16–13). The available food at a given location is usually used up in half an hour, so the

heart urchins are constantly on the move creating new chambers and associated features. Each move involves a distance of about 15 cm (6 in.), about twice the body length, and is accomplished by the short, fine spines that serve primarily as a means of transporting the heart urchin through the sand.

Meiofauna   Living in the spaces between sediment particles are tiny organisms ranging in size from 0.1 to 2 mm (0.004–0.08 in) in length (figure 16–14). They feed primarily on bacteria removed from the surface of sediment particles. The meiofauna population, composed primarily of polychaetes, mollusks, arthropods, and nematodes, are found in sediment from the intertidal zone to the deep ocean trenches.

Intertidal Zonation   Although it requires more investigation to observe it, there is a characteristic faunal distribution across the intertidal range of sediment-covered shores similar to that observed on rocky shores. Although the species of animals found in each of the corresponding intertidal zones are appropriately different, the distribution of the life forms across the intertidal rocky and sediment-covered shores have two characteristics in common: The maximum number of species and the greatest biomass are found near the low tide shoreline, and both decrease toward the high tide shoreline. This zonation, shown in figure 16–15, is best developed on steeply sloping, coarse sand beaches and is less readily identified on the gentler sloping, fine sand beaches. The tiny clay-sized particles of the mud flat produce a deposit with essentially no slope, which eliminates the possibility of zonation in this protected, low-energy environment.

A.

B.

C.

FIGURE 16–14
Meiofauna. Scanning electron micrographs of *A,* nematode head (×804). The projections and pit on right side are possibly sensory structures. *B,* amphipod (×20). This organism builds a burrow of cemented sand grains. *C,* polychaete (×55) with its proboscis extended. Courtesy of Howard J. Spero, University of South Carolina.

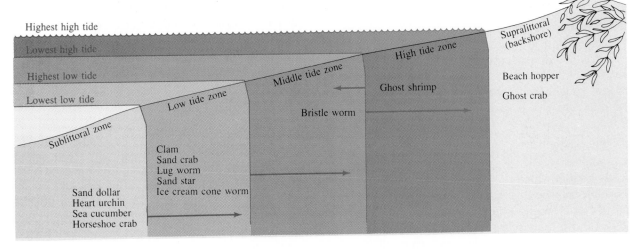

FIGURE 16–15

Intertidal Zonation on the Sediment-Covered Shore. Zonation is best displayed on coarse sand beaches with steep slopes. As the sediment becomes finer and the beach slope decreases, zonation becomes less distinct. It disappears entirely on flat mud flats.

The Mud Flat

Two widely distributed plants associated with the mud flat are eel-grass, *Zostera,* and turtle-grass, *Thalassia,* which occupies the low tide zone and adjacent shallow sublittoral regions bordering the flats. Numerous openings at the surface of mud flats attest to a large population of bivalve mollusks and other invertebrates.

Quite visible and interesting inhabitants of the mud flats are the fiddler crabs, *Uca,* living in burrows that may be more than 1 m (3 ft) deep. Relatives of the shore crabs, they usually measure no more than 2 cm (0.8 in.) across the carapace. Fiddler crabs get their name from one out-sized claw, up to 4 cm (1.6 in.) long, on the male. This large claw is waved around in such a manner the crab seems to be fiddling. The females have two normal-sized claws. The large claw of the male is used to court females and fight off other competing males. They feed by extracting organic matter from the mud and show a degree of dislike for the sea by jumping into their burrows and closing the mud door when the tide rises to cover the mud flat (figure 16–16).

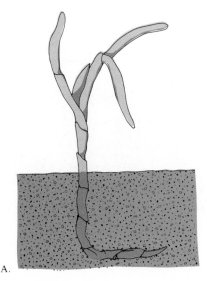

FIGURE 16–16

Life of the Mud Flat. *A, Zostera,* eelgrass. *B, Uca,* fiddler crab.

THE SHALLOW OFFSHORE OCEAN FLOOR

Extending from the spring low-tide shoreline to the seaward edge of the continental shelf is an environment that is mainly sediment-covered, although bare rock exposures may be found locally near shore. This region conforms roughly with what has been described as the sublittoral zone.

The Rocky Bottom (Sublittoral)

A rocky bottom within the shallow inner sublittoral region will usually be covered with algae. Along the Pacific coast of North America, the giant bladder kelp, *Macrocystis,* is known to attach to rocks at depths as great as 30 m (100 ft) if the water is clear enough to allow sunlight to support plant growth at this depth. The giant bladder kelp and another fast-growing kelp, *Nereocystis,* often form bands of kelp forest along the Pacific coast. Smaller tufts of red and brown algae are found on the bottom and living epiphytically (one plant living nonparasitically on another plant) on the kelp fronds. Epifauna commonly found growing along with the algal

tufts on the fronds are hydroids and bryozoan colonies. All these smaller life forms serve as food for many of the animals found living within the kelp forest community. Nudibranch are important as predators on the hydroids and bryozoan colonies. Surprisingly, very few animals feed directly on the living kelp plant. Among those that do are the large sea hare, *Aplysia,* and sea urchins (figure 16–17).

Varieties of barnacles, annelid worms, and bivalve mollusks are known to bore into the rocky substrate. Among the bivalves are the date mussels, which bear a striking resemblance to date fruit and can be found embedded in soft shale and sandstone. They possess no apparent means of mechanically boring into rock and probably secrete a dissolving chemical. Piddocks belonging to the genus *Pholas* are able to bore into most sedimentary rock after settling onto it in larval form. Developing a filelike rasping surface on the anterior end of their valves, they mechanically create a small hole that extends deeper into the rock and increases in diameter as the animal grows. Housed inside the rock, the piddock is protected from predators. It feeds by extending siphons into the water and filtering food much like the

FIGURE 16–17

Life of the Sublittoral Zone. *A, Macrocystis. B,* sea hare (*Aplysia*) in a kelp forest. *C,* sea urchin, *Strongylocentrotus.*

clams that bury themselves in soft sediment (figure 16–18).

Large crustaceans called *lobsters* are common to rocky bottoms. They are a somewhat varied group with robust external skeletons. The chelaless, or spiny lobsters, are named for their spiny carapace and are characterized by two very large and spiny antennae that have noise-making devices near their base. The genus *Palinurus* is found at depths greater than 20 m (65 ft) along the coast of Europe, reaches lengths up to 50 cm (20 in.), and is a delicacy. The species *Panulirus argus* is found in the Caribbean and is sometimes observed to migrate single file for unknown reasons over distances of several kilometers. *Panulirus interruptus* is the spiny lobster of the American west coast. Other related species are found in the Juan Fernandez Islands and along the coasts of New Zealand, Australia, and South Africa. All the spiny lobsters are taken for food, but none are as highly regarded as the so-called true lobsters belonging to the genus *Homarus*. Although they are scavengers like their spiny relatives, the true lobsters—which include the American lobster, *Homarus americanus*—also feed on live animals, including mollusks, crustaceans, and members of their own species. Migrating into deep water during winter and returning to nearshore waters in summer, the American lobster is nocturnal. Related species are found along the European coast from Norway to the Mediterranean and the coast of South Africa, and the American lobster is found from Labrador to Cape Hatteras (figure 16–19).

Oysters are sessile bivalve mollusks found in estuarine environments. They prefer locations where there is a steady flow of clean water to provide plankton and oxygen. Having great commercial importance through-

SPINY LOBSTER (*Panulirus* sp.)

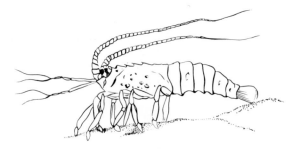

AMERICAN LOBSTER (*Homarus americanus*)

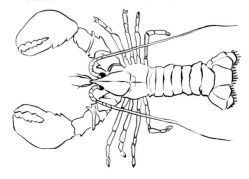

FIGURE 16–19
Spiny and American Lobsters.

out the world, they have been closely studied. Oyster beds consist of the empty shells of generation after generation cemented to rock bottom or one another, with the living generation on top. Each female produces many millions of eggs each year, which become planktonic larvae when fertilized. After a few weeks as plankton, the larvae settle to attach themselves to the bottom. The larvae prefer live oyster shells, dead oyster shells, and rock, in that order, as a substrate for attachment. Only a small percentage of the larval forms survive to the attachment stage, and after attachment, many oysters perish as a result of predation and competition for food and space before reaching maturity in 1 to 5 years. A great variety of attached forms of sponges, coelenterates, annelids, crustaceans, and tunicates are common to oyster beds; like oysters, most live by filtering plankton from the water. Oysters serve as prey for a variety of sea stars, fishes, crabs, and boring snails that drill through the shell and rasp away the soft tissue of the oyster (figure 16–20).

Coral Reefs

Although coral distribution throughout the ocean floor is rather general, coral accumulations that might be classified as reefs are restricted to the warmer water regions where the average monthly temperature exceeds 18°C (64°F) throughout the year (figure 16–21). Such temper-

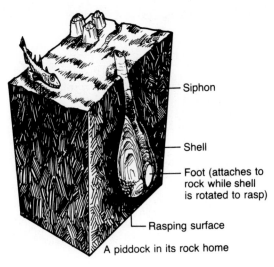

Siphon

Shell

Foot (attaches to rock while shell is rotated to rasp)

Rasping surface

A piddock in its rock home

FIGURE 16–18
Rock Borer, Piddock.

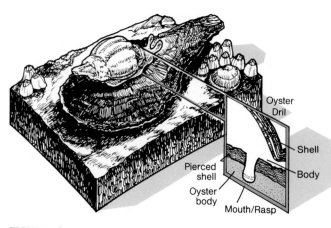

FIGURE 16–20

An Oyster Drill Feeding on an Oyster.

ature conditions are found primarily between the tropics, although reefs grow to latitudes approaching 35°N and S on the western margins of ocean basins where warm water masses move into the high latitude areas and raise average temperatures.

Not only coral but algae, mollusks, and foraminifera make important contributions to the reef structure. Reef corals are found only in the shallow water surrounding the continents and islands of the open ocean because they are *hermatypic*. This means they have a symbiotic relationship with the green alga, *Zooxanthella,* that lives within the tissue of the reef-building coral. The alga contributes to the calcification capability of corals by extracting carbon dioxide from the animals' body fluids, thus increasing the concentration of the carbonate ion needed for the precipitation of calcium carbonate. This relationship is representative of *mutualism,* in which both parties benefit. The coral contribute to the rela-

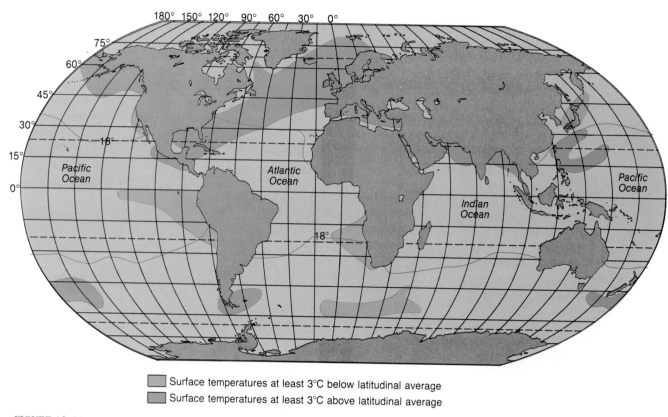

☐ Surface temperatures at least 3°C below latitudinal average
☐ Surface temperatures at least 3°C above latitudinal average

FIGURE 16–21

Coral Reef Distribution. Minimum water temperatures of 18°C (64°F) in the surface waters of the Northern and Southern hemispheres occur in February and August, respectively. Coral reef development is restricted to the low-latitude area between the two 18°C (64°F) temperature lines shown on the map. Observe that in each ocean basin, the width of the coral reef belt is wider on the west side because here boundary currents carry warm water away from the equator, while cold currents converge on the equator on the east side of the oceans.

tionship by providing a supply of nutrients to the *Zooxanthella.*

Reef growth also requires that the water have a relatively normal salinity and that it be free from particulate matter. Therefore, we see very little coral reef growth near the mouths of large rivers that lower the salinity and carry large quantities of suspended material that would choke the reef colony. Maximum reef growth will occur where a constant flow of water carries food to the waiting tentacles of the polyps, which extrude and feed primarily at night. Such conditions are optimum in areas where water circulation through current activity is relatively good.

Reefs are widely developed throughout the Pacific and Indian oceans on the flanks of the many volcanic islands rising above the ocean surface from the deep-ocean floor. Consisting primarily of the skeletal remains of hermatypic corals and calcareous algae, reefs occur around the margins of all islands and continents where the proper conditions for their existence are found.

Coral reefs actually contain up to three times as much plant as animal biomass. The *Zooxanthellae* account for less than 5 percent of the reef's plant mass; most of the rest is filamentous green algae. The concentration of phytoplankton in the waters of coral reefs is many times greater than in the remarkably nonproductive tropical waters of the open ocean. This is because the reef structure serves to concentrate and hold the nutrients required for plants to prosper.

Because of changes in wave energy, salinity, water depth, temperature, and other less obvious factors, there is a well-developed vertical and horizontal zonation of the reef. These zones are readily identified by the assemblages of plant and animal life found associated with them (figure 16–22).

The greatest depth to which active coral growth extends is 150 m (492 ft), below which there is not enough sunlight to support hermatypic mutualism. Water motions are not great at these depths, so relatively delicate varieties can live on the outer slope of the reef from 150 (492 ft) to about 50 m (165 ft). From 50 m (165 ft) to about 20 m (66 ft), the strength of water motion associated with breaking waves increases on the side of the reef facing into the prevailing current flow. Correspondingly, the mass of coral growth and the strength of the coral structure supporting it increases toward the top of this zone. The genus *Acropora* is representative of the coral found here. The buttress zone extends from 20 m (66 ft) to the low tide line; within it more massive varieties of coral and encrusting algae such as the red algae, *Lithothamnion,* withstand the crashing waves. The waves cut surge channels across the *Lithothamnion* ridge, which is inhabited by only a few animals—such as snails, limpets, and the slate pencil urchin, *Heterocentrotus*—that can withstand the constant beating of the surf. The surge channels extend down the reef slope as debris channels that carry the products of wave erosion. Small reef fish find protection from larger

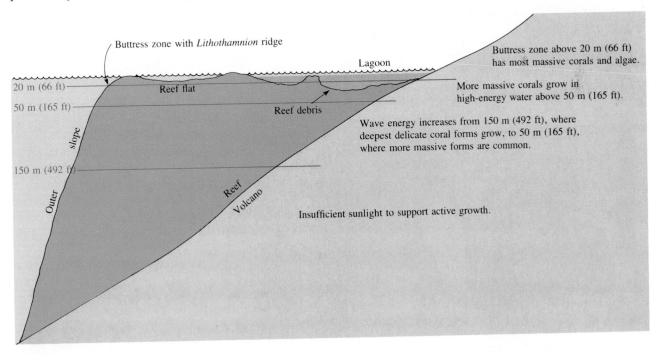

Buttress zone with *Lithothamnion* ridge

Lagoon

Buttress zone above 20 m (66 ft) has most massive corals and algae.

20 m (66 ft)

Reef flat

More massive corals grow in high-energy water above 50 m (165 ft).

50 m (165 ft)

Reef debris

Wave energy increases from 150 m (492 ft), where deepest delicate coral forms grow, to 50 m (165 ft), where more massive forms are common.

slope

150 m (492 ft)

Reef

Outer

Volcano

Insufficient sunlight to support active growth.

FIGURE 16–22
Coral Reef Zonation.

predators such as sharks, barracuda, and jacks in the debris grooves cutting between the buttress ridges. The reef flat extends across the lagoons of atolls and to the shoreline of the island protected by fringing or barrier reefs. Here the reef may be under a few centimeters to a few meters of water at low tide. A variety of beautiful reef fish swim through this shallow water (figure 16–23). The sand of reef debris and foraminifera tests fills in the deeper holes and provides a home for sea cucumbers, worms, and a variety of mollusks. In the protected water behind the *Lithothamnion* ridge lagoonal reefs may form, where species of *Porites* and *Acropora* grow into beautiful large colonies. Gorgonian coral, anemones, crustaceans, mollusks, and echinoderms of great variety also will be found in the lagoon reef.

An obvious example of *commensalism* (one organism benefits while the other is unaffected) is observed in the behavior of the shrimpfish that swims head down among the long slender spines of the reef sea urchins.

A.

B.

FIGURE 16–23
Fishes Found in the Coral Reef Lagoon. *A,* moray eel. *B,* Moorish idol.

FIGURE 16–24
Clown Fish and Sea Anemone.

The spines are a significant deterrent to any predator, and the sea urchin is neither hindered nor aided by the presence of the little fish. The clown fish receives a similar protection by swimming among the tentacles of two species of sea anemones. This relationship is believed to be mutual, since the anemones benefit by the clown fish serving as bait to draw other fish within reach of anemone tentacles and actually carrying food to them (figure 16–24). A variety of cleaner shrimp and fishes that set up cleaning stations on coral reefs are known to be essential to the well-being of many of the larger fishes on the reef because they remove parasites and infected tissue.

Sediment-Covered Bottom

Most of the continental shelf is covered with sand or mud deposits. These may be replaced by pebbles, cobbles, and boulders of glacial origin in the higher latitudes. Generally similar physical environments may extend over greater distances than in the intertidal region. Although one might expect to find finer sediment near the seaward edge of the shelf than near the shore, this is not always the case. It is not uncommon to find relict beach deposits left behind during periods of glacial advance near the edge of the continental shelf. During these periods when sea level was 150 m (492 ft) or more lower than today, beaches were created that remain uncovered by later deposits of sediment. In general, light availability, temperature, and water movement decrease with increasing depth across the continental shelf.

Below the low-tide line, some of the same species described as burrowing near the low-tide line in the intertidal zone will be found in the sediment. The burrowers usually form similar communities. Although the

PLATE 27

The identity of this fish, photographed at a depth of 130 ft, remains one of the sea's secrets.

A

Narwhals are unusual toothed whales that may reach 6 m (20 ft) in length. Narwhals possess no dorsal fin and live in the Arctic. They are rarely found south of 65°N latitude. The male has a modified tooth that projects forward from the upper lip as a long tusk with a left-hand spiral. Living in schools of 15 to 20, narwhals feed on cuttlefish, fish, and crustaceans. These narwhals were found in the Beaufort Sea north of Alaska.

B

No creature in the sea captures the human imagination like the whales, huge mammals that returned to the sea about 60 million years ago.

Two suborders are present in today's oceans, the toothed whales (Odontoceti) and baleen whales (Mysticeti). Represented on this plate are members of each

PLATE 28

group drawn to scale.

The toothed whales—bottle-nosed dolphin, narwhal, killer whale, and sperm whale—are active predators and may track down their prey by use of echolocation.

The baleen whales—humpback whale, right whale, and blue whale—leisurely fill their mouths with water

ATLANTIC RIGHT WHALE (Eubalaena glacialis)
This cold-water whale was the first recorded target of whaling by Basque seamen during the middle ages. Similar species are found near Japan and in high southern latitudes. Length is up to 18 m (60 ft).

SPERM WHALE (Physeter catodon)
Found mostly in tropical waters, this deep diver has a huge snout that contains a large amount of oil. Length of male is to 19 m (63 ft). Length of female is to 10.5 m (35 ft).

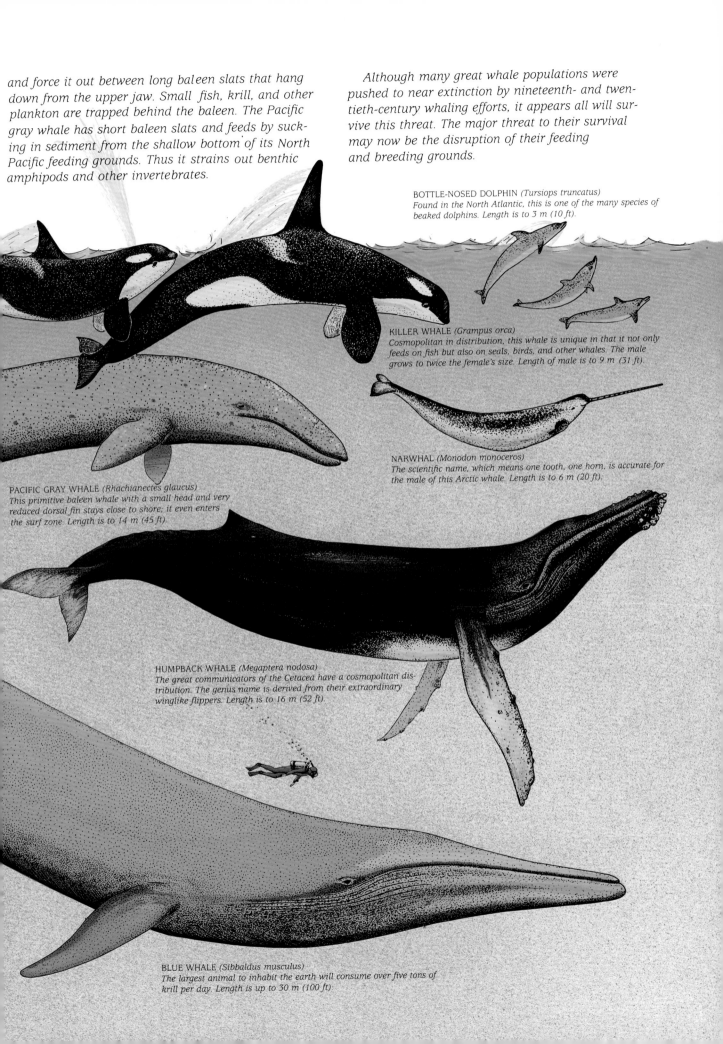

and force it out between long baleen slats that hang
down from the upper jaw. Small fish, krill, and other
plankton are trapped behind the baleen. The Pacific
gray whale has short baleen slats and feeds by suck-
ing in sediment from the shallow bottom of its North
Pacific feeding grounds. Thus it strains out benthic
amphipods and other invertebrates.

Although many great whale populations were
pushed to near extinction by nineteenth- and twen-
tieth-century whaling efforts, it appears all will survive this threat. The major threat to their survival
may now be the disruption of their feeding
and breeding grounds.

BOTTLE-NOSED DOLPHIN (Tursiops truncatus)
Found in the North Atlantic, this is one of the many species of
beaked dolphins. Length is to 3 m (10 ft).

KILLER WHALE (Grampus orca)
Cosmopolitan in distribution, this whale is unique in that it not only
feeds on fish but also on seals, birds, and other whales. The male
grows to twice the female's size. Length of male is to 9 m (31 ft).

NARWHAL (Monodon monoceros)
The scientific name, which means one tooth, one horn, is accurate for
the male of this Arctic whale. Length is to 6 m (20 ft).

PACIFIC GRAY WHALE (Rhachianectes glaucus)
This primitive baleen whale with a small head and very
reduced dorsal fin stays close to shore; it even enters
the surf zone. Length is to 14 m (45 ft).

HUMPBACK WHALE (Megaptera nodosa)
The great communicators of the Cetacea have a cosmopolitan distribution. The genus name is derived from their extraordinary
winglike flippers. Length is to 16 m (52 ft).

BLUE WHALE (Sibbaldus musculus)
The largest animal to inhabit the earth will consume over five tons of
krill per day. Length is up to 30 m (100 ft).

A

B

Killer whales frolic in the icy waters of the north Pacific. Leaping whale displays the bold white undermarkings. The erect triangular dorsal fin of a mature male projects from the water. Male dorsal fin may reach a height of more than 1.5 m (5 ft).

The remains of thirteen porpoises and fourteen seals were found in the stomach of one killer whale and give evidence of the animal's voracious appetite.

A view of the underside of the throat of a Bryde's whale reveals the numerous longitudinal grooves that allow the throat to expand to immense proportions when they gulp large volumes of water that contain their primary foods, krill and small fish. The Bryde's whale is a smaller, close relative of the blue whale. This Bryde's whale is "side feeding" on shrimplike krill near the north end of the Sea of Cortez.

PLATE 29

species may differ throughout the world's temperate and cold water environments, they are similar. The sandy environments are dominated by bivalve mollusks with a few annelid worms. As the sediment becomes muddier, the bivalves become less varied, and the community contains a greater variety of burrowing annelids. In tropical waters there are many different invertebrate inhabitants of the sediment-covered bottom, and rarely is the community dominated by one or even a few species.

Interesting inhabitants of some shores are horseshoe crabs, arthropods that have not changed substantially from fossilized forms as old as 360 million years. They spend most of their lives in the sandy and muddy offshore waters where they feed on worms, mollusks, and algae. Their common name comes from their horseshoe-shaped carapace, although an equally striking feature is their swordlike tail that aids in moving through the sediment. Of the three living species, one is found along the Atlantic coast of North America, and the others inhabit the western Pacific shores. The crabs come ashore to breed. The female deposits the eggs in a shallow nest near the spring high-tide shoreline on sandy beaches, and the male fertilizes them. With the next spring tide 2 weeks later, the eggs are ready to hatch. The newly hatched crabs quickly enter the ocean, where they undergo a series of molts before achieving adult form (figure 16–25).

DEEP-OCEAN FLOOR

The Physical Environment

Beyond the edge of the continental shelf, the bathyal zone extends to a depth of 4000 m and includes all the continental slope in most parts of the ocean. The conti-

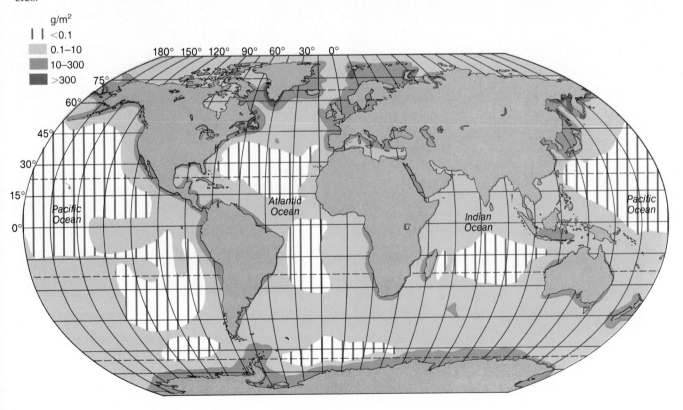

FIGURE 16–26
Benthic Biomass. Distribution of benthic macrofauna (g/fresh weight/m²) in the world ocean. (Redrawn from Zenkevitch et al., 1971.)

nental rise and deep-ocean floor—between 4000 and 6000 m (13,100–19,700 ft) deep, the abyssal zone—represents over 80 percent of the benthic environment. The hadal zone of the deep trenches exceeding 6000 m (19,700 ft) represents the most restricted environment within the benthic habitat. No plant life exists throughout this deep-ocean region. Light is present in only the lowest concentrations above 1000 m (3280 ft) and is absent below this depth. Everywhere the temperature is low, rarely exceeding 3°C (37.4°F) and falling as low as −1.8°C (30.7°F) in the high latitudes. Pressure exceeds 200 atmospheres on the oceanic ridges, ranges between 300 and 500 atmospheres on the deep-ocean abyssal plains, and climbs to over 1000 atmospheres in the deepest trenches. (The pressure is 1 atmosphere at the ocean surface and increases by one atmosphere for each 10 m (33 ft) of increased depth.) Most of the deep-ocean floor is covered by at least a thin layer of sediment ranging from the mudlike abyssal clay deposits of the abyssal plains and deep trenches, through the oozes of the oceanic ridges and rises, to some coarse sediment de-

posited as turbidites on the continental rise. Occasionally, near the crest of the oceanic ridges and rises and down the slopes of seamounts and oceanic islands, basaltic ocean crust is the substrate.

The Deep Fauna

Although the deep-ocean communities have not been extensively sampled, it is becoming apparent that the diversity of deep-ocean benthos is much greater than initially thought. The overriding limiting factor for life on the deep-ocean floor must be the availability of food. Little is known of the nature and availability of food in the deep ocean, but some is apparently cycled through a food chain that exists in the deep environments. Claude Zobell has estimated that 30 to 40 percent of the organic matter reaching the ocean floor is first used by benthic bacteria at the ocean-sediment interface. The bacteria are then ingested by deposit-feeding infauna. The smaller bacteria-feeding infauna are then preyed upon by larger bottom-dwellers. The carcasses of large

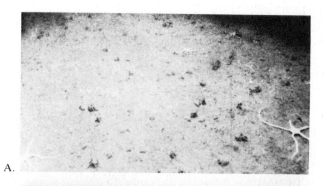

A.

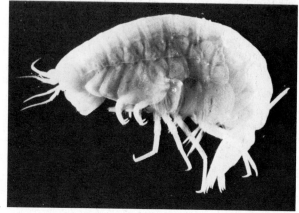

B.

C.

FIGURE 16–27

Deep-Ocean Benthos. *A,* although serpent stars such as these are frequently found on the deep-ocean floor, their variety and abundance are considerably less there than in shallow water. *B,* close view of a preserved amphipod recovered in a trap from the floor of the Marianas Trench below 10,400 m (6.46 mi). *C,* amphipods are more abundant on the deep-ocean floor. Here they are attracted to bait in the Philippine Trench at a depth of 9600 m (5.96 mi). (Photos from A. A. Yayanos, courtesy Scripps Institution of Oceanograph, University of California, San Diego.)

FIGURE 16–28

A, locations of two well-studied hydrothermal vent communities in the Pacific Ocean and two more recently discovered communities in the Atlantic Ocean. The first discovered is in the Galapagos Rift. *B, Alvin* explores a hydrothermal vent field and biocommunity typical of that observed on the East Pacific Rise (see also Plate 2).

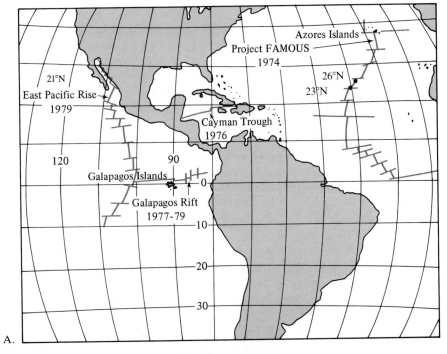

A.

B.

fish and mammals may provide food on the bottom, as they would sink rather rapidly once they have died.

The pattern of distribution of benthic biomass is similar to that of photosynthetic primary productivity in the surface waters of the world ocean (compare figure 16–26 with figure 14–8). This suggests that benthic productivity is largely supported by the photosynthetic productivity of the ocean's surface waters.

Although most of the marine phyla may be represented in the deep-ocean fauna, there appears to be a characteristic deep-sea fauna that begins to appear on the lower continental slope. However, the organisms are also frequently found on the continental shelves of high-latitude regions—an indication that temperature is a major factor limiting their distribution. In general, deep-water fauna is differentiated from the shallow-water benthos by a decrease in the variety of sea stars and mollusks and an increase in the diversity of other echinoderms, isopod and amphipod crustaceans, polychaete worms, pycnogonids, and pogonophorans (figure 16–27).

An exciting discovery in 1977 was the very specialized biological community found in association with warm water vents of the Galapagos Rift spreading center. The first view of the community was observed in pictures taken by the camera sled ANGUS (Acoustically Navigated Underwater Survey System). It showed hundreds of large clams and mussels at depths below 2500 m (8200 ft). The immediate question was, How do they survive? Clams and mussels normally feed by filtering phytoplankton out of the water, and since there could be no phytoplankton at this depth, an alternate food supply, such as bacteria, had to be present. Subsequent investigations found that the organic content of the water, in the form of bacteria, was 500 times greater than normal for this bottom environment and 4 times greater than in productive surface waters. The bacteria are chemosynthetic forms that obtain their energy for the synthesis of organic matter by oxidizing hydrogen sulfide contained in the hydrothermal emissions (figure 16–28).

Continued study of this area and another discovered off the tip of Baja California in 1979 revealed a recognizable zonation of organisms from the 2°C (35.6°F) ambient temperature on the outer edge of the community toward the warmer water near the hydrothermal springs. Spaghetti-like acorn worms and benthic siphonophores were found on the outer edge. The siphonophores are relatives of the Portuguese man-of-war and resemble dandelions gone to seed. At intermediate distances, clams, mussels small anemones, and serpulid worms were found. Thriving in the warm 10–15°C (50–59°F) water near the vent openings were tube-dwelling pogonophoran worms that, like the clams and mussels,

have symbiotic sulfur oxidizing bacteria living in their tissue that directly supply their energy needs. The worms have red hemoglobin-filled tentacles that probably have a great capacity for dissolving oxygen extracted from the surrounding water. Living in plasticlike tubes over 2 m (6 ft) long, these worms have no mouth or gut. Similar communities have been found in association with subsequently discovered hydrothermal vents.

During 1985 and 1986, active hydrothermal vents were found on the Mid-Atlantic Ridge. Associated with two of these vent areas at 23°N and 26°N are biocommunities that differ from those in the Pacific Ocean. The predominant animals at the Atlantic Ocean vents are particulate-feeding shrimp (figure 16–29).

A relict vent area found on the Galapagos spreading center indicates such communities may have a relatively short life span. This inactive vent was identified by an accumulation of dead clams. It appears that when the vent becomes inactive and the hydrogen sulfide that serves as the source of energy for the community is no longer available, the community dies.

FIGURE 16–29

Hydrothermal Vent Communities. Swarm of particulate-feeding shrimp, the predominant animals observed at hydrothermal vents near 23°N and 26°N on the Mid-Atlantic Ridge. This swarm was photographed at 26°N by Peter A. Rona, NOAA.

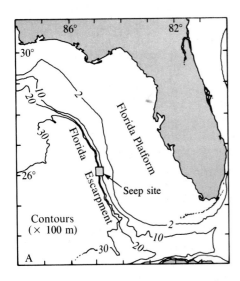

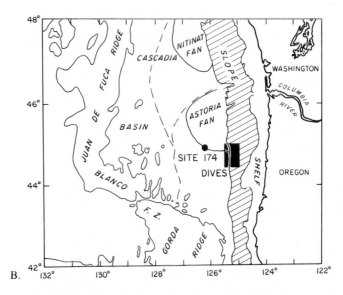

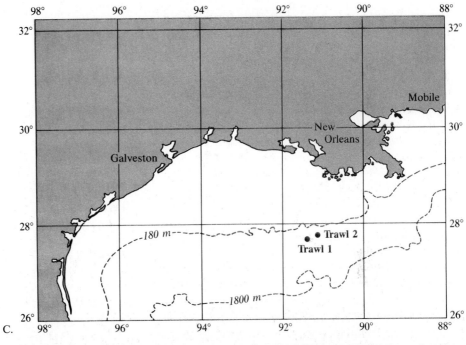

FIGURE 16–30

Low-Temperature Seep Biocommunities *A,* location of hypersaline seep biocommunity at base of Florida Escarpment. *B,* location of vent communities off the coast of Oregon. *C,* locations of trawls that recovered members of hydrocarbon seep biocommunities in the Gulf of Mexico. *D,* at the contact of the abyssal Gulf of Mexico clay-rich sediments and the fractured limestones of the Florida Escarpment, a biological community similar to those of the hydrothermal vents is found. The most obvious organisms are mussels with many small white gastropods on them. At lower right are tube worms heavily encrusted with hydrozoans and a galatheid crab. (Photo by Charles K. Paull, Scripps Institute of Oceanography, University of California, San Diego.)

During 1984, three new settings for spring-supported deep-sea benthic communities similar to those of the hydrothermal vents were discovered. Figure 16–30 shows the locations of the initial discoveries of these communities.

At the base of the broad Florida Escarpment, which marks the seaward edge of the Florida Platform, hot, salty water that percolates down through the limestone platform bedrock is trapped by deep-sea clay deposits that abut the escarpment. This water is dammed up behind the clay and seeps out of the fractured limestone at the sediment surface at a depth of 3266 m (10,700 ft) in the Gulf of Mexico. This salty, hydrogen sulfide-rich water supports a community of sulfur-oxidizing bacteria that support a biological community that is in many respects similar to those of the hydrothermal vents (figure 16–30D).

Also in the Gulf of Mexico, at depths between 600 and 700 m (1968 and 2296 ft), oil and gas seeps support communities in which large tube worms and clams appear to be supported by sulfur-oxidizing chemosynthetic bacteria, while a mussel population depends on symbiotic methane using bacteria, as do mussels at the base of the Florida Escarpment.

Finally, in the subduction zone of the Juan de Fuca Plate off the coast of Oregon, the third community was found. At a depth of 2036 m (6678 ft) along the base of the continental slope, methane-rich water seeps out of sediments that are being squeezed by compression. This community may derive most of its energy from methane using bacterial symbionts. Subsequent investigations have identified similar communities associated with the Japan Trench and the Peru-Chile Trench.

SUMMARY

Over 98 percent of the 200,000 species of marine animals live on the ocean floor. Most of these live on the continental shelf, with only 400-odd species found in deep-ocean trenches. The importance of temperature on the development of species diversity is reflected in the fact that there are three times as many species of benthos along the European coast warmed by the Gulf Stream than along a similar length of North American coast cooled by the Labrador Current.

Because of tidal motions, the **littoral** (intertidal) zone can be divided into the **high-tide zone** (mostly dry), **middle-tide zone** (equally wet and dry), and the **low-tide zone** (mostly wet). The littoral is bounded by the **supralittoral** (covered only by storm waves) and the **sublittoral**, which extends below the low tide shoreline.

The supralittoral zone along rocky coasts is characterized by the presence of the **periwinkle snail**, the **rock louse**, and the **limpet**. (Rock lice stay well above the high-tide line, but snails and limpets may extend down into the high-tide zone.)

Within the intertidal zone are found the **barnacles**, crustaceans that attach themselves to the rocks and comb plankton from the water. The barnacles most characteristic of the high-tide zone are the tiny buckshot barnacles. The most conspicuous plants of the high tide zone are *Fucus* along colder shores and *Pelvetia* in warmer latitudes. These plants become more abundant in the middle tide zone, and the diversity and abundance of the flora and fauna in general increase toward the lower intertidal zone.

Larger species of acorn barnacles are found in the middle tide zone. A common assemblage of rocky middle tide zones includes the goose barnacle, the **mussels**, and **sea stars**. Joining the sea stars as predators on the barnacles and mussels are carnivorous snails.

Tide pools within the middle tide zone commonly house **sea anemones**, fishes such as the **opaleye**, **woolly sculpin**, and **rockpool blenny**, and **hermit crabs**. At the lower limit of the middle tide zone, **sea urchins** become numerous.

The dominance of animals in the upper and middle tide zone is ended in the low-tide zone, where plants become dominant. The encrusting *Lithothamnion* red algae that is present in the middle tide zone becomes very abundant in tide pools. The low tide zone in temperate latitudes is characterized by a variety of moderately sized red and brown algae providing a drooping canopy for the animal life.

Scampering across the entire intertidal zone are the scavenging **shore crabs**.

The only sediments usually found along rocky shores are a few cobbles and boulders. Moving into more protected segments of the shore, lower levels of wave energy allow deposition of deposits of sand and mud. Due to the higher **permeability** of sand, sand deposits are usually well oxygenated, and mud deposits are **anaerobic** below a thin surface layer. Hydrogen sulfide (H_2S), a product of anaerobic decomposition of organic matter, gives mud deposits a smell of rotten eggs.

Compared to the rocky shore, the **diversity of species** is reduced on and in beach deposits and is quite restricted in mud deposits. This does not mean that the abundance of life is reduced. Although life is less visible, up to 8000 burrowing clams have been recovered from 1 m^2 (10.8 ft^2) of mud flat. **Suspension feeders** (filter feeders) characteristic of the rocky shore are still found in the sediment, but there is a great increase in the relative abundance of **deposit-feeders** that ingest sediment and detritus.

Bivalve mollusks are well suited for sediment-covered shores. Clams that burrow into the sediment pump water through siphons and extract food and oxygen. The **lug-worm** is a deposit-feeder that ingests sand and extracts whatever organic content is available.

Amphipods called *beach hoppers* can be found feeding on the kelp deposited high on the beach by storm waves. **Sand crabs** filter food from the water with their long, curved antennae while buried in the sand near the shoreline.

Echinoderms are represented in the sandy beach by the **sand star**, which feeds on buried invertebrates, and the **heart urchin**, which scrapes organic matter from sand grains.

As is true for the rocky shore, the diversity of species and abundance of life on the sediment-covered shore increases toward the low-tide shoreline. Although it is more difficult to observe on sediment-covered shores, the high-, middle- and low-tide zonation is also present. It is best developed on steep, coarse sand beaches and is probably missing on mud flats with little or no slope.

Although many burrowing forms inhabit the **mud flats**, the more visible life forms are the **eelgrass**, *Zostera,* and **turtle-grass**, *Thalassia,* found at its low tide margin and the **fiddler crabs**, *Uca,* which leave their burrows at low tide.

Attached to the **rocky sublittoral** bottom just beyond the shoreline is a band of algae including large **kelp** plants. Growing on the large fronds of the kelp are small varieties of algae, **hydroids**, and **bryozoans**. **Nudibranchs** feed on the small hydroids and bryozoans, and the **sea hare** and **sea urchin** feed on the kelp and other algae. **Piddocks** and **date mussels** bore into the rocky bottom. **Spiny lobsters** are common to rocky bottoms in the Caribbean and along the West Coast, and the **American lobster** is found from Labrador to Cape Hatteras.

Oyster beds found in estuarine environments consist of individuals that attach themselves to the bottom or the empty shells of previous generations. Living within these beds is a community consisting of suspension feeders such as **sponges, coelenterates, annelids, crustaceans**, and **tunicates**. Sea stars, fishes, crabs, and boring snails prey on the oysters.

Above a depth of 150 m (492 ft) in tropical waters, living coral reef can be found off the shores of islands and continents. Delicate varieties are found at 150 m (492 ft), and they become more massive near the surface, where wave energy is higher. The top 20 m (66 ft), the buttress zone, is reinforced with calcium carbonate deposited by algae such as *Lithothamnion,* producing the **Lithothamnion ridge**. Waves cut **surge channels** across the ridge, and **debris channels** extend down the reef front below the surge channels. Many varieties of **commensalism** and **mutualism** are found within the coral reef biological community.

On sediment-covered bottoms, the dominant form of animal life in shallow water is the burrowing bivalve mollusk. As water depth increases and sediment texture becomes finer, annelid worms become dominant. Off the Atlantic coast of North America and some western Pacific shores, species of **horseshoe crabs** that date back more than 360 million years feed on burrowing mollusks and annelids.

Although little is known of the **deep-ocean benthos**, it is clear that it is much more varied than previously thought. There seems to be a decreased variety of sea stars and mollusks and increased diversity of other **echinoderms, isopods, amphipods, polychaetes, pycnogonids**, and **pogonophorans**. Patterns of **benthic biomass distribution** are similar to patterns of biological productivity in surface waters, suggesting that benthos depend greatly on this productivity as a source of food.

An important discovery of the **hydrothermal vent communities** made in 1977 at the Galapagos Rift has shown that, at least locally, **chemosynthesis** is an important means of primary productivity in the ocean. Subsequent discoveries have identified hydrothermal vent biocommunities in the eastern and western Pacific Ocean and on the Mid-Atlantic Ridge. Similar communities have been found in association with lower-temperature sulfide- and methane-rich seeps in the Gulf of Mexico and in subduction zones.

QUESTIONS AND EXERCISES

1. Discuss the general distribution of life in the ocean. Include species diversity (variety of types of organisms) between the pelagic and benthic environments and within the benthic environment.

2. Diagram the intertidal zones of the rocky shore and list characteristic organisms of each zone.

3. Describe the dominant feature of the middle tide zone along rocky coasts, the mussel bed. Include a discussion of other organisms found in association with the mussels.

4. List and describe crabs found within the rocky shore intertidal zone.

5. Discuss how sediment stability and permeability of sandy and muddy shores differ.

6. Other than predation, discuss the two types of feeding styles that are characteristic of the rocky, sandy, and muddy shores. One of these feeding styles is rather well represented in all these environments; name it and give an example of an organism that uses it in each environment.

7. How does the diversity of species on sediment-covered shores compare with that of the rocky shore? Can you think of any reasons why this should be so? If you can, discuss them.

8. In which intertidal zone of a steeply sloping coarse sand beach would you find the following: clams, beach hoppers, ghost shrimp, sand crabs, and heart urchins?

9. Why do lug-worms not need to move frequently, yet heart urchins, which are also deposit feeders, must move about every half-hour? Consider the specific source of the sediment they ingest.

10. What relationships exist among intertidal zonation, sediment particle size, and beach slope?

11. Discuss the dominant species of kelp, their epifauna, and animals that feed on kelp in the Pacific coast kelp forest.

12. Where in the world are spiny lobsters and American lobsters (and their related species) found?

13. Discuss the preferred environment, reproduction, and threats to survival of larval and adult forms of oysters.

14. Describe the environment suited to development of coral reefs.

15. What physical factors contribute to vertical zonation of the seaward edge of a coral reef?

16. Why are horseshoe crabs called living fossils?

17. As one moves from the shoreline to the deep ocean floor, what changes in the physical nature of the ocean floor can be expected?

18. Describe the process Claude Zobell thinks may be important in distributing organic matter throughout the animal populations of the deep ocean floor.

19. What do you think is the most significant new knowledge gained from the discovery of the hydrothermal vent biological communities and similar communities associated with lower temperature seeps of water and hydrocarbons?

REFERENCES

Childress, J. J., Fisher, C. R., Brooks, J. M., Kennicutt, M. C. II, Bidigare, R., and Anderson, A. E. 1986. A methanotrophic marine molluscan (*Bivalvia, Mytilidae*) symbiosis: Mussels fueled by gas. *Science* 233:4770, 1306–08.

Fagerstrom, J. A. 1987. *The evolution of reef communities.* New York: John Wiley.

George, D., and George, J. 1979. *Marine life: An illustrated encyclopedia of invertebrates in the sea.* New York: Wiley-Interscience.

Hessler, R. R., Ingram, C. L., Yayanos, A. A., and Burnett, B. R. 1978. Scavenging amphipods from the floor of the Philippine Trench. *Deep-Sea Research* 25:1029–47.

MacGinitie, G. E., and MacGinitie, N. 1968. *Natural history of marine animals,* 2d ed. New York: McGraw-Hill.

Parsons, T. R., Takahashi, M., and Hargrave B. 1984. *Biological oceanographic processes,* 3d ed. New York: Pergamon Press.

Ricketts, E. F., Calvin, J., and Hedgpeth, J. 1968. *Between Pacific tides.* Stanford, Calif.: Stanford University Press.

Rona, P. A., Klinkhammer, G., Nelson, T. A., Trefry, J. H., and Elderfield, H. 1986. Black smokers, massive sulfides and vent biota at the Mid-Atlantic Ridge. *Nature* 321:6065, 33–37.

Thorson, G. 1971. *Life in the sea.* New York: McGraw-Hill.

Zenkevitch, L. A., Filatove, A., Belyaev, G. M., Lukyanove, T. S., and Suetove, I. A. 1971. Quantitative distribution of zoobenthos in the world ocean. *Bulletin der Moskauer Gen. der Naturforscher, Abt. Biol.* 76, 27–33.

SUGGESTED READING

Sea Frontiers

Bellomy, M. D. 1973. Blossoms in the sea. 19:1, 2–13.
An informative discussion of the variety, distribution, and life style of sea anemones.

Coleman, N. 1974. Shell-less molluscs. 20:6, 338–42.
A description of nudibranchs, gastropods without shells.

George, J. D. 1970. The curious bristle-worms. 16:5, 291–300.
The variety of worms belonging to the class Polychaeta of phylum Annelida is described. The discussion includes locomotion, feeding, and reproductive habits of the various members of the class.

Gibson, M. E. 1981. The plight of *Allopora.* 27:4, 211–18.
The reason the author believes the unusual California hydrocoral is headed for the endangered species list is the topic of this article.

Harman, A. 1984. CASM's deep dive finds chasm of life. 30:4, 240–47.
Recounts the findings of the Canadian-American Seamount expedition to the sea floor along the Juan de Fuca Ridge.

Heston, T. B. 1987. Living history: The horseshoe crab. 33:3, 195–99.
Describes the natural history of this arthropod that has been around for about 300,000,000 years.

Shinn, E. A. 1981. Time capsules in the sea. 27:6, 364–74.
The method by which geologists determine past climatic and environmental conditions by studying coral reefs is discussed.

Scientific American

Alkon, Daniel L. 1983. Learning in a marine snail. 249:1, 70–85.
Neural mechanisms of associative learning in the snail, *Hermissenda,* may be similar to those of the human brain.

Caldwell, R. L., and Dingle, H. 1976. Stomatopods. 234:1, 80–89.
Presents the ecology of these interesting crustaceans that have appendages specialized for spearing and smashing prey.

Cameron, James N. 1985. Molting in the blue crab. 252:5, 102–9.
New knowledge is offered concerning the chemistry of this process.

Feder, H. A. 1972. Escape responses in marine invertebrates. 227:1, 92–100.
Discusses the surprisingly rapid movements and other responses made by invertebrates to the presence of predators. Some interesting photographs accompany the text that describes the escape responses of limpets, snails, clams, scallops, sea urchins, and sea anemones.

Richardson, J. R. 1986. Brachiopods. 255:3, 100–107.
Common in the fossil record, brachiopods were thought to be less widely distributed in today's oceans. They appear to have a much wider distribution than previously thought.

Wicksten, M. K. 1980. Decorator crabs. 242:2, 146–57.
Describes how species of spider crabs use materials from their environment to camouflage themselves.

Yonge, C. M. 1975. Giant clams. 232:4, 96–105.
The distribution and general ecology of the tridacnid clams, some of which grow to lengths well in excess of 1 m, are investigated.

17
FOOD FROM THE SEA

A catch of Peruvian anchovy during the early 1960s when the fishery was undergoing rapid development. These small fish represented one of the world's major fisheries until the collapse of 1972. Photograph courtesy of FAO.

Since well before the beginning of recorded history, humans have used the sea as a source of food. Although food from the sea has in recent years provided only about 5 percent of the protein consumed by the world's population, this seemingly small percentage has meant the difference between starvation and an adequate diet for millions of the world's inhabitants.

In this chapter, the problems and prospects for the future of the world's natural marine fisheries and mariculture (marine aquaculture) industries will be discussed.

FISHERIES

It is important that we understand not only the feeding and other ecological relationships that exist among the ocean's organisms but the effects of our attempts to exploit the biological resources of the oceans as a source of human food. While we seek to increase the food resources we remove from the oceans, it is clear that the greatest negative impact humans have had on the ecology of the oceans is through overfishing.

It is very disturbing that the fishery of fish, shellfish, and seaweeds from the ocean has increased only from 59.7 metric tons (mt) in 1970 to 73.1 mt in 1984, the latest year for which complete information is available—despite a significantly increased fishery effort. It is also certain that the cost per unit of production is rapidly increasing (table 17–1).

Fishing peoples have always understood that fish must be permitted to reproduce if a fishery is to be maintained. However, it has become increasingly difficult to regulate fisheries in a manner that will assure sufficient reproduction to maintain the fishery. This has been primarily the result of the great demand that accompanies an increasing human population. Fisheries such as the anchovy, cod, flounder, haddock, herring, and sardine apparently are suffering from overfishing; that is, they are being harvested at a rate above the natural rate of production in such numbers that their population is driven below the size necessary to produce maximum yield. International efforts are now

TABLE 17-1
World Commercial Fish Catch, 1951–1984 (million metric tons).*

Year	Fresh Water	Peruvian Anchovy	Other Marine	Total Marine	Total
1951	2.6	NA	20.9	20.9	23.5
1952	2.8	NA	22.3	22.3	25.1
1953	3.0	NA	22.9	22.9	25.9
1954	3.2	NA	24.4	24.4	27.6
1955	3.4	NA	25.5	24.5	28.9
1956	3.5	.1	27.2	27.3	30.8
1957	3.9	.3	27.5	27.8	31.7
1958	4.5	.8	28.0	28.8	33.3
1959	5.1	2.0	29.8	31.8	36.9
1960	5.6	3.5	31.1	34.6	40.2
1961	5.7	5.3	32.6	37.9	43.6
1962	5.8	7.1	31.9	39.0	44.8
1963	5.9	7.2	33.5	40.7	46.6
1964	6.2	9.8	35.9	45.7	51.9
1965	7.0	7.7	38.5	46.2	53.2
1966	7.3	9.6	40.4	50.0	57.3
1967	7.2	10.5	42.7	53.2	60.4
1968	7.4	11.3	45.2	56.5	63.9
1969	7.6	9.7	45.4	55.1	62.7
1970	8.4	13.1	46.6	59.7	68.1
1971	9.0	11.2	48.3	59.5	68.5
1972	5.7	4.8	53.7	58.5	64.2
1973	5.8	1.7	55.3	57.0	62.8
1974	5.8	4.0	56.8	60.7	66.5
1975	6.2	3.3	57.0	60.2	66.4
1976	5.9	4.3	59.7	63.9	69.8
1977	6.1	.8	62.3	62.8	68.9
1978	5.8	1.4	63.3	64.6	70.4
1979	5.9	1.4	63.8	65.2	71.1
1980	6.2	.8	65.4	65.8	72.0
1981	6.6	1.5	66.7	68.2	74.8
1982	6.8	1.8	67.9	69.7	76.5
1983	7.2	.1	69.2	69.3	76.5
1984	9.7	.1	73.0	73.1	82.8

NA—Not available

Source: U.S. Department of Commerce, N.O.A.A. publication. Fisheries of the United States, 1984.

being made to provide worldwide ocean fisheries management.

Early History of Fisheries Assessment and Management in the United States

Federal research into the nature of fisheries has been significant since the founding of the U. S. Fish Commission in 1871. However, regulation of North American fisheries dates to the prohibition of seining for mackerel by the Plymouth Colony in 1684.

The first major U. S. fishing ground was Georges Bank, where a cod fishery was being conducted by 1721 (figure 17–1). Mackerel and halibut fisheries were added by 1831. Although cod held up under heavy fishing, halibut began to decline in 1850. The *purse seine* was introduced out of Gloucester in 1825, and by 1850 the mackerel fishery began to decline. The mackerel, which salted well and were fished only during spring and summer, were protected in 1887 by the U. S. Congress. They prohibited the landing of purse seined mackerel from March to May. This action did not improve the state of the mackerel population, and the salt mackerel fishery continued to decline. Within a few years, it collapsed.

This early failure in fishery management is but one in a series of failures. The details may vary, but the results are always similar. Of all the human endeavors, no activity has been so consistently fruitless as fisheries management.

Fisheries and Marine Biology

Over the years, huge fluctuations in fish stocks that were not clearly related to fishing pressures convinced many involved in fisheries management that the study of fisheries ecology was needed.

Recruitment, the addition of young adult fish to the fishery, was found to depend on survival at critical stages in the life of a fish. For eggs and larvae, the mortality rate is very high. Mortality decreases in juvenile fish and reaches its lowest level in adults.

Larval survival After the larvae use up their yolk sacs, they must have appropriate food available or die. For instance, anchovy larvae require at least 20 phytoplankton cells per milliliter of water within 2 ½ days, and the minimum cell size must be at least 40 μm (0.00016 in.) in diameter. Another factor that can influence survival of fish larvae is competition for available food from other larval fishes. Information is gradually becoming available on these factors for plaice, herring, sardines, and anchovy.

Juvenile survival Once a fish survives the larval stage, it still must survive a juvenile existence before it can be recruited into the fishery. Juvenile fish may die from predation, disease, parasitism, genetic defects, and the effects of pollution. Availability of adequate food is still important because rapid growth rate is the best insurance against predators. If the juvenile fish grows faster than its predator (and under good conditions, it does),

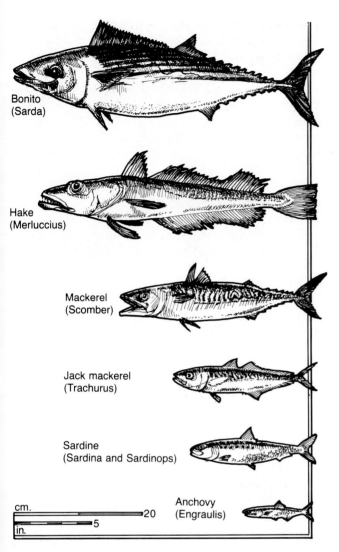

Bonito
(Sarda)

Hake
(Merluccius)

Mackerel
(Scomber)

Jack mackerel
(Trachurus)

Sardine
(Sardina and Sardinops)

Anchovy
(Engraulis)

cm.
20
5
in.

FIGURE 17–5
Fishes Common to All Four Subtropical Eastern Boundary Current Upwelling Ecosystems. (Courtesy of National Marine Fishery Service.)

in knowledge. Dr. Paul Smith of the National Marine Fisheries Service is participating in the study, and he believes that observing the variables that occur in the four areas over a 5-year span will allow as much new knowledge to be gained as would be obtained from a 20-year study of only one upwelling area.

New Technologies Useful in Studying Fisheries Ecology

The use of *side-directed sonar* has become increasingly effective in assessing the masses of populations within the ecosystem associated with fisheries. When these signals are reflected off schools of fish, squid, or other animals, the return signals can be analyzed to determine the volume and mass of the aggregations. Careful anal-

ysis of the frequencies that return with the greater resonance can tell investigators the species and even the age of organisms in the school.

Using a version of this sonar equipment attached to the manned submersible *Johnson Sea Link,* investigators added significantly to the knowledge of the ecosystem that supports the important fisheries of Georges Bank off the coast of New England.

It had long been a problem to explain how the benthic community of Georges Bank could support the large demersal fish (fish that live on or near the bottom) and squid fisheries there. The benthic and demersal zooplankton communities did not appear to be large enough to support these fisheries.

Sonar investigations in the submarine canyons along the southeast flank of Georges Bank (figure 17–1) revealed that large masses of the krill *Meganyctiphanes norvegica* (figure 17–6) are present on or near the bottom day and night. In the 50m (165 ft) adjacent to the bottom, densities of krill average 813 animals per cubic meter. It is known that these krill are an important component of the diet of squid, cod, flounder, haddock, hake, pollock, and redfish. It is clear that if these fishes descend into the canyons and deep waters surrounding Georges Bank to feed on this previously unknown food supply, the krill can easily provide the needed support for the existence of the large fisheries represented by these animals.

Fishermen and Marine Science

Such activities as we have just discussed fall into the realm of the fisheries scientist. The fisherman is much less concerned with these findings. For instance, fishermen have always fished where the greatest concentrations of salable fish could be found. They began by finding fish in the nearshore waters. Technology made it possible for them to go greater distances to catch fish and get them back to market in good enough shape to sell. The distant-water fleets of the 1960s and 1970s steadily increased the world catch. Now that the world fishery has leveled off at more than 70 million metric tons, the economy of the distant-water fleets is failing. They are becoming the victims of their own success—a common occurrence. As fishery scientists learn more about fishery ecology, this information will be of great interest to those assigned the task of managing the fisheries, but this increased knowledge will probably continue to be of little interest to fishermen, who are primarily concerned with finding and catching the fish. Although many of those involved in fishing management are pessimistic about those who exploit the fisheries ever developing a significant interest in fishery ecology, let us hope they are wrong.

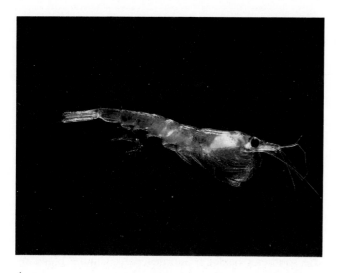

A.

B.

FIGURE 17–6

High-Density Near-Bottom Krill Aggregations in Submarine Canyons off Georges Bank. *A,* close-up view of the krill *Meganyctiphanes norvegica.* It is about 3.8 cm (1.5 in.) long. (Photo by Pam Blades-Eckelbarger, Harbor Branch Oceanographic Institution, Inc.) *B,* A large aggregation of the krill observed in the submarine canyons off the southeast flank of Georges Bank (see figure 17–1). (Photo by Marsh Youngbluth, Harbor Branch Oceanographic Institution, Inc.)

The Future of Fisheries Management

A disturbing trend in fisheries management developed during the recent past that reduced the priority of preserving the ecology of the fishery in making management decisions: Increased emphasis was given to social benefits such as occupational opportunity. Since it is the ecosystem that makes the social benefits possible, it was surely shortsighted to relegate the welfare of the ecosystem to any position less than the number one priority in fisheries management. The fact that this philosophy is being abandoned will surely increase the likelihood of successful fisheries management in the future.

With all the effort being directed at gathering data upon which to base the management of fisheries, it is still often said that the fisheries of the world are unmanaged. One reason for this assessment of fisheries management is the method commonly used to control catches. Most regulating bodies now set a *total allowable catch (TAC)* for each species. However, with the fishing methods in use, many species are recovered as a "by catch" along with the fish at which the fishery is primarily directed. To cope with this inevitable occurrence, a total allowable catch is established to cover all species affected by the fishery. This value is always less than the total of catch allowances for all the species that were set on an individual species basis. The setting of these total allowable catch values is in itself difficult but not nearly so difficult as assuring accurate reporting of catches.

Fisheries scientists are an optimistic lot, and in spite of the dismal record of management in the past, most believe they have at least begun to appreciate the nature of the problem. What they need now is experimental evidence that they are correct in their understanding. This evidence may well be 40 or 50 years in coming. The success of the future will require large-scale studies involving both physical oceanographers and marine biologists in an ecosystem approach to understanding fisheries. Studies such as that of the eastern boundary current system upwellings previously discussed will be the types of studies fishery scientists will be conducting in the future. Although the world catch may not be much larger than it is today, the technology and knowledge that will be available to fisheries managers of the future will likely make it possible to set and enforce catch limits that will assure the sustained well being of most fisheries.

Regulation

The effects of inadequate management can be seen in the history of the fisheries in the northwest Atlantic. In this area regulated by the International Commission for the Northwest Atlantic Fisheries, the fishing capacity of the international fleet increased 500 percent from 1966 to 1976; the total catch, however, rose by only 15 percent. This indicated a significant decrease in the catch

equator. This southbound surface flow of warm equatorial water seldom reaches more than 2° or 3° south of the equator, except during the summer, when it reaches its maximum strength. Beneath the surface current is the southward-flowing Peru Undercurrent.

Biological Conditions

Biologically, the significance of the cold water and nutrients provided by upwelling is great. During the years 1964–1970 fully 20 percent of the world fishery was represented by anchoveta taken off the coast of Peru.

The food web starts with diatoms and other phytoplankton that support a vast zooplankton population of copepods, arrowworms, fish larvae, and other small animals. Energy is passed on through a series of predators, but most stops with the anchoveta, which once attained a biomass of 20 million metric tons. Natural predators that feed on the anchoveta are marine birds—boobies, cormorants, and pelicans—and large fishes and squid.

A Natural Hazard

Periodically during the summer, the southeasterly trade winds weaken. The northward-flowing current and upwelling slows down, and the Peru Countercurrent may move as far as 1000 km (621 mi) south of the equator, laying down a warm low-salinity layer over the coastal water. Since this event commonly occurs around Christmas, it is called El Niño, The Child.

The increase in water temperature and decrease in nutrients brought about by the slowdown or complete halt of upwelling cause a severe decrease in the anchoveta population, the remnants of which seek deeper, cooler water. Marine birds that feed primarily on the anchoveta and produce the enormous guano (excrement rich in phosphorus and nitrogen compounds) deposits on the islands off the Peruvian coast have their populations cut severely by the decrease in the anchoveta.

In 1972 a severe El Niño occurred. The total biomass of anchoveta decreased from an estimated 20 million mt in 1971 to an estimated 2 million mt in 1973. But before blaming this decrease solely on the natural event of El Niño, with which the anchoveta have contended for thousands of years, another factor must be evaluated.

An Artificial Hazard

The Peruvian anchoveta fishery began to develop into a major industry in 1957. In 1960 the Instituto del Mar del Peru was established with the aid of the United Nations to study the fishery and recommend a management program to the government. It was hoped that early study would prevent the fishery from going the way of the Japan herring and California sardine fisheries that had collapsed as a result of overexploitation.

Biologists estimated that the maximum annual take that the anchoveta could sustain was 10 million metric tons. Figure 17–10B shows that this limit was exceeded in 1968, 1970, and 1971. The 1970 catch of 12.3 million metric tons probably reached 14 million metric tons if processing losses and spoilage were included.

Dr. Paul Smith, who is studying the anchoveta fishery problem for the National Marine Fisheries Service in the Southwest Fisheries Center at La Jolla, California, believes, as do other biologists who are familiar with the problem, that the yearly anchoveta catches may have been much larger than the official figures indicate. This belief is based on the fact that many catches contained large numbers of *pelladilla,* which are young anchoveta that are low in protein. This underreporting would have resulted when buyers reduced the total tonnage of a catch they purchased by a percentage determined by the proportion of the catch that was accounted for by high-water-content and low-oil-content pelladilla. For example, a 100-mt catch may have been recorded as a 70-mt catch because it contained a large number of pelladilla.

Dr. Smith thinks the crisis may have resulted from a combination of the effects of El Niño and the industry's lack of understanding of the anchoveta fishery. This is reflected in the fact that the January 1972 allotment was set at 1.2 million metric tons. This figure was reached easily by the industry with about half the expected effort. Because the fishing vessels were able to meet their allotment so easily without going far from shore, the fishery was thought to have developed handsomely, and the February allotment was raised to 1.8 million metric tons. Much more effort was required to meet this allotment, and in March the fishery collapsed. What had happened?

Apparently, El Niño, or the Peru Countercurrent, had come south in January, but it had remained offshore. The anchovy were concentrated in the nearshore portion of the Peru Current that had not been affected (figure 17–10A). In January the nets were filled easily near shore, and it was natural to think the high concentration of anchovy extended over the normal fishery area as well. This one event cannot explain the collapse of the fishery, but it is a symptom common to fisheries that are poorly understood.

It may be that even with upwelling, ammonia (NH_3) from decomposition of anchoveta excrement is required to support enough phytoplankton to allow the anchoveta population to expand. Since anchoveta populations are now small and scattered, they may swim through great expanses of water without adding much nutrient enrichment. When the population was larger, the whole

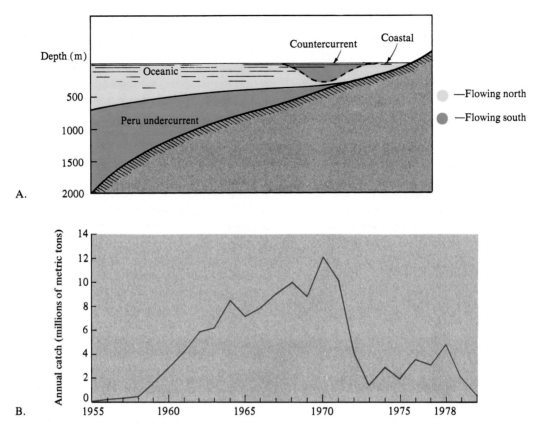

FIGURE 17–10

The Peruvian Anchoveta Fishery—The Loss of a Fishery. *A,* as the Peru Current flows north along the Peruvian coast, it is periodically split into oceanic and coastal branches by the Peru Countercurrent. The Peru Countercurrent is a warm water mass that moves south from the equator during the Southern Hemisphere summer season. Although it usually moves south no more than 2° or 3° below the equator, it periodically moves well down the Peruvian coast in an event called the El Niño. An active fish meal industry based on anchoveta made Peru, for a while, the world's leading fishing nation. *B,* annual anchoveta catch from 1955 through 1980. After a peak catch of 12.3 million mt in 1970, the fishery rapidly declined and experienced a major collapse in 1972.

region received it. It may require a long time, with a gradual increase in the population, until this nutrient supplement is made available to the phytoplankton throughout the region.

The anchoveta are gradually being replaced by sardines, a more valuable variety of fish. Whereas the entire anchoveta catch was converted to fish meal and exported as poultry and hog feed, the sardine fishery will be directed at producing canned and frozen products for direct human consumption.

With anchoveta fishing banned since 1980 and an emphasis on species conservation in Peru's future fishing activity, the anchoveta fish-meal fishery may be viable in the future. However, it will never reach previous levels. Although it is much smaller in biomass, the sardine fishery may eventually develop into a major source of revenue due to the higher value of the products that can be produced from it. To improve the present state of affairs, the Ministry of Fisheries will have to conduct a

prudent program of fishery assessment and management.

MARICULTURE

Marine aquaculture, or *mariculture,* has been conducted for years throughout the Far East, making major contributions to the available food supply in this part of the world. Most attempts at commercial mariculture in the United States have not been successful because most of the projects were based on intensive production tainted with overambition and impatience.

Organisms chosen for mariculture should be popular marine products that command a high price, are easy and inexpensive to grow, and reach marketable size within a year or less. Candidates should be hardy and resistant to disease and parasites. They should also be capable of feeding with a high degree of growth per unit of intake. Severe economic problems can result if the chosen organism is not able to reproduce or cannot be brought to sexual maturity in captivity.

Algae

Algae, in the form of certain seaweeds, are grown and marketed as luxury foods. The life cycle of most of these algae is very complicated, and most of the successful operations are conducted in Japan, where spore-producing plants are cultivated in laboratories. When the spore-producing plants release their spores into the water, local growers come to the laboratories to dip their nets, ropes, or whatever devices they use for attachment, into the water and allow the spores to attach themselves to these devices. They are next submerged into estuarine waters where the spores grow into the desired seaweed form using naturally available nutrients.

Bivalves

Cultivation of oysters and mussels is probably the most successful form of mariculture. Commercially, no bivalves have been reared from the larval stage to a marketing size in a controlled environment because of the problems related to producing sufficient phytoplankton to feed them. Usually, hatchery-produced juveniles are reared to maturity in natural environments such as bays, estuaries, and other protected coastal waters. They may be set on the bottom or in trays or other suspension devices above the bottom (figure 17–11). This simple procedure results in yields ranging from 10 to over 1000 mt/acre of edible meat per year. There is no attempt at artificial feeding, and the mollusks are fed by the phytoplankton carried to them by the tides and other water movements.

An interesting outcome of placing oil platforms in the coastal ocean has developed off the Santa Barbara, California, coast. Mussels grew so rapidly on the substructure of the platforms that it had cost thousands of dollars per year to remove them from each platform.

In a deal that has worked out well for Bob Meek and his Ecomar Marine Consulting firm, the oil companies now get free cleaning of their platforms, and Ecomar harvests 3500 lb of mussels per day. The mussels are sold along the West Coast and as far inland as Chicago. The main market is the restaurant industry, which claims that these mussels are as fine as can be found. Ecomar also cultures up to 50,000 scallops and oysters annually on trays that are suspended from the platforms.

Crustaceans

Crustaceans such as shrimp and lobster have not yet been successfully raised in the United States, even though the price for such organisms is relatively high (figure 17–12). The problem with most shrimp operations is that the shrimp must be fed extraneously, and so far suitable artificial food has not been developed. The brightest spot in shrimp mariculture in the western hemisphere is Ecuador, where a very profitable operation is producing 45,000 tn per year. This success in Ecuador, where the cost of labor and dealing with government regulation is low, contrasts with high costs in the United States, where a similar operation may have to cope with over 35 regulating agencies, along with high labor costs.

American lobsters (*Homarus americanus*) can easily be carried through the complete life cycle in captivity, but they have a cannibalistic nature. To be grown successfully, these lobsters would have to be compartmentalized and maintained at an optimum temperature of about 20°C (68°F) for 2 years. Artificial feeding would also be necessary. Encouraging results have been achieved in rearing American lobsters in the warm effluent from an electrical generating plant in Bodega Bay, California. Similar projects will be tried in the coastal waters off Maine. The spiny lobster is not so easily reared through its long, complicated larva development, and less effort has been exerted toward developing commercial mariculture with these animals.

Fish

Mariculture with fish is also difficult, and only in Japan and Norway have significant marine programs been developed. Salmon, yellowtail, tuna, puffers, and a few

FIGURE 17–11

Bamboo Raft Used in Culturing Oysters. Oysters are suspended from bamboo rafts in nutrient-rich coastal waters off South Korea. This method of culturing off the bottom helps protect oysters from predators such as boring snails and sea stars. (Photo from M. Grant Gross.)

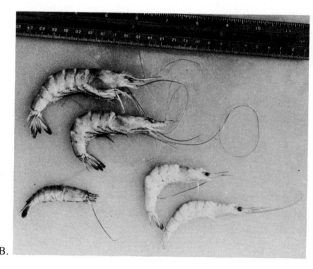

A.

B.

FIGURE 17–12

Shrimp Research. *A,* a European biologist and the Philippine students to whom he is teaching research methods examine shrimp kept in concrete ponds to study their molting process. (FAO photo by S. Bunnag.) *B,* varieties of shrimp, a popular sea food, that are being studied as possible objects of mariculture. (FAO photo by Pat Morin.)

other fishes are being raised in Japan on an experimental basis.

Norway has pioneered the rearing of Atlantic salmon in suspended cages with success far greater than that achieved throughout the rest of the world in salmon-ranching efforts. Norwegian salmon production reached 100,000 mt in 1990, and Norwegian technology is now being used in the Pacific Northwest. In Washington, Oregon, and British Columbia it has replaced the failed salmon ranching, which released young salmon into the ocean in hope of their return. The returns were very poor. This new activity is controversial in the Pacific Northwest because of the introduction of Atlantic salmon into Pacific waters. Although the technology does not involve releasing the salmon from the cages, some have escaped and created a worrisome situation.

Good returns of salmon released by the Alaska Fish and Game hatcheries are being reported. In Canada, the addition of nitrates and phosphates to lakes provided a larger biomass of plankton for young salmon to feed on before going to sea. This rather simple modification, which had no ill effects on the ecology of the lake, increased salmon production in the lake, and salmon returns from the open ocean. It is hoped that when the management of commercial operations is refined as knowledge is gained, we will be hearing success stories from the Pacific Northwest.

Some estuary fish farming involving mullet and milkfish has been practiced successfully in the Far East for centuries. These fishes are hardy and tolerate salinity ranging from that of fresh water to full seawater, and they do not have to be artificially fed. However, since they cannot be spawned and raised to sexual maturity in an artificial environment, the fry must be collected from a natural nursery environment. The costs of operating the farms are relatively low, and yields of about 1 mt/acre are common (figure 17–13).

WHAT LIES AHEAD?

Mariculture requiring intensive artificial feeding and high labor costs cannot be expected to help alleviate the world shortage of animal protein. Actually, more food is consumed than is produced by such systems, which are typical of the enterprises that have been attempted in the United States. However, those farming enterprises that use natural feeding in estuarine environments have good potential for helping increase the food supply. Most of these efforts require juvenile animals that are obtained from natural nursery areas, which represent the greatest costs to the marine farmer. Potential for increased productivity using the 1 billion acres or so of coastal wetlands in the world would be quite significant. If only 10 percent of the wetlands were to be put into production, it could result in 100 million metric tn of fish based on the productivity that has been established for mullet and milkfish.

Many supporters of mariculture believe that this industry, which now accounts for less than 10 percent of the present food taken from the sea, can grow to a business of 60 million metric tons per year. Many scientists who have studied the field of mariculture are not so optimistic about the ability of the industry to significantly increase the world food supply. The new projects

FIGURE 17–13

Milkfish Ponds. Milkfish are harvested using a gill net. Brackish water fish culture techniques applied to ponds such as these in the Philippines are proving successful. (FAO photo by P. Boonserm.)

being developed in the United States are directed at luxury foods that will certainly be too expensive for the world's hungry poor. They think that these projects may generate a profit but not much food.

The Food and Agriculture Organization (FAO) of the United Nations has estimated that, based on traditional fisheries, the maximum sustainable yield that might be expected from the oceans is about 120 million tons. This estimate may be considered conservative and could well be exceeded, particularly with the addition of new species to the significant fisheries. A candidate for addition to the list of major fisheries is the Antarctic krill. Grenadiers and lantern fishes found in deeper waters and the pelagic red crab of the Eastern Pacific and subantarctic regions may also be added in the near future. Another significant addition would be cephalopods such as squid, cuttlefish, and octopus, which yield about 80 percent edible flesh—compared with approximately 20 to 50 percent for the most common representatives of our present fishery.

Surely, improved management of existing fisheries, progress in mariculture, and the addition of new species to the world fishery will lead to a much needed increase in the supply of marine food.

SUMMARY

Increased effort in some of the world's fisheries has not significantly increased catches. In some cases, the result has been a dramatic decrease in takes, indicating that **overexploitation** of these fisheries has resulted in decreased catch per unit of effort.

Fisheries management in coastal U. S. waters began with the prohibition against seining for mackerel imposed by the Plymouth Colony in 1684. Attempts at regulation of fisheries have not been successful to date. This failure may in large part be due to the fact that the ecology of most fisheries is not well understood. Present studies of the ecology of marine fisheries attempt to assess the factors that affect survival of fish eggs, larvae, and juveniles. Understanding these important factors helps fishery scientists make better predictions of **recruitment** of young adults to the fishery. A promising method of **fishery assessment** involves an **ecosystem approach**. This method requires knowledge of the annual **nutrient flux** into the ecosystem, **size grades** of organisms, **standing stocks**, and **size- doubling times** of each component of the food chain.

About 50 percent of the world fishery comes from about 0.1 percent of the ocean surface area where **upwelling** occurs. Therefore, studies of **duration** and **rate of upwelling** in these regions have been undertaken. They reveal that moderate rates of upwelling over long periods of time—or even short durations—produce larger fisheries than high or low rates of upwelling.

Studies of the four major subtropical eastern boundary current fisheries supported by upwelling reveal that they have much in common. They are dominated by four species of pelagic fishes: **sardine, anchovy, jack mackerel**, and **hake**. Spawning occurs mostly in areas of moderate upwelling where food conditions are optimum for fish larvae. The **Peruvian anchoveta** are an exception that spawn where upwelling rates are high. There is as yet no explanation for this behavior.

There has been an agreement to extend coastal nations' control of the ocean to 200 mi (322 km) beyond their coasts to protect coastal fisheries. The incidental killing of **dolphins** while conducting the **tuna fishery** in the eastern Pacific has produced serious problems for the U.S. tuna fleet and in the field of international relations. Such problems will surely increase until much more international cooperation is developed in the conduct of open-ocean fisheries.

An example of the combined effects of natural phenomena and human activities on a marine population is the sharp decline in the population of Peruvian anchoveta that occurred in 1972. This reduction seems to have been the result of a severe **El Niño** condition and a poor understanding of the fishery in the coastal waters of Peru. A less euphemistic statement of the cause would be that the fishery was destroyed by failure of regulators to enforce recommended procedures.

Mariculture projects around the world have, in some cases, indicated potential for increasing the food taken from the ocean. The most successful mariculture to date has been directed at algae, **oysters**, **mussels**, **salmon**, **mullet**, and **milkfish**. These activities, combined with adding new species such as **krill**, **grenadiers**, **lantern fish**, **red crabs**, and **cephalopods** to our open-ocean fisheries, may help increase the amount of food we take from the ocean.

QUESTIONS AND EXERCISES

1. Describe the nature of and the results of the first attempt at fisheries management in U. S. coastal waters.

2. What are the two critical survival stages that must occur before young adults can be recruited to a fishery, and what factor is most important in increasing the chance of survival at each stage?

3. Explain the relationship between the influx of nutrients (nitrogen) into an ecosystem and the amount of the fishery that can safely be removed from the ecosystem each year.

4. What characteristics of an upwelling system are related to the magnitude of the fishery it can produce? What combination of these factors appears to produce the maximum environment for the development of a large fishery?

5. What are the characteristics that the fisheries of the four major eastern boundary current upwellings of the subtropical gyres have in common?

6. Discuss how the discovery of large demersal aggregations of krill found in submarine canyons off the flank of Georges Bank relate to the ecology of the fisheries on the bank.

7. List and discuss some positive and negative effects that you think may result from the establishment of the 200-m (322-km) Exclusive Economic Zones.

8. Discuss the natural and artificial conditions that may have combined to cause the catastrophic decline of the Peruvian anchoveta fishery in 1972.

9. What characteristics are desirable in an organism chosen for a mariculture project?

10. What do the cultivation of algae and bivalves have in common that helps make them good choices for mariculture operations?

11. Discuss the positive and troublesome aspects of mariculture of crustaceans.

12. Discuss how the mariculture of mullet and milkfish is similar to that of bivalves.

13. From which mariculture and natural fisheries do we have the greatest hope of obtaining an increased quantity of food from the sea?

REFERENCES

Backus, R. H., and Bourne, D. W., eds. 1987. *Georges Bank.* Cambridge, Mass.: The MIT Press.

Borgese, E. M., and Ginsburg, N. 1988. *Ocean yearbook 7.* Chicago: University of Chicago Press.

Greene, C. H., Wiebe, P. H., Burczynxki, J., and Youngbluth, M. J. 1988. Acoustical detection of high-density krill demersal layers in the submarine canyons off Georges Bank. *Science* 241:4863, 359–61.

Knauss, J. A. 1974. Marine science and the 1974 Law of the Sea Conference: Science faces a difficult future in changing Law of the Sea. *Science* 184:1335–41.

Longhurst, A. R., and Pauly, D. 1987. *Ecology of tropical oceans.* San Diego: Academic Press.

Parsons, T. R., Takahashi, M., and Hargrave, B. 1984. *Biological oceanographic processes,* 3rd ed. New York: Pergamon Press.

Parrish, R., Bakun, A., Husby, D. M., and Nelson, C. S. 1983. Camparative climatology of selected environmental processes in relation to eastern boundary current pelagic fish reproduction. Unpublished F. A. O. study.

Royce, W. F. 1987. *Fishery development*. San Diego: Academic Press.

Smith, P. E. 1969. The horizontal dimensions and abundance of fish schools in the upper mixed layer as measured by sonar. La Jolla, Calif.: Bureau of Commercial Fisheries.

SUGGESTED READING

Sea Frontiers

Brown, M. 1984. Counting dolphins: New answers to an important question. 30:2, 68–75.
Describes a new method used to study the number of dolphins in the tropical Pacific Ocean.

Cooper, L. 1981. Seaweed with potential. 27:1, 24–27.
Describes how an increased demand for seaweeds as a source of food and natural chemical products has stimulated interest in the mariculture of macroalgae.

Cuyvers, L. 1984. Milkfish: Southeast Asia's protein machine. 30:3, 173–79.
The importance of milkfish mariculture operations to the food supply in Southeast Asia is discussed.

Iverson, E. S., 1984. Mariculture: Promise made, debt unpaid. 30:1, 20–29.
Progress in mariculture in North America is considered.

Iverson, E. S. and Jory, D. E. 1986. Shrimp culture in Ecuador: Farmers without seed. 32:6, 442–53.
The rapidly growing shrimp mariculture industry of Ecuador encounters some growing pains.

Katz, A. S. 1984. Herring fisheries: Comparing the two stories. 30:6, 335–41.
A discussion of how the San Francisco herring fishery management might use the history of the collapsed North Sea herring fishery to prevent their fishery from being overfished.

Maranto, G. 1988. Caught in conflict: Managers trapped in fisheries dilemma. 34:3, 144–51.
With the heavy pressure on fisheries and the conflict between recreational and commercial fishermen, fisheries managers are hard pressed to effectively regulate fisheries.

Nicol, S. 1987. Krill: Food of the future? 33:1, 12–17.
Before krill can be considered as a viable fishery, many questions about the cost of the fishery and processing as well as the fishery's effect on krill-eating mammal populations must be answered.

Scientific American

Holt, S. J. 1969. The food resources of the ocean. 221:3, 178–97.
A review of world fisheries and their potential for the future is presented. The critical problem of international regulation of fisheries is also explored.

Pinchot, G. B. 1970. Marine farming. 223:6, 14–21.
Presents a brief history and description of efforts and potential for future marine farming, with an emphasis on the means by which nutrients may be economically introduced into low-nutrient waters.

18
MARINE POLLUTION

As society provides for an ever-increasing population, the role played by the oceans in meeting these needs is also increased. One of these roles has been disposal for a growing amount of waste materials from sewage sludge to radioactive waste. An additional threat is the accidental contamination from transportation and production accidents associated with the efforts to meet present energy needs with petroleum.

The ocean is an important recreational resource in the United States, since 80 percent of its population live within easy access of the coasts of either the oceans or the Great Lakes. Well over 100 million Americans participate in marine recreational activities, spending almost $150 billion annually. In the past it has not been unusual to see local areas of the shore closed to recreational activities such as fishing, boating, and swimming because of pollution.

We have little difficulty understanding the meaning of *pollution*. The World Health Organization defines pollution of the marine environment as

> The introduction by man, directly or indirectly, of substances or energy into the marine environment, including estuaries, which results or is likely to result in such deleterious effects as harm to living resources and marine life, hazards to human health, hindrance to marine activities, including fishing and other legitimate uses of the sea, impairment of quality for use of sea water and reduction of amenities.

But as far as the marine environment is concerned, there is great difficulty in establishing the degree to which pollution is occurring. In most cases, the ocean has not been studied sufficiently prior to society's introduction of pollutants. Therefore, we cannot tell how the marine environment has been altered by these activities.

CAPACITY OF THE OCEANS FOR ACCEPTING SOCIETY'S WASTES

Although, at present, the consensus developing in the United States and in many other nations is that we should not propose to dispose of any materials or en-

The *Exxon Valdez* tanker aground in Prince William Sound off the coast of Valdez, Alaska. Photo from N. Y. Bureau/Wide World Photos.

ergy in the ocean, a more realistic view is that we will have to consider the oceans as an option in disposing of some waste materials. An even more realistic view of this problem is that regardless of what we say we are going to do, polluting materials will continue to enter the oceans for the foreseeable future. In light of this probability, research into the capacity of the oceans to absorb various substances that could lead to their becoming polluted continues.

One of the present goals of marine pollution research is to determine an illusive entity, the *assimilative capacity* of the oceans—the amount of a given material that can be contained within the ocean without causing an unacceptable degree of harm to the living or nonliving resources of the oceans. Determining at what concentrations an unacceptable degree of harm is done to the ocean's resources requires long-term monitoring of disposal activities. Although some researchers believe we have enough data to establish values for the assimilative capacity of the coastal ocean for certain substances, all agree that what we do not know greatly exceeds what we do know. It seems likely that establishing dependable limits of input rates for the broad range of substances that are dumped in the ocean is many years away.

PREDICTING THE EFFECTS OF POLLUTION ON MARINE ORGANISMS

To date, the most widely used technique for determining the level of pollutant concentration that negatively affects the living resources of the ocean is the standard laboratory *bioassay*. Regulatory agencies such as the Environmental Protection Agency use a bioassay that determines the concentration of a pollutant that causes 50 percent mortality among the test organisms to set allowable concentration limits of pollutants in coastal water effluents. The shortcoming of the bioassay is that it does not predict the long-term effect of sublethal doses on the organisms.

It is desirable to be able to predict the effects of a given concentration of pollutants at four different levels of biological organization, some of which can be detected in a few minutes, whereas others may require decades of observation. These levels of biological response, the time spans over which they occur, and examples of each follow.

1. *Biochemical-cellular:* Minutes to hours. Exposure to organic pollutants such as hydrocarbons and PCBs produces dysfunction in metabolic processes, causing death.
2. *Organism:* Hours to months. Pollution stress may create changes in physiological processes such as

metabolism or digestive efficiency that will severely reduce the energy available for growth and reproduction.

3. *Population dynamics:* Months to decades. Although it is difficult to tie long-term effects to a certain pollutant, short-term or chronic effects may be tied to events such as oil spills or malfunctions at sewage plants. These responses occur as changes in the abundance and distribution of a species, population structure on an age-class basis, growth rates within age classes, fecundity, and incidence of disease.
4. *Community dynamics and structure:* Years to decades. Communities, as assemblages of populations, become less diverse, and more resistant populations increase in numbers under the stress of pollutants.

Much progress is needed before the identification of early warning signs of stress at each level of biological response is possible. Since the degree of system complexity and the time needed to measure a response increase exponentially from the biochemical-cellular level to the community level, the predictive difficulties also increase at each level.

CEPEX The Controlled Ecosystem Pollution Experiment (CEPEX) used plastic experimental bags (figure 18–1) to simulate communities of plankton ranging in size from 68 to 1700 m³ (2400–60,000 ft³) to investigate the effects of pollutants. Observations of the recovery rates of bacteria, phytoplankton, and zooplankton to pollution by heavy metals and hydrocarbons showed that the bacteria and phytoplankton, with their short generation times, recovered much more rapidly than the larger and more complex zooplankton.

MARINE POLLUTION CONTROL IN THE UNITED STATES

The United States has long been concerned with the legal protection of the marine environment. The first Rivers and Harbors Act of 1899 provided a straightforward prohibition against dumping any kind of refuse into the navigable waters of the United States. Most current regulations on waste disposal in the coastal ocean, the 200-mi (322-km) Exclusive Economic Zone, are contained in two major statutes: The Federal Water Pollution Control Act (Clean Water Act) of 1948 and its amendments deal primarily with problems of point sources of municipal and industrial waste and spills of oil and hazardous materials. The Marine Protection, Research, and Sanctuaries Act (Ocean Dumping Act) of 1972 controls dumping of wastes at sea, research, and the establishment of marine sanctuaries.

Although these regulations are quite complex, they basically (1) prohibit the dumping of materials known to be harmful, and (2) specify the criteria under which other materials may be dumped. The basic premise of each seems to be that land disposal is preferable to marine disposal. If it cannot be proved to the satisfaction of the regulating agencies that dumping of a material will not adversely affect human health, it will not be allowed in the ocean.

Despite the strong language of the laws, the coastal ocean has been adversely affected by the dumping of waste materials. Conversely, some waste materials that could possibly have been safely disposed of in the coastal ocean were not because of the difficulty in proving it would not have been harmful to dump them. It is very easy to prove the harmful effects of dangerous materials, but it is almost impossible to prove that other seemingly benign substances will not create a long-term harmful condition.

Because some scientists say that we can probably better predict the paths of materials dumped into the ocean than on land, the option of ocean disposal has received renewed attention. In the past, uncertainty led to a decision to prohibit dumping; then support grew for plac-

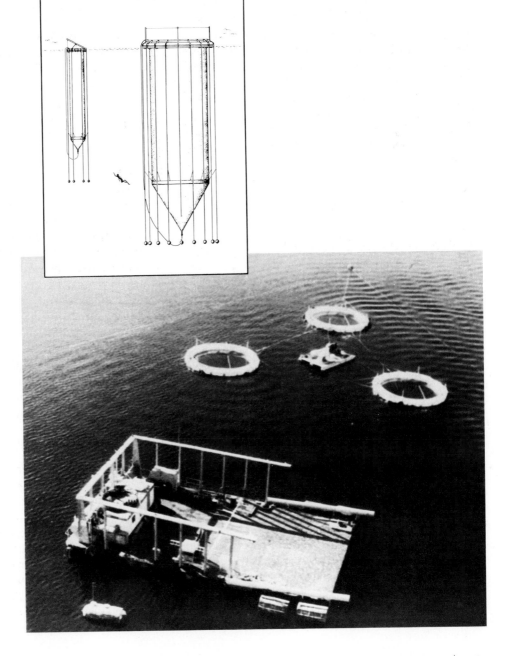

FIGURE 18–1

Controlled Ecosystem Pollution Experiment (CEPEX). Views of CEPEX bags used in experiments to investigate the effects of pollutants on plankton communities. A scuba diver is shown for scale. (Courtesy of the National Science Foundation.)

ing the obligation of proving harmful effect on the regulating agency instead of requiring the dumping applicant to prove that no harmful effect would occur. After the public outcry that resulted from the littering of northeastern U.S. beaches with medical debris during the summer of 1988, legislation to prohibit marine dumping was quickly passed. In light of recent advances in scientific knowledge, it is frustrating to have to admit the broad scope of our ignorance of the marine environment. Considering the great cost and amount of time required to remove uncertainty from the field of marine waste disposal in the near future, it would seem prudent to continue exercising a great deal of caution in pursuit of this option. Considering the rapid escalation of the cost of land-based disposal, short-term political priorities point to stopping ocean dumping, and economic reality tends to make ocean dumping more attractive. Clearly, politicians and their constituents need to have the best possible data available to proceed rationally. Some very difficult decisions will have to be made in the near future.

AREAS OF CONCERN

Plastics

During the last 30 years, the use of plastics has increased at a tremendous rate. The problem of disposing of this "throwaway" component of Western culture has already strained the capacity of our land-based disposal systems. It is also an increasingly abundant component of oceanic flotsam.

Small pellets that are used in the production of essentially all plastic products are transported in bulk aboard commercial vessels. Found throughout the oceans, the pellets probably find their way into the oceans as a result of spillage at loading terminals. In coastal waters, plastic products used in fishing and thrown overboard by recreational and commercial vessels are common. They also find their way into the open ocean from careless dumping by commercial vessels.

The best documentation of the negative effects of plastics in the ocean on marine organisms is found in the strangulation of seals and birds caught in plastic netting and packing straps (figure 18–2). Marine turtles are known to mistake plastic bags for jellyfish or transparent plankton on which they typically feed.

It is believed that all plastics that enter the ocean may eventually be removed by the shorelines of islands and continents that "filter" out the particles as ocean currents wash against their shores. Beaches throughout the world are probably increasing their plastic pellet content as a result of the filtering process. Some Bermuda beaches have up to 10,000 pellets/m². The beaches of

FIGURE 18–2

Elephant Seal with Plastic Packing Strap Tight Around Its Neck. This animal was found on a beach of San Clemente Island off the Southern California coast. It was saved from strangulation by the removal of the plastic packing strap by scientists from the National Marine Fisheries Service. (Photo by Wayne Perryman, NMFS.)

Menemsha Harbor on Martha's Vineyard yielded 16,000 plastic spherules/m². Thin plastic film products will eventually be broken into smaller particles by photochemical degradation, and others may sink as they become denser because of photochemical degradation and encrustation by epifauna such as bryozoans and hydroids. These data are from a survey conducted between 1984 and 1987, which also found more than 10,000 plastic pieces and 1500 pellets/km² in the northern Sargasso Sea between 28° and 40°N latitude. Compared with a survey conducted in 1972, the 1987 survey found that the concentration of plastic pellets had doubled.

The U.S. Congress is considering a bill to ban disposal of plastics in the 200-mi-wide Exclusive Economic Zone. Internationally, the Convention for the Prevention of Pollution from Ships forbids disposal of all plastics into the oceans. The acceptance of both of these regulations would aid greatly in decreasing the concentration of plastics in the world's oceans.

Petroleum

There have been a number of major oil spills since the *Torrey Canyon* ran aground at Seven Stones Rocks off the Cornwall coast on March 18, 1967. They have resulted from additional tanker accidents and the blowout of oil wells being drilled or produced in the coastal ocean. Although it is important to know how much oil enters the oceans from what sources, our primary concern is the effect of hydrocarbon pollution on marine organisms and the marine environment in general.

Since hydrocarbons are organic substances biodegradable by microorganisms, they are considered by

FIGURE 18–5
***Argo Merchant* Oil Spill, December 15, 1976.** *A*, location of the grounding site and oiled area of the ocean. *B, Argo Merchant,* after breaking up and spilling much of its cargo into the Atlantic Ocean southeast of Nantucket. (*B,* photo courtesy of U.S. Environmental Protection Agency.)

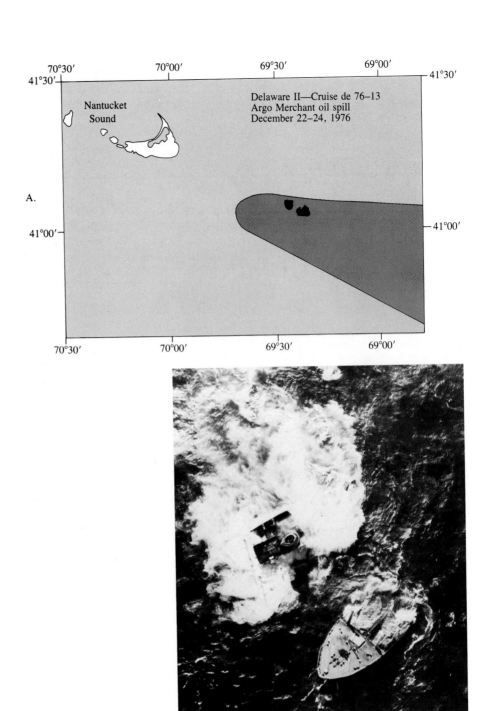

A.

Delaware II—Cruise de 76–13
Argo Merchant oil spill
December 22–24, 1976

Nantucket Sound

There were, however, numerous oiled birds that washed ashore at Nantucket and Martha's Vineyard.

Because pollock females spawn about 225,000 and cod about 1 million eggs each season over a range extending from New Jersey to Greenland, it is unlikely that this single occurrence will have a major effect on those fisheries. However, it is easy to see why the fishing communities of the Northwest Atlantic are so happy about the marginal results of oil exploration on Georges Bank. It may be that a fishery already severely stressed by overexploitation could not survive even the lowest level of pollution brought about by the establishment of oil production on the Bank.

Exxon Valdez The first major oil spill resulting from the development of the North Slope of Alaska petroleum reserves occurred March 23, 1989, when the California-bound *Exxon Valdez* went aground on rocks 40 km (25 mi) out of Valdez, Alaska. The 43,000 tons of oil spilled makes this the largest spill to occur in U.S. waters. This spill in the pristine waters of Prince William Sound resulted in a large number of dead birds and sea otters.

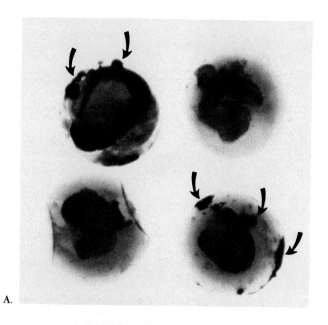

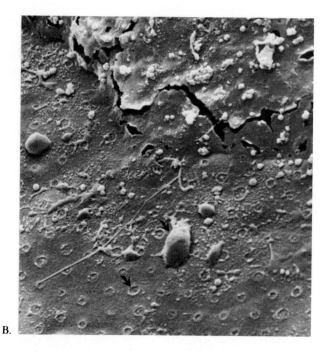

FIGURE 18–6

Pollock Eggs Affected by the *Argo Merchant* Spill. *A,* these eggs, in the tail-bud and tail-free embryo stages, were taken from the edge of the oil slick. The outer membranes of the eggs at the upper left and lower right are contaminated with a tarlike oil; arrows point to some of the oil masses. The uncontaminated egg at the upper right has a malformed embryo; that at the lower left is collapsed and also has an abnormal embryo. The actual size of pollock eggs is about 1 mm (0.04 in.); of cod eggs, about 1.5 mm (0.06 in.). *B,* a portion of the surface of an oil-contaminated egg. The upper arrow points to one of many oil droplets; the lower arrow to one of the membrane pores. (Scanning electron microscope; about 5000 X.) (Courtesy of A. Crosby Longwell.)

The total death toll will never be known, but it is predicted that these Alaskan waters may have a long, slow recovery as a result of the low water temperatures which will slow natural cleanup processes.

Oil spills are a problem that will be with us for many years, and they may become more common as the petroleum reserves underlying the continental shelves of the world are increasingly exploited. The largest oil spill on record occurred June 3, 1979, as a result of this exploitation. The Petroleus Mexicanos IXTOC #1 blew out and caught fire. Before it was capped on March 24, 1980, it spewed 468,000 tons of oil into the Gulf of Mexico (figure 18–7).

Oil can be broken down by microorganisms such as bacteria, fungi, and yeast. Certain of these organisms are effective in breaking down only a particular variety of hydrocarbon, and none is effective against all forms. In 1980 Dr. A. M. Chakrabarty, a microbiologist at the General Electric Research and Development Center in Schenectady, New York, produced a microorganism capable of breaking down nearly two-thirds of the hydrocarbons in most crude oil spills. By manipulating special rings of DNA called *plasmids,* he has been able to combine the consumptive characteristics of numerous natural strains

of bacteria into what might be considered a "superbug" that could be very effective in cleaning oil spills. However, before this superstrain can be used against actual oil spills, much testing is needed to ensure that it does not have adverse effects on the environment.

Sewage

Over 500,000 mt of sewage sludge containing a mixture of human waste, oil, zinc, copper, lead, silver, mercury, PCBs, and pesticides per year have been dumped through sewage outfalls into the coastal waters of southern California. A more massive 8 million metric tons have been dumped into the New York Bight each year. Although the Clean Water Act of 1972 prohibited dumping of sewage into the ocean after 1981, the high cost of treating and disposing of it on land resulted in extended waivers being granted to these areas. Ironically, the non-biodegradable debris that washed up on Atlantic coast beaches during the summer of 1988 and hurt the tourist business probably was responsible for legislation again being passed to terminate disposing of sewage into the ocean. The cutoff date is now January, 1992. There is little likelihood that this debris is associated with sew-

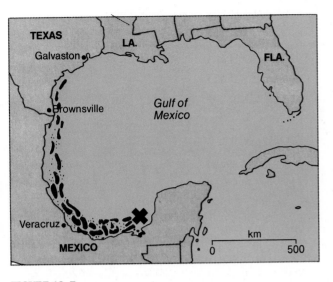

FIGURE 18–7

IXTOC #1 Blowout, June 3, 1979. Location of the Gulf of Campeche blowout and oil slick that threatened the Texas coast during the summer of 1979.

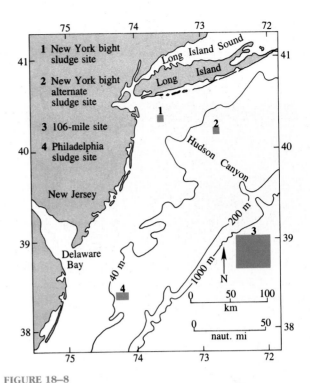

FIGURE 18–8

Previous and Present Atlantic Sewage Sludge Disposal Sites.

age. It is more likely the result of plastic debris being carried to the ocean by storm drain systems after heavy rains.

It now appears that the political pressure to clean up the coastal water is sufficient to assure that the more expensive land treatment and disposal will be the primary disposal procedure of the future. For the present, Los Angeles has stopped pumping sewage into the sea. If it is allowed to resume this practice, the sewage sludge pumped through the lines will have had its toxic chemicals and pathogens removed.

Off the east coast over 8 million metric tons of sewage sludge was dumped by barge over dump sites totalling 150 km² (58 mi²) each year during the past few years. These dumps occurred at the New York Bight sludge site and the Philadelphia sludge site shown on figure 18–8. The New York Bight water depth is about 29 m (95 ft), and Philadelphia's site is 40 m (130 ft) deep. In such shallow water, the water column displays relatively uniform characteristics from top to bottom; even the smallest sludge particles reach the bottom without having undergone much horizontal transport, and the ecology of the dump site can be totally destroyed. At the very least, the concentration of organic and inorganic nutrients seriously disrupt the biogeochemical cycling. Greatly reduced species diversity results, and in some locations the environment becomes anoxic.

The shallow-water sites have been abandoned, and sewage is now being transported to a deep-water site 106 miles out. (figure 18–8). At the deep-water site beyond the shelf break, a well-developed density gradient usually separates low-density, warmer surface water from high-density, colder deep water. Internal waves moving along this density gradient can retard the sinking rate of particles and may allow a horizontal transport rate 100 times greater than the sinking rate. It was believed that this process would greatly reduce the degradation of the sea-floor ecology.

The local fishing community began complaining about the adverse effect of the deep-water dumping on their fisheries soon after it began. As it stands now, this East Coast dumping program for sewage is also doomed, and municipalities will have to find the money for land-based processing.

Radioactive Waste

As with other waste materials, a case can be made that oceans can accept a given amount of radioactive waste without doing significant harm to the marine environment. Since 1944, artificial *radionuclides* (radioactive atoms) have been reaching the oceans as *fallout* from the production and testing of nuclear weapons. This input has been about four times greater in the Northern Hemisphere than in the Southern Hemisphere. Because of the nature of the fallout process, no high concentrations of any radionuclides have resulted. Table 18–1 shows the amount of radiation in thousands of curies that has reached the oceans via fallout. (A *curie* (ci) is a

TABLE 18–1

Artificial Radionuclides in the Oceans and Their Sources.

	Plutonium-239, -240 (kCi)	Cesium-137 (kCi)	Strontium-90 (kCi)	Carbon-14 (kCi)	Tritium (kCi)
Total worldwide fallout by early 70s	320	16,700	11,500	6,000	3,000,000
North Atlantic Ocean— early 70s	63	3,300	2,300		650,000
Windscale, 1957–78 discharge	14	830	130		
	Total α-emitters	Total β/γ-emitters (other than tritium)			
The NEA dumpsite (1967–79)	8.3	258			

quantity of any radioactive nuclide in which exactly 3.7 $\times 10^{10}$ disintegrations occur per second.)

The second largest source of radionuclides is the *nuclear fuel cycle*. The release from this source occurs in local concentrations. Nuclear power generating plants release little radiation to the oceans, but nuclear fuel reprocessing plants contribute a significant quantity. Such plants are found at Windscale and Dounreay in Great Britain and Cape de la Hague, France.

Another significant source of radioactive waste is a dumping operation carried out over the past three decades under international agreement at various locations. Since 1967, the only significant dumping operation has been conducted in the Northeast Atlantic under the coordination of the Nuclear Energy Agency of the Organization for Economic Cooperation and Development. The input of this operation is shown in table 18–1.

In order to determine safe input rates for various radionuclides, oceanographers must determine the average concentration of a radionuclide if mixed totally in the ocean. This is of primary importance for radionuclides with half-lives much greater than the mixing time* of the ocean; it is a major consideration for materials like plutonium, which has a half-life of 24,400 years. A second consideration is to determine the transfer mechanisms to critical populations; this is important for all radionuclides in areas near a dumping locality, regardless of the length of their half-lives.

The International Atomic Energy Agency has determined maximum dumping rates on the basis of these considerations. As an example, the dumping rate for one of the most dangerous radionuclides, plutonium-239 is placed at 10,000 Ci/year. This limit means that the plu-

tonium concentration would reach the safe dose limit in 40,000 years. Although there is presently much resistance to development of the industry, we may assume that nuclear fission generation of energy will be a viable source of energy for a few hundred years. If so, there would seem to be little likelihood of seriously damaging the marine environment by the dumping of plutonium at this rate.

Although research has been done on deep-sea disposal of high-level nuclear waste, it has now been terminated. The Nuclear Energy Agency has concluded that if this waste is to be disposed of in the thick sediments of abyssal plains, it will not occur for at least 50 years. Land-based disposal sites are now the top priority of research toward the solution of this problem.

Halogenated Hydrocarbons (DDT and PCBs)

DDT (dichlorodiphenyltrichlorethane) and PCBs (polychlorinated biphenyls) are found throughout the marine environment. They are persistent, biologically active chemicals that have been put into the oceans entirely as a result of human activities.

At the time of a near-total ban on U.S. production of the pesticide DDT in 1971, 2.0×10^{12} g had been manufactured, most of it by the United States. Its use in the Northern Hemisphere since 1972 has virtually ceased. The danger of excessive use of DDT and other similar pesticides was first manifested in the marine environment in the effects on marine bird populations. During the 1960s, there was a serious decline in the brown pelican population of Anacapa Island off the coast of California; high concentrations of DDT in the fish eaten by the birds had caused the production of excessively thin eggshells. A decline in the osprey population of Long Island Sound that began in the late 1950s and continued throughout the 1960s was also caused by thinning eggshells brought on by DDT contamination. Studies

---

Mixing time is the time required to distribute evenly a substance added to the ocean. The scientific community's best guess is that the mixing time of the world ocean is several thousand years; that of the Atlantic Ocean may be less than 1000 years.

FIGURE 18–9
Brown Pelicans. Pelican populations in the Channel Islands off Southern California are increasing, following a serious decline in the 1960s caused by DDT toxicity.

showed a 1 percent increase in brown pelican and osprey eggshell thickness from 1970 to 1976, while the concentration of pesticide residue decreased. Since there was an increased egg hatching rate associated with these changes, it was concluded that DDT was the cause of the decline in the bird populations during the 1960s (figure 18–9).

PCBs are industrial chemicals found in a variety of products from paints to plastics. They have been indicated as causes of spontaneous abortions in sea lions and the death of shrimp in Escambia Bay, Florida.

The main route by which DDT and PCBs enter the ocean is the atmosphere. They are concentrated initially in the thin surface slick of organic chemicals at the ocean surface, until they gradually sink to the bottom attached to sinking particles. A study off the coast of Scotland indicated that open-ocean concentrations of DDT and PCBs are 10 and 12 times less, respectively, than in coastal waters. Long-term studies have shown that DDT residue content in mollusks along the U.S. coasts reached a peak in 1968. The pervasiveness of DDT and PCBs in the marine environment can best be demonstrated by the fact that Antarctic marine organisms contain measurable quantities of halogenated hydrocarbons. Obviously there has been no agriculture or industry on that continent to explain this presence; these substances have been transported from far-distant sources by winds and ocean currents.

Mercury

The stage was set for the first tragic occurrence of mercury poisoning with the establishment of a chemical factory on Minamata Bay, Japan, in 1938. One of the products of this plant was acetaldehyde, the production of which requires mercury as a catalyst. The first ecological changes in Minamata Bay were reported in 1950, effects on humans were noted in 1953, and minamata disease became epidemic in 1956. It was not until 1968 that the Japanese government declared mercury the cause of the disease, which involves a breakdown of the nervous system. The plant was immediately shut down, but by 1969 over 100 people were known to suffer from the disease. Almost half of these victims died. A second occurrence of mercury poisoning resulted from pollution by an acetaldehyde factory, shut down in 1965, that was located in Niigata, Japan. Between 1965 and 1970, 47 fishing families contracted the disease.

During the 1960s and 1970s much attention was given to the problem of mercury contamination in seafood. Studies done on the amount of seafood consumed by various populations of humans have led to the establishment of safe levels of mercury content in fish to be marketed. To establish these levels for a given population, three variables must be considered:

1. Fish consumption rate of the human population under consideration
2. Mercury concentration of the fish being consumed by that population
3. Minimum ingestion rate that induces symptoms of disease

If dependable data on these variables are available and a safety factor of 10 is applied to the determination, a maximum concentration level allowable for the contaminant (in this case mercury) can be established with a high degree of confidence that it will protect the health of the human population.

Figure 18–10 shows how such data can be used to arrive at a safe level of mercury concentration in fish to be consumed by a population. For the general populations of Japan, Sweden, and the United States, the average individual fish consumption rates are 84, 56, and 17 g/day, respectively. By comparison, the members of the Minamata fishing community averaged 286 and 410 g/day during winter and summer seasons, respectively. Based on the minimum level of mercury consumption that is known to bring on symptoms of poisoning as determined by Swedish scientists using Japanese data— 0.3 mg/day over a 200-day period—the three populations shown in figure 18–10 would begin to show symptoms if they ate fish with the following mercury concentrations:

Japan	4 ppm
Sweden	6 ppm
United States	20 ppm

FIGURE 18–10

Mercury Concentrations in Fish vs. Consumption Rates for Various Populations. Plotted curves show the safe ingestion level of mercury until the first signs of Minamata disease can be expected (0.3 mg/day), and the safe ingestion level of mercury after the safety factor of 10 is included (0.03 mg/day). For the United States, it appears that tuna and swordfish are safe to consume, although some swordfish will be banned because their mercury concentrations are above the present 1.0 mg/kg (ppm) FDA limit.

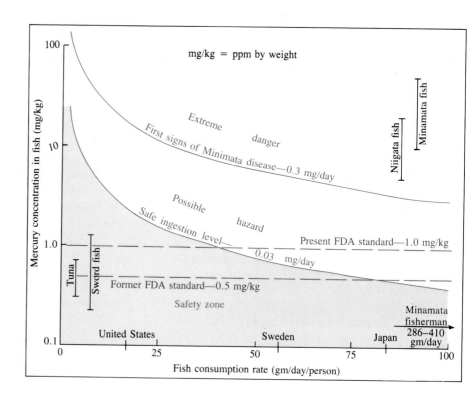

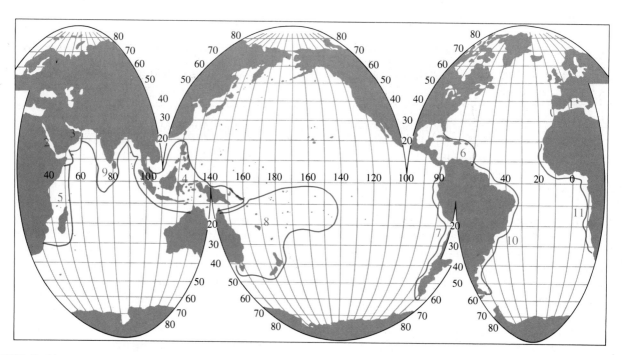

FIGURE 18–11

Areas Covered by UNEP's Regional Seas Programme. According to the United Nations Environment Programme, the definition of boundaries is the responsibility of the governments concerned, and those shown here are only illustrative. Environmental units with action plans in effect or in preparation include (1) Mediterranean Sea, (2) Red Sea and Gulf of Aden, (3) Kuwait region, (4) East Asian Seas, (5) East Africa, (6) Caribbean Sea, (7) Southeast Pacific, (8) Southwest Pacific, (9) South Asian Seas, (10) Southwest Atlantic, and (11) West and Central Africa.

If a safety factor of 10 is applied, the maximum concentration of mercury concentration in fish that could be safely consumed by these populations would be

Japan	0.4 ppm
Sweden	0.6 ppm
United States	2.0 ppm

Although the U.S. Food and Drug Administration (FDA) initially established an extremely cautious limit of 0.5 mg/kg (ppm), the present limit of 1 mg/kg adequately protects the health of U.S. citizens. Essentially all tuna falls below this concentration, and most swordfish is acceptable. The present limit amounts to the use of a safety factor of 20 instead of 10, so unless one eats an unusually large amount of tuna or swordfish, there should be little concern about consuming these fishes.

INTERNATIONAL EFFORTS TO PROTECT THE MARINE ENVIRONMENT

Beginning with the Action Plan for the Human Environment established by the United Nations in 1972, a comprehensive plan for international environmental protection has evolved. It involves procedures for assessment of problem areas and monitoring activities that are potential sources of pollution.

One of the most critically polluted areas identified was the Mediterranean Sea, polluted in many coastal regions from essentially every conceivable source—domestic sewage, industrial discharge, pesticides, and petroleum. To aid its recovery, an action plan was initiated in 1975, was completed in 1976, and has been ratified by 13 of the 18 nations with Mediterranean coasts.

Other regions identified as environmental units with action plans in effect are the Red Sea, Gulf of Aden, East Asian Seas, East Africa and Kuwait (figure 18–11). Areas with action plans in preparation are the Caribbean, South Asian Seas, West and Central Africa, Southeast Pacific, and Southwest Pacific.

If such efforts proceed effectively, they can reverse the trend toward progressive degradation of the environment that accelerated throughout the 1960s. Recent efforts appear to have slowed this process and may well have reversed it in local coastal areas. The prognosis for the future health of the marine environment is much improved from a decade ago; this change serves as testimony to the human will to preserve the quality of the earth's environment. There is still much to do, and a continuing effort will be required if we are to succeed.

SUMMARY

Pollution is the introduction of substances or energy into the marine environment that results in harm to the living resources of the oceans or humans that use these resources. Scientists are trying to determine the ocean's **assimilative capacity,** the amount of a substance it can absorb without becoming polluted.

The U.S. government has been concerned about coastal pollution since it passed the **Rivers and Harbors Act** in 1899. Later laws were passed to prohibit the dumping of materials known to be harmful and to specify criteria under which other materials may be dumped.

The amount of **plastic** accumulating in the oceans has increased dramatically. Certain forms of plastic are known to be lethal to marine mammals and turtles, and national and international legislation is being considered to ban the disposal of plastic in the oceans.

Study of areas such as **Wild Harbor,** Massachusetts, and **Chedabucto Bay,** Nova Scotia, have shown that oil pollution reduces species diversity and persists longer on muddy bottoms than on sandy or rocky bottoms. Oil spills that do not come ashore, like that of the *Argo Merchant,* do considerably less environmental damage than those that do. The greatest damage resulting from the *Argo Merchant* spill was probably to plankton, especially fish eggs. The *Exxon Valdez* spill on March 23, 1989, in Prince William Sound, Alaska, released 43,000 mt and is the worst tanker spill in U.S. history.

Although 1972 legislation required an end to dumping of **sewage** in the coastal ocean by 1981, exceptions continued to be made. Especially off the **southern California** coast and in **New York Bight,** large amounts of sewage and sewage sludge continued to be dumped. Increased public concern resulted in new legislation that sets January 1992 as the deadline for stopping sewage dumping in the ocean. Some scientists expected that the transfer of dump sites along the northeast Atlantic coast to deeper water farther offshore would make it safe to continue to dump sewage from Philadelphia and the New York area into the coastal ocean. However, people who fish have complained that the deep-water dumping has damaged fisheries less than 1 year after it began.

Most of the **radioactive waste** in the ocean comes from **fallout** resulting from the testing of nuclear explosives since 1944. However, the areas where the greatest concentration of radionuclides is entering the ocean are near the fuel reprocessing plants in Great Britain and France.

DDT pollution produced a decline in the **Long Island osprey** population in the 1950s and the **brown pelican** population of the California coast in the 1960s. Virtual cessation of DDT use in the Northern Hemisphere in 1972 allowed the recovery of both populations. The DDT had thinned the eggshells and reduced the number of successful hatchings.

The first major human disaster resulting from ocean pollution reached epidemic proportions in 1956 within **Minamata Bay, Japan**. More than 50 people died, and another 100 were affected by **mercury poisoning**. A similar episode occurred during the late 1960s at Niigata, Japan. Mercury contamination levels of mercury have been set based on the fish consumption of populations throughout the world, and they appear to have been effective in preventing further poisonings.

With its Action Plan for the Human Environment initiated in 1972, the **United Nations Environment Programme** established 11 environmental units within which assessment and monitoring have begun or are planned. They are the Mediterranean Sea, Red Sea, Kuwait, Caribbean, West and Central Africa, East Asian Seas, Southeast Pacific, Southwest Pacific, East Africa, South Asian Seas, and Southwest Atlantic.

QUESTIONS AND EXERCISES

1. Discuss why it is difficult to determine the assimilative capacity of the oceans for accepting our waste materials.

2. List the acts of the U.S. Congress that have protected water quality, and describe restrictions contained in each.

3. Do you think it is a better policy (1) not to allow dumping into the ocean of any substance that cannot be proven to be harmless to the ocean or (2) to consider the effects of disposing of the substance on land and in the ocean before deciding where to dispose of it? Discuss the reasons for your choice.

4. Discuss the means by which plastics are thought to enter the ocean, their known negative effects on organisms, and legislation being considered to outlaw dumping of plastics in the ocean.

5. Describe the effect of oil spills on species diversity and recovery in the benthos of Wild Harbor and Chedabucto Bay.

6. Discuss whether oil spills that wash ashore or those that do not, like that from the *Argo Merchant,* are more destructive to marine life.

7. How might dumping sewage in deeper water of the East Coast help reduce the degradation of the ocean bottom?

8. What components of the nuclear energy generation industry introduce the most radioactive wastes into the oceans?

9. Which group of animals, from what we have been able to determine, suffered the greatest threat from DDT?

10. Refer to figure 18–10 and discuss the desirability of setting a mercury contamination level for fish at the 1.0 mg/kg level for the citizens of the United States, Sweden, and Japan.

REFERENCES

Bascom, W. 1978. Quantifying man's influence on coastal waters. *Journal of Ocean Engineering* 3:4.

Borgese, E. M., and Ginsburg, N., eds. 1988. *Ocean Yearbook 7.* Chicago: University of Chicago Press.

Comptroller General of the United States. 1977. *Problems and progress in regulating ocean dumping of sewage sludge and industrial wastes.*

Frey, R. W., Howard, J. D., and Dörjes, J. 1989. Coastal sequences, eastern Buzzards Bay, Massachusetts: Negligible record of an oil spill. *Geology* 17:5, 461–65.

International Atomic Energy Agency. 1980. *International symposium on the impacts of radionuclide releases into the marine environment.*

Krendle, P. A., ed. 1975. *Heavy metals in the aquatic environment.* New York: Pergamon Press.

Lawrence, W. W. 1979. *Of acceptable risk: Science and the determination of safety.* Los Altos, Calif.: William Kaufman, Inc.

Mearns, A. J., and Word, J. Q. 1981. Forecasting effects of sewage solids on marine benthic communities. In *Ecological effects of environmental stress,* ed. J. O'Connor and G. Mayer. Estuarine Research Foundation, special publication.

National Academy of Sciences, National Research Council. 1979. *Report on a multimedium management of municipal sludge,* Vol. 9.

National Advisory Committee on Oceans and Atmosphere. 1980. *The role of the ocean in a waste management strategy*. A Special Report to the President and the Congress.

National Oceanic and Atmospheric Administration. 1979. *Assimilative capacity of U.S. coastal waters for pollutants*.

Proceedings of Workshop at Crystal Mountain, Washington, July 29-August 4, 1979.

Oil spills and spills of hazardous substances. 1975. Washington, D.C.: U.S. Environmental Protection Agency.

Wilbur, R. J. 1987. Plastic in the North Atlantic. *Oceanus* 30:3, 61–68.

SUGGESTED READING

Sea Frontiers

Driessen, P. K. 1987. Oil rigs and sea life: A shotgun marriage that works. 33:5, 362–72.
Despite resistance to offshore drilling because of aesthetic and environmental concerns, there is evidence that offshore drilling rigs enhance the ecology of many areas.

Gruber, M. 1971. The great ocean sweepstakes. 17:3, 146–59.
The problems related to oil pollution of the ocean are discussed. Devices and methods designed to aid in confining and cleaning up oil spills are also covered.

Hull, E. W. S. 1978. Oil spills: The causes and the cures. 24:6, 360–69.
A general discussion of oil spills, their frequency, and methods to clean them up.

Idyll, C. P. 1971. Mercury and fish. 17:4, 230–40.
An article written just after the mercury scare of 1970 discusses the effects of mercury on marine life.

Nowadnick, J. 1977. There is but one ocean. 23:3, 130–40.
Problems related to exploitation of the ocean in terms of environmental impact and international politics are discussed.

Parry, J. 1983. Nations unite to fight pollution. 29:3, 143–50.
The efforts of the United Nations Environment Programme are described.

Reynolds, J. E., and Wilcox, J. R. 1987. People, power plants, and manatees. 33:4, 263–69.
The effects of water heating by power plants on the distribution of manatees is discussed.

White, I. C. 1985. An organization to help combat oil spills. 31:1, 15–21.
The efforts of the International Tanker Owners Pollution Federation Limited to increase the efficiency of cleanup operations and reduce the negative effects of oil spills have resulted in more intelligent responses to oil spills.

Scientific American

Bascom, W. 1974. Disposal of waste in the ocean. 231:2, 16–25.
An informative overview of people's misconceptions concerning the disposal of waste in the oceans and the methods that may be used to safely use the oceans for such a purpose.

Butler, J. N. 1975. Pelagic tar. 232:6, 90–97.
Deals with the origin and distribution of lumps of tar found floating at the ocean surface and their possible implications to marine biology.

APPENDIX I
Scientific Notation

To simplify writing very large and very small numbers, scientists indicate the number of zeros by scientific notation, in which one integer is placed to the left of the decimal and a multiplication times a power of 10 tells which direction and how far the decimal would be moved to write the number out in its long form. For example:

$$2.13 \times 10^5 = 213,000$$
or
$$2.13 \times 10^{-5} = 0.0000213$$

Further examples showing numbers that are powers of 10 are:

$$
\begin{aligned}
1,000,000,000 &= 1.0 \times 10^9 \ \text{ or } 10^9 \\
1,000,000 &= 1.0 \times 10^6 \ \text{ or } 10^6 \\
1,000 &= 1.0 \times 10^3 \ \text{ or } 10^3 \\
100 &= 1.0 \times 10^2 \ \text{ or } 10^2 \\
10 &= 1.0 \times 10^1 \ \text{ or } 10^1 \\
1 &= 1.0 \times 10^0 \ \text{ or } 10^0 \\
0.1 &= 1.0 \times 10^{-1} \text{ or } 10^{-1} \\
0.01 &= 1.0 \times 10^{-2} \text{ or } 10^{-2} \\
0.001 &= 1.0 \times 10^{-3} \text{ or } 10^{-3} \\
0.000001 &= 1.0 \times 10^{-6} \text{ or } 10^{-6} \\
0.000000001 &= 1.0 \times 10^{-9} \text{ or } 10^{-9}
\end{aligned}
$$

To add or subtract numbers written as powers of 10, they must be converted to the same power:

Addition

$$
\begin{array}{r}
2.1 \times 10^3 \\
+1.0 \times 10^5 =
\end{array}
\begin{array}{r}
0.021 \times 10^5 \\
+1.000 \times 10^5 \\
\hline
1.021 \times 10^5
\end{array}
$$

Subtraction

$$
\begin{array}{r}
3.4 \times 10^4 \\
-2.0 \times 10^3 =
\end{array}
\begin{array}{r}
3.4 \times 10^4 \\
-0.2 \times 10^4 \\
\hline
3.2 \times 10^4
\end{array}
$$

To multiply or divide numbers written as powers of 10, the exponents are added or subtracted:

Multiplication

$$
\begin{array}{r}
6.04 \times 10^2 \\
\times 2.1 \times 10^4 \\
\hline
12.684 \times 10^6
\end{array}
$$

Division

$$
\begin{array}{r}
3.0 \times 10^3 \\
\div 1.5 \times 10^2 \\
\hline
2.0 \times 10^1
\end{array}
$$

APPENDIX II
The Metric System and Conversion Factors

Units of Length   1 micrometer (μm) = 10^{-6} m = 0.0000394 in.
1 millimeter (mm) = 10^{-3} m = 0.0394 in. = 10^3 μm
1 centimeter (cm) = 10^{-2} m = 0.394 in. = 10^4 μm
1 meter (m) = 10^2 cm = 39.4 in = 3.28 ft = 1.09 yd = 0.547 fath
1 kilometer (km) = 10^3 m = 0.621 statute mi = 0.540 nautical mi

Units of Volume   1 liter = 10^3 cm^3 = 1.0567 liquid qt = 0.264 U.S. gal
1 cubic meter (m^3) = 10^6 cm^3 = 10^3 I = 35.3 ft^3 = 264 U.S. gal
1 cubic kilometer (km^3) = 10^9 m^3 = 10^{15} cm^3 = 0.24 satute mi^3

Units of Area   1 square centimeter (cm^2) = 0.155 in.^{2}
1 square meter (m^2) = 10.7 ft^2
1 square kilometer (km^2) = 0.292 nautical mi^2 = 0.386 statute mi^2

Units of Time   1 day = 8.64 $\times$ 10^4 s (mean solar day)
1 year = 8765.8 h = 3.156 $\times$ 10^7 s

Units of Mass   1 gram (g) = 0.035 oz
1 kilogram (kg) = 10^3 g = 2.205 lb
1 metric ton (mt) = 10^6 g = 2205 lb

Units of Speed   1 centimeter per second (cm/s) = 0.0328 ft/s
1 meter per second (m/s) = 2.24 statute mi/h = 1.94 kt
1 kilometer per hour (km/h) = 27.8 cm/s = 0.55 kt
1 knot (1 nautical mile per hour, kt) = 1.15 statute mi/h = 0.51 m/s

Units of Temperature

Celsius (°C)	Fahrenheit (°F)	Kelvin (K)	
−237.2	−459.7	0	Absolute zero (lowest possible temperature)
0	32	273.2	Freezing point of water
100	212	373.2	Boiling point of water

Conversions: °F = (1.8 × °C) + 32   °C = (°F − 32)/1.8

APPENDIX III
Periodic Table of the Elements

Atomic number — 1
Symbol of element — H
Atomic weight — 1.0080

Radioactive state at 1 atmos. and 20°C (68°F)

Inert gas ☐
Gas ▨
Liquid ◯
Solid—all others

Light Metals

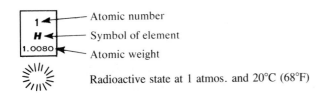

Transitional Elements

Nonmetals

Heavy Metals

	I A	II A	III B	IV B	V B	VI B	VII B		VIII B			I B	II B	III A	IV A	V A	VI A	VII A	VIII A
1	1 H 1.0080																		2 He 4.003
2	3 Li 6.939	4 Be 9.012												5 B 10.81	6 C 12.011	7 N 14.007	8 O 15.9994	9 F 18.998	10 Ne 20.183
3	11 Na 22.990	12 Mg 24.31												13 Al 26.98	14 Si 28.09	15 P 30.974	16 S 32.064	17 Cl 35.453	18 Ar 39.948
4	19 K 39.102	20 Ca 40.08	21 Sc 44.96	22 Ti 47.90	23 V 50.94	24 Cr 52.00	25 Mn 54.94	26 Fe 55.85	27 Co 58.93	28 Ni 58.71	29 Cu 63.54	30 Zn 65.37	31 (Ga) 69.72	32 Ge 72.59	33 As 74.92	34 Se 78.96	35 (Br) 79.909	36 Kr 83.80
5	37 Rb 85.47	38 Sr 87.62	39 Y 88.91	40 Zr 91.22	41 Nb 92.91	42 Mo 95.94	43 Tc (99)	44 Ru 101.1	45 Rh 102.90	46 Pd 106.4	47 Ag 107.870	48 Cd 112.40	49 In 114.82	50 Sn 118.69	51 Sb 121.75	52 Te 127.60	53 I 126.90	54 Xe 131.30
6	55 (Cs) 132.91	56 Ba 137.34	57 TO 71	72 Hf 178.49	73 Ta 180.95	74 W 183.85	75 Re 186.2	76 Os 190.2	77 Ir 192.2	78 Pt 195.09	79 Au 197.0	80 (Hg) 200.59	81 Tl 204.37	82 Pb 207.19	83 Bi 208.98	84 Po (210)	85 At (210)	86 Rn (222)
7	87 (Fr) (223)	88 Ra 226.05	89 TO 103															

Rare Earth Elements

Lanthanide series	57 La 138.91	58 Ce 140.12	59 Pr 140.91	60 Nd 144.24	61 Pm (147)	62 Sm 150.35	63 Eu 151.96	64 Gd 157.25	65 Tb 158.92	66 Dy 162.50	67 Ho 164.93	68 Er 167.26	69 Tm 168.93	70 Yb 173.04	71 Lu 174.97
Actinide series	89 Ac (227)	90 Th 232.04	91 Pa (231)	92 U 238.03	93 Np (237)	94 Pu (242)	95 Am (243)	96 Cm (247)	97 Bk (249)	98 Cf (251)	99 Es (254)	100 Fm (253)	101 Md (256)	102 No (254)	103 Lw (257)

367

APPENDIX IV
The Geologic Time Table

Era	Period	Epoch & Age 10^6 Yr		Development of Life and Orogenic Events	
Cenozoic	Quaternary	Recent			
		Pleistocene		Pacific Coast Ranges form	
		Pliocene	2 —		
	Tertiary	Miocene	7 —		
		Oligocene	25 —		
		Eocene	38 —		
		Paleocene	55 —		Rocky Mts. form
Mesozoic	Cretaceous		65 —	First primates	
	Jurassic		136 —		
	Triassic		190 —	First birds	Sierra Nevada form
			225 —	First mammals	
Paleozoic	Permian		280 —	First flowering plants	
	Pennsylvanian			First reptiles	
	Mississippian		310 —	First insects	Appalachian Mts. form
	Devonian		345 —		
	Silurian		400 —	First amphibians	
	Ordovician		430 —	First land plants	
	Cambrian		500 —	First vertebrates (fish)	
	Eocambrian		570 —	First abundant animal fossils	
Precambrian			640 —	First algae, bacteria	
			4500 —		

368

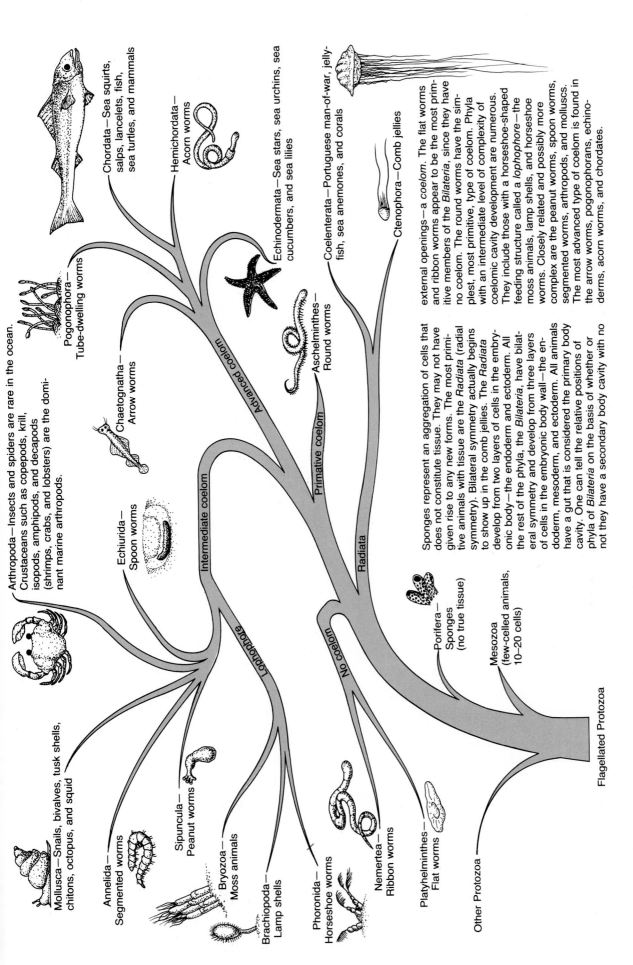

Mollusca—Snails, bivalves, tusk shells, chitons, octopus, and squid

Annelida—Segmented worms

Sipuncula—Peanut worms

Bryozoa—Moss animals

Brachiopoda—Lamp shells

Phoronida—Horseshoe worms

Nemertea—Ribbon worms

Platyhelminthes—Flat worms

Other Protozoa

Arthropoda—Insects and spiders are rare in the ocean. Crustaceans such as copepods, krill, isopods, amphipods, and decapods (shrimps, crabs, and lobsters) are the dominant marine arthropods.

Pogonophora—Tube-dwelling worms

Echiurida—Spoon worms

Chaetognatha—Arrow worms

Lophophore

Advanced coelom

Intermediate coelom

Primative coelom

Radiata

No coelom

Mesozoa (few-celled animals, 10–20 cells)

Porifera—Sponges (no true tissue)

Chordata—Sea squirts, salps, lancelets, fish, sea turtles, and mammals

Hemichordata—Acorn worms

Echinodermata—Sea stars, sea urchins, sea cucumbers, and sea lilies

Coelenterata—Portuguese man-of-war, jelly-fish, sea anemones, and corals

Ctenophora—Comb jellies

Aschelminthes—Round worms

Flagellated Protozoa

Sponges represent an aggregation of cells that does not constitute tissue. They may not have given rise to any new forms. The most primitive animals with tissue are the *Radiata* (radial symmetry). Bilateral symmetry actually begins to show up in the comb jellies. The *Radiata* develop from two layers of cells in the embryonic body—the endoderm and ectoderm. All the rest of the phyla, the *Bilateria*, have bilateral symmetry and develop from three layers of cells in the embryonic body wall—the endoderm, mesoderm, and ectoderm. All animals have a gut that is considered the primary body cavity. One can tell the relative positions of phyla of *Bilateria* on the basis of whether or not they have a secondary body cavity with no

external openings—a *coelom*. The flat worms and ribbon worms appear to be the most primitive members of the *Bilateria*, since they have no coelom. The round worms have the simplest, most primitive, type of coelom. Phyla with an intermediate level of complexity of coelomic cavity development are numerous. They include those with a horseshoe-shaped feeding structure called a *lophophore*—the moss animals, lamp shells, and horseshoe worms. Closely related and possibly more complex are the peanut worms, spoon worms, segmented worms, arthropods, and molluscs. The most advanced type of coelom is found in the arrow worms, pogonophorans, echinoderms, acorn worms, and chordates.

APPENDIX V The Phylogenetic Tree

Taxonomic Classification of Common Marine Organisms

Kingdom Monera

Organisms without nuclear membranes; nuclear material is spread throughout cell; predominantly unicellular.

Phylum Schizophyta Smallest known cells; bacteria (1500 species).

Phylum Cyanophyta Blue-green algae; chlorophyll *a*, carotene and phycobilin pigments (200 species).

Kingdom Protista

Organisms with nuclear material confined to nucleus by a membrane.

Phylum Chrysophyta Golden-brown algae; includes diatoms, coccolithophores and silicoflagellates; chlorophyll *a* and *c*, xanthophyll and carotene pigments (6000$^+$ species).

Phylum Pyrrophyta Dinoflagellate algae; chlorophyll *a* and *c*, xanthophyll and carotene pigments (1100 species).

Phylum Chlorophyta Green algae, chlorophyll *a* and *b* and carotene pigments (7000 species).

Phylum Phaeophyta Brown algae; chlorophyll *a* and *c* xanthophyll and carotene pigments (1500 species)

Phylum Rhodophyta Red algae; chlorophyll *a*, carotene and phycobilin pigments (4000 species).

Phylum Protozoa Nonphotosynthetic, heterotrophic protists (27,400 species).
Class Mastigophora Flagelleted; dinoflagellates (5200 species).
Class Sarcodina Ameboid; foraminiferans and radiolarians (11,500 species).
Class Ciliophora Ciliated (6000 species).

Kingdom Fungi

Phylum Mycophyta Fungi, lichens; most fungi are decomposers found on the ocean floor, whereas lichens inhabit the upper intertidal zones (3000 species of fungi, 160 species of lichen).

Kingdom Metaphyta

Multicellular, complex plants.

Phylum Tracheophyta Vascular plants with roots, stems, and leaves that are serviced by special cells that carry food and fluids (287,200 species).
Class Angiospermae Flowering plants with seeds contained in a closed vessel (275,000 species).

Kingdom Metazoa

Multicellular animals.

Phylum Porifera Sponges; spicules are the only hard parts in these sessile animals that do not possess tissue (10,000) species.
Class Calcarea Calcium carbonate spicules (50 species).
Class Desmospongiae Skeleton may be composed of siliceous spicules or spongin fibers or be nonexistent (9500) species).
Class Sclerospongiae Coralline sponges; massive skeleton composed of calcium carbonate, siliceous spicules and organic fibers (7 species).
Class Hexactinellida Glass sponges; six-rayed with siliceous spicules (450 species).

Phylum Cnidaria (Coelenterata) Radially symmetrical, two-cell, layered body wall with one opening to gut cavity; polyp (asexual, sexual, benthic) and medusa (sexual, pelagic) body forms (10,000 species).
Class Hydrozoa Polypoid colonies such as pelagic Portuguese man-of-war and benthic Obelia common; me-

dusa present in reproductive cycle but reduced in size (3000 species).

Class Scyphozoa Jellyfish; medusa up to 1 m in diameter is dominant form; polyp is small if present (250 species).

Class Anthozoa Corals and anemones possessing only polypoid body form and reproducing asexually and sexually (6500 species).

Phylum Ctenophora Predominantly planktonic comb jellies; basic eight-sided radial symmetry modified by secondary bilateral symmetry (80 species).

Phylum Platyhelminthes Flatworms; bilateral symmetry; hermaphroditic (25,000 species).

Phylum Nemertea Ribbon worms; as long as 30 m; benthic and pelagic (800 species).

Phylum Nematoda Roundworms; marine forms are primarily free-living and benthic; most 1 to 3 mm in length (5000 marine species).

Phylum Rotifera Ciliated, unsegmented forms less than 2 mm in length (1500 species, only a few marine).

Phylum Bryozoa (Ectoprocta) Moss animals; benthic, branching or encrusting colonies; lophophore feeding structure (4500 species).

Phylum Branchiopoda Lamp shells; lophophorate benthic bivalves (300 species).

Phylum Phoronida Horseshoe worms; 24-cm-long lophorate tubeworms that live in sediment of shallow and temperate shallow waters (15 species).

Phylum Sipuncula Peanut worms, benthic (325 species).

Phylum Echiura Spoon worms; sausage shaped with spoon-shaped proboscis; burrow in sediment or live under rocks (130 species).

Phylum Pogonophora Tube-dwelling, gutless worms 5 to 80 cm long; absorb organic matter through body wall (100 species).

Phylum Tardigrada Marine meiofauna that have the ability to survive long periods in a cryptobiotic state (diversity is poorly known).

Phylum Mollusca Soft bodies possessing a muscular foot and mantle that usually secretes calcium carbonate shell (75,000 species).

Class Monoplacophora Rare trench-dwelling forms with segmented bodies and limpetlike shells (10 species).

Class Polyplacophora Chintons; oval, flattened body covered by eight overlapping plates (600 species).

Class Gastropoda Large, diverse group of snails and their relatives; shell spiral if present (64,500 species).

Class Bivalvia Bivalves; includes mostly filter-feeding clams, mussels, oysters and scallops (7500 species).

Class Aplacophora Tusk shells; sand-burrowing organisms that feed on small animals living in sand deposits (350 species).

Class Cephalopoda Octopus, squid and cuttlefish which possess no external shell except in the genus *Nautilus* (600 species).

Phylum Annelida Segmented worms in which musculature, circulatory, nervous, excretory, and reproductive systems may be repeated in many segments; mostly benthic (10,000 marine species).

Phylum Arthropoda Jointed-legged animals with segmented body covered by an exoskeleton (30,000 marine species).

Subphylum Crustacea Calcareous exoskeleton, two pairs of antennae; cephlon, thorax and abdomen body parts; includes copepods, ostracods, barnacles, shrimp, lobsters and crabs (26,000 species).

Subphylum Chelicerata Horseshoe crabs (4 species).
Class Merostomata
Class Pycnogonida Sea spiders.

Subphylum Uniramia Insects; genus *Halobites* is the only truly marine insect.

Phylum Chaetognatha Arrowworms; mostly planktonic, transparent and slender; up to 10 cm long (50 species).

Phylum Echinodermata Spiny-skinned animals; benthic animals with secondary radial symmetry and water vascular system (6000 species).

Class Asteroidea Starfishes; free-living, flattened body with five or more rays with tube feet used for locomotion; mouth down (1600 species).

Class Ophiuroidea Brittle stars and basket stars; prominent central disc with slender rays; tube feet used for feeding; mouth down (200 species).

Class Echinoidea Sea urchins, sand dollars, heart urchins; free-living forms without rays; calcium carbonate test; mouth down or forward (860 species).

Class Holothuroidea Sea cucumbers; soft bodies with radial symmetry obscured; mouth forward (900 species).

Class Crinoidea Sea lillies, feather stars; cup-shaped body attached to bottom by a jointed stalk or appendages; mouth up (630 species).

Phylum Hemichordata Acorn worms and pterobranchs; primitive nerve chord; gill slits; benthic (90 species).

Phylum Chordata Notochord; dorsal nerve chord and gills or gill slits (55,000 species).

Subphylum Urochordata Tunicates; chordate characteristics in larval stage only; benthic sea squirts and planktonic thaliaceans and larvaceans (1375 species).

Subphylum Cephalochordata Amphioxus or lancelets; live in coarse temperate and tropical sediment (25 species).

Subphylum Vertebrata Internal skeleton; spinal column of vertebrae; brain (52,000 species).

Class Agnatha Lampreys and hagfishes; most primitive vertebrates with cartilaginous skeleton, no jaws, and no scales (50 species).

Class Chondrichthyes Sharks, skates and rays; cartilaginous skeleton; 5 to 7 gill openings; placoid scales (625 species).

Class Osteichthyes Bony fishes; cycloid scales; covered gill opening; swim bladder common (30,000 species).

Class Amphibia Frogs, toads, and salamanders; Asian mud flat frogs are the only amphibians that tolerate marine water (2600 species).

Class Reptilia Snakes, turtles, lizards, and alligators; orders Squamata (snakes) and Chelonia (turtles) are major marine groups (6500 species).

Class Aves Birds; many live on and in the ocean but all must return to land to breed (8600 species).

Class Mammalia Warm-blooded; hair; mammary glands; bear live young; marine representatives found in the orders Sirenia (sea cows, dugong), Cetacea (whales), and Carnivora (sea otter, pennipeds) (4100 species).

GLOSSARY

Abiotic environment The nonliving components of an ecosystem.

Abyssal clay Deep-ocean (oceanic) deposits containing less than 30 percent biogenous sediment.

Abyssal hill Volcanic peaks rising less than 1 km (0.621 mi) above the ocean floor.

Abyssal plain A flat depositional surface extending seaward from the continental rise or oceanic trenches.

Abyssal zone The benthic environment between 4000 and 6000 m (13,120 and 19,680 ft).

Abyssopelagic Open-ocean (oceanic) environment below 4000 m depth.

Algae One-celled or many-celled plants that have no root, stem, or leaf systems. Simple plants.

Amino acid One of more than 20 naturally occurring compounds that contain NH_2 and $COOH$ groups. They combine to form proteins.

Amphidromic point A nodal or no-tide point in the ocean or sea around which the crest of the tide wave rotates during one tidal period.

Amphineura Class of molluscs with eight dorsal calcareous plates. Chitons.

Amphipoda Crustacean order containing laterally compressed members such as the "sand hoppers."

Anaerobic respiration Respiration carried on in the absence of free oxygen (O_2). Some bacteria and protozoans carry on respiration this way.

Annelida Phylum of elongated segmented worms.

Antarctic bottom water A water mass that forms in the Weddell Sea, sinks to the ocean floor, and spreads across the bottom of all oceans.

Antinode Zone of maximum vertical particle movement in standing waves where crest and trough formation alternate.

Aphelion The point in the orbit of a planet or comet where it is farthest from the sun.

Aphotic zone Without light. The ocean is generally in this state below 1000 meters (3280 ft).

Apogee The point in the orbit of the moon or an artificial satellite that is farthest from the earth.

Aschelminthes Phylum of wormlike pseudocoelomates.

Assimilative capacity The level of concentration to which the ocean can accumulate foreign substances without harming marine life or humans that use the ocean or its resources.

Asthenosphere A plastic layer in the upper mantle 80–200 km deep which may allow lateral movement of lithospheric plates and isostatic adjustments.

Atlantic-type margin The passive trailing edge of a continent that is subsiding due to lithospheric cooling and increasing sediment load.

Atom The smallest particle of an element that can combine with similar particles of other elements to produce compounds.

Autolytic decomposition Decomposition of organic matter that is achieved by enzymes present in the tissue. The enzymes are triggered to begin their work at death of the tissue.

Autotroph Plants and bacteria that can synthesize organic compounds from inorganic nutrients.

Backshore The inner portion of the shore, lying landward of the mean spring tide high water line. Acted upon by the ocean only during exceptionally high tides and storms.

Barnacle *See* Cirripedia.

Barrier island A long, narrow wave-built island separated from the mainland by a lagoon.

Barrier reef A coral reef separated from the nearby landmass by open water.

Basalt A dark-colored volcanic rock characteristic of the ocean crust. Contains minerals with relatively high iron and magnesium content.

Bathyal zone The benthic environment between the depths of 200 and 4000 (656 and 13,120 ft). It includes mainly the continental slope and the oceanic ridges and rises.

Bathymetry The study of ocean depth.

Bathypelagic zone The pelagic environment between the depths of 1000 and 4000 m (3,280 and 13,120 ft).

Bay barrier A marine deposit attached to the mainland at both ends and extending entirely across the mouth of a bay, separating the bay from the open water.

Beach Sediment seaward of the coastline through the surf zone that is in transport along the shore and within the surf zone.

Benthic Pertaining to the ocean bottom.

Benthos The forms of marine life that live on the ocean bottom.

Biogenous sediment Sediment containing material produced by plants or animals, e.g., coral reefs, shell fragments, and housing of diatoms, radiolarians, foraminifera, and coccolithophores.

Biogeochemical cycles The natural cycling of compounds among the living and nonliving components of an ecosystem.

Bioluminescence Light produced by chemical reaction. Found in bacteria, phytoplankton, and metazoans.

Biomass The total mass of a defined organism or group of organisms in a particular community or the ocean as a whole.

Biotic community The living organisms that inhabit an ecosystem.

Bore A steep-fronted tide crest that moves up a river in association with high tide.

Boundary current The northward- or southward-flowing currents that form the western and eastern boundaries, respectively, of the subtropical circulation gyres.

Buoyancy The ability or tendency to float or rise in a liquid.

Breaker zone Region where waves break at the seaward margin of the surf zone.

Bryozoa Phylum of colonial animals that often share one coelomic cavity. Encrusting and branching forms secrete a protective housing (zooecium) of calcium carbonate or chitinous material. Possess lophophore feeding structure.

Calcareous Containing calcium carbonate.

Calcium carbonate, CaCO$_3$ A chalklike substance secreted by many organisms in the form of coverings or skeletal structures.

Calorie Unit of heat defined as the amount of heat required to raise the temperature of 1 g of water 1°C.

Capillary wave Ocean waves whose wavelength is less than 1.74 cm. The dominant restoring force for such waves is surface tension.

Carapace Chitinous or calcareous shield that covers the cephalothorax of some crustaceans. Dorsal portion of a turtle shell.

Carbohydrates An organic compound containing the elements carbon, hydrogen, and oxygen with the general formula (CH$_2$O)n.

Carnivore An animal that depends on other animals solely or chiefly for its food supply.

Carotin A red to yellow pigment found in plants.

Celsius temperature scale 0°C = 273.16 K; 0°C = freezing point of water; 100°C = boiling point of water.

Centrifugal force A fictional force that seems to make an object move away from the center of a curved path it is following. It results from the application of a centripetal force that acts against the inertia of the object.

Centripetal force A center-seeking force that tends to make rotating bodies move toward the center of rotation.

Cephalopoda A class of the phylum Mollusca with a well-developed pair of eyes and a ring of tentacles surrounding the mouth. The shell is absent or internal on most members. The class includes the squid, octopus, and nautilus.

Cephalothorax The head and midbody region of crustaceans.

Cetacea An order of marine mammals that includes the whales.

Chaetognatha A phylum of elongate transparent wormlike pelagic animals commonly called *arrowworms*.

Chela Arthropod appendage modified to form a pincer.

Chemical energy A form of potential energy stored in the chemical bonds of compounds.

Chemosynthesis A process by which bacteria synthesize organic molecules from inorganic nutrients using chemical energy released from the bonds of some chemical compound by oxidation.

Chiton Common name for any member of the class of molluscs, Amphineura, with eight dorsal plates.

Chloride ion, Cl$^-$ A chlorine atom that has become negatively charged by gaining one electron.

Chlorinity The amount of chloride ion and ions of other halogens in ocean water expressed in parts per thousand by weight, ‰.

Chlorophyll A group of green pigments that make it possible for plants to carry on photosynthesis.

Chlorophyta Green Algae. Characterized by the presence of chlorophyll and other pigments.

Chrysophyta An important phylum of planktonic algae, including the diatoms. The presence of chlorophyll is masked by the pigment carotin that gives the plants a golden color.

Cilium Short hairlike structures common on lower animals. Beating in unison, they may create water currents that carry food toward the mouth of an animal or be used for locomotion.

Circumpolar Current Eastward-flowing current that extends from the surface to the ocean floor and encircles Antarctica.

Cirripedia An order of crustaceans with up to six pairs of thoracic appendages that strain food from the water. They are the barnacles that attach themselves to the substrate and secrete an external calcareous housing.

Clastic A rock or sediment composed of broken fragments of preexisting rocks. Two common examples are beach deposits and sandstone.

Clay A term relating to particle size between silt and colloid. Clay minerals are hydrous aluminum silicates with plastic, expansive, and cation exchange properties.

Cnidoblast Stinging cell of phylum Coelenterata that contains the stinging mechanism (nematocyst) used in defense and capturing prey.

Coast A strip of land that extends inland from the coastline as far as marine influence is evidenced in the landforms.

Coastal upwelling The movement of deeper nutrient-rich water into the surface water mass as a result of windblown surface water moving offshore.

Coastline Landward limit of the effect of the highest storm waves on the shore.

Coccolithophore A microscopic planktonic form of algae, encased by a covering composed of calcareous discs (coccoliths).

Coelenterata Phylum of radially symmetrical animals that includes two basic body forms, the medusa and the polyp.

Includes jellyfish (medusoid) and sea anemones (polypoid).

Coelom Secondary body cavity (gut cavity is primary cavity). Forms within the mesoderm in higher animals, is lined with peritoneum, and contains vital organs.

Colonial animals Animals that live in groups of attached or separate individuals. Groups of individuals may serve special functions.

Comb jelly Common name for members of the phylum Ctenophora. (*See* Ctenophora.)

Commensalism A symbiotic relationship in which one party benefits and the other is unaffected.

Compensation depth, CaCO₃ The depth at which the amount of $CaCO_3$ produced by the organisms in the overlying water column is equal to the amount of $CaCO_3$ the water column can dissolve. There will be no $CaCO_3$ deposition below this depth.

Compensation depth, O₂ The depth where the oxygen produced by photosynthesis is equal to the oxygen requirements of plant respiration. A plant population cannot be sustained below this depth, which will be greater in the open ocean than near the shore due to the relatively deeper light penetration in the open ocean.

Compound A substance containing two or more elements combined in fixed proportions.

Condensation The conversion of water from the vapor to the liquid state. When it occurs, the energy required to vaporize the water is released into the atmosphere. This is about 585 cal/g of water at 20°C.

Conduction The transmission of heat by the passage of energy from particle to particle.

Conservative property A property of ocean water that the water attains at the surface and is changed only by mixing and diffusion after the water sinks below the surface.

Consumers The animal populations within an ecosystem that consume the organic mass produced by the producers.

Continent About one-third of the earth's surface that rises above the deep-ocean floor to be exposed above sea level. Continents are composed primarily of granite, an igneous rock of lower density than the basaltic oceanic crust.

Continental borderland A highly irregular portion of the continental margin that is submerged beneath the ocean and is characterized by depths greater than those characteristic of the continental shelf.

Continental drift A term applied to early theories supporting the possibility the continents are in motion over the earth's surface.

Continental rise A gently sloping depositional surface at the base of the continental slope.

Continental shelf A gently sloping depositional surface extending from the low water line to the depth of a marked increase in slope around the margin of a continent or island.

Continental slope A relatively steeply sloping surface lying seaward of the continental shelf.

Convection In a fluid being heated unevenly, the warmer part of the mass will rise and the cooler portions will sink. If the heat source is stationary, cells may develop as the rising warm water cools and sinks in regions on either side of the axis of rising.

Convergence There are polar, tropical, and subtropical regions of the oceans where water masses with different characteristics come together. Along these lines of convergence, the denser mass will sink beneath the others.

Copepoda An order of microscopic to nearly microscopic crustaceans that are important members of the zooplankton in temperate and subpolar waters.

Coral A group of benthic anthozoans that exist as individuals or in colonies and secrete $CaCO_3$ external skeletons. Under the proper conditions corals may produce reefs composed of their external skeletons and the $CaCO_3$ material secreted by varieties of algae associated with the reefs.

Coriolis effect A fictional force resulting from the earth's rotation causes particles in motion to be deflected to the right in the Northern Hemisphere and to the left in the Southern Hemisphere.

Cosmogenous sediment All sediment derived from outer space.

Cotidal lines Lines connecting points where high tide occurs simultaneously.

Crust Unit of earth's structure that is composed of basaltic ocean crust and granitic continental crust. The total thickness of the crustal units may range from 5 km beneath the ocean to 50 km beneath the continents.

Crustacea A class of phylum Arthropoda that includes barnacles, copepods, lobsters, crabs, and shrimp.

Crystalline rock Igneous or metamorphic rocks. These rocks are made up of crystalline particles with orderly molecular structures.

Ctenophora A phylum of gelatinous organisms that are more or less spheroidal with biradial symmetry. These exclusively marine animals have eight rows of ciliated combs for locomotion and most have two tentacles for capturing prey.

Current A horizontal movement of water.

Cypris Advanced free-swimming larval stage of barnacles. After attaching to substrate, it metamorphoses into adult.

Decapoda 1. An order of crustaceans with five pairs of thoracic "walking legs." Includes crabs, shrimp, and lobsters. 2. Suborder of cephalopod molluscs with 10 arms that includes squids and cuttlefish.

Decomposers Primarily bacteria that break down nonliving organic material, extract some of the products of decomposition for their own needs, and make available the compounds needed for plant production.

Deep boundary current Relatively strong deep currents flowing across the continental rise along the western margin of ocean basins.

Deep scattering layer A layer of marine organisms in the open ocean that scatter signals from an echo sounder. It migrates daily from depths of slightly over 100 m at night to more than 800 m during the day.

Deep-sea system Includes all benthic environments beneath the littoral (sublittoral, bathyal, abyssal, and hadal).

Deep water The water beneath the permanent thermocline (pycnocline) that has a uniformly low temperature.

Deep-water wave Ocean wave traveling in water that has a depth greater than one-half the average wave length. Its speed is independent of water depth.

Delta A low-lying deposit at the mouth of a river.

Density Mass per unit volume of a substance. Usually expressed as *grams per cubic centimeter*. For ocean water with a salinity of 35‰ at 0°C, the density is 1.028 g/cm³. (*See* sigma-T.)

Denitrifying bacteria Bacteria that reduce oxides of nitrogen to produce free nitrogen (N_2).

Detritus Any loose material produced directly from rock disintegration. (Organic: material resulting from the disintegration of dead organic remains.)

Diatom Member of the class Bacillariophyceae of algae that possesses a wall of overlapping silica valves.

Diffraction Any bending of a wave around an obstacle that cannot be interpreted as refraction or reflection.

Diffusion A process by which fluids move through other fluids from areas of high concentration to areas in which they are in lower concentrations by random molecular movement.

Dinoflagellates Single-celled microscopic organisms that may possess chlorophyll and belong to the plant phylum Pyrrophyta (autotrophic) or may ingest food and belong to the class Mastigophora of the animal phylum Protozoa (heterotrophic).

Discontinuity An abrupt change in a property such as temperature or salinity at a line or surface.

Disphotic zone The dimly lit zone, corresponding approximately to the mesopelagic, in which there is not enough light to carry on photosynthesis; sometimes called the twilight zone.

Dissolved oxygen Oxygen that is dissolved in ocean water.

Distributary A small stream flowing away from a main stream. Such streams are characteristic of deltas.

Diurnal inequality The difference in the heights of two successive high waters or two successive low waters during a lunar (tidal) day.

Diurnal tide A tide with one high water and one low water during a tidal day.

Divergence A horizontal flow of water from a central region, as occurs in upwelling.

Doldrums A belt of light variable winds within 10°–15° of the equator, resulting from the vertical flow of low density air within this equatorial belt.

Dolphin 1. A brilliantly colored fish of the genus *Coryphaena*. 2. The name applied to the small, beaked members of the cetacean family, Delphinidae.

Dorsal Pertaining to the back or upper surface of most animals.

Drifts Thick sediment deposits on the continental rise produced where the deep boundary current slows and loses sediment when it changes direction to follow the base of the continental slope.

Dynamic topography A surface configuration resulting from the geopotential difference between a given surface and a reference surface of no motion. A contour map of this surface is useful in estimating the nature of geostrophic currents.

Earthquake A sudden motion or trembling in the earth caused by the sudden release of slowly accumulated strain by faulting (movement along a fracture in the earth's crust) or volcanic activity.

Ebb current During a decrease in the height of the tide, the ebb current flows seaward.

Echinodermata Phylum of animals that have bilateral symmetry in larval forms and usually a five-sided radial symmetry as adults. Benthic and possessing rigid or articulating exoskeletons of calcium carbonate with spines, this phylum includes sea stars, brittle stars, sea urchins, sand dollars, sea cucumbers, and sea lilies.

Echo sounding Determining the depth of water by measuring the time required for a sonic or ultrasonic signal to travel to the bottom and back to the ship that emitted the signal.

Ecological efficiency Efficiency with which energy is transferred from one trophic level to the next. The ratio of the amount of protoplasm added to a trophic level to the amount of food required to produce it.

Ecosystem All the organisms in a biotic community and the abiotic environmental factors with which they interact.

Ectoderm Outermost layer of cells in an animal embryo. In vertebrates it gives rise to skin, nervous system, sense organs, etc.

Eddy A current of any fluid forming on the side of a main current. It usually moves in a circular path and develops where currents encounter obstacles or flow past one another.

Ekman spiral A theoretical consideration of the effect of a steady wind blowing over an ocean of unlimited depth and breadth and of uniform viscosity. The result is a surface flow at 45° to the right of the wind in the Northern Hemisphere. Water at increasing depth with drift in directions increasingly to the right until at about 100 m depth it is moving in a direction opposite to that of the wind. The net water transport is 90° to the wind, and speed decreases with depth.

Ekman transport The net transport of surface water set in motion by wind. Due to the Ekman spiral phenomenon, it is theoretically in a direction 90° to the right and 90° to the left of the wind direction in the Northern Hemisphere and Southern Hemisphere, respectively.

Electromagnetic energy Energy that travels as waves or particles with the speed of light. Different kinds possess different properties based on wavelength. The longest wavelengths belong to radio waves, up to 10 km in length. At the other end of the spectrum are cosmic rays with greater penetrating power and wavelengths of less than 0.000001 μm.

Electromagnetic spectrum   The spectrum of radiant energy emitted from stars and ranging between cosmic rays with wavelengths less than $10-11$ cm and very long waves with wavelengths in excess of 100 km (62 mi).

Element   One of a number of substances, each of which is composed entirely of like particles, atoms, that cannot be broken into smaller particles by chemical means.

El Niño   A southerly flowing warm current that generally develops off the coast of Ecuador shortly after Christmas. Occasionally it will move father south into Peruvian coastal waters and cause the widespread death of plankton and fish.

Emergent shoreline   A shoreline resulting from the emergence of the ocean floor relative to the ocean surface. It is usually rather straight and characterized by marine features usually found at some depth.

Endoderm   Innermost cell layer of an embryo. Develops into digestive and excretory system and forms lining for respiratory system, etc., in vertebrates.

Endothermic reaction   A chemical reaction that absorbs energy. For example, energy is stored in the organic products of the chemical reaction photosynthesis.

Entropy   A thermodynamic quantity that reflects the degree of randomness in a system. It increases in all natural systems.

Environment   The sum of all physical, chemical, and biological factors to which an organism or community is subjected.

Epicenter   The point on the earth's surface that is directly above the focus of an earthquake.

Epifauna   Animals that live on the ocean bottom, either attached or moving freely over it.

Epipelagic zone   A subdivision of the oceanic province that extends from the surface to a depth of 200 m (656 ft).

Equatorial tide   A twice monthly tide occurring when the moon is over the equator. It displays a minimal diurnal inequality.

Equatorial upwelling   The surfacing of deeper, nutrient-rich water into the equatorial surface mass that results when wind-driven surface water moves to higher latitudes.

Equilibrium tide theory   A tidal hypothesis that considers the ocean to be of uniform nature and depth throughout the earth's surface. It is further assumed that this ocean will respond instantly to the gravitational forces of the sun and moon.

Equinox   The times when the sun is over the equator. Day and night are of equal length throughout the earth. *Vernal equinox* occurs about March 21 as the sun is moving into the Northern Hemisphere. *Autumnal equinox* occurs about September 21 as the sun is moving into the Southern Hemisphere.

Estuarine circulation   Circulation characteristic of an estuary and other bodies of water having restricted circulation with the ocean that results from an excess of runoff and precipitation as compared to evaporation. Surface flow is to the ocean with a subsurface counter flow.

Estuary   The mouth of a river valley where marine influence is manifested as tidal effects and increased salinity of the river water.

Euphotic zone   The surface layer of the ocean that receives enough light to support photosynthesis. The bottom of this zone, which marks the compensation depth, varies and reaches a maximum value of around 150 m in the very clearest open ocean water.

Euryhaline   Pertains to the ability of a marine organism to tolerate a wide range of salinity.

Eurythermal   Pertains to the ability of a marine organism to tolerate a wide range of temperature.

Eutrophism   A condition in which waters are nutrient-rich and support a high level of biological productivity.

Evaporation   The physical process by which a liquid is converted to a gas. Commonly considered to occur at a temperature below the boiling point of the liquid.

Exothermic reaction   A chemical reaction that liberates energy. For example, the energy stored in the products of photosynthesis is released by the chemical reaction respiration.

Fahrenheit temperature scale (°F)   Freezing point of water is 32°; boiling point of water is 212°.

Fat   An organic compound formed from alcohol glycerol and one or more fatty acids; a lipid, it is a solid at atmospheric temperatures

Fan   A gently sloping, fan-shaped feature normally located near the lower end of a canyon.

Fast ice   Sea ice that is attached to the shore and therefore remains stationary.

Fathom   A unit of depth in the ocean, commonly used in countries using the English system of units. It is equal to 1.83 m, or 6 ft.

Fault   A fracture or fracture zone in the earth's crust along which displacement has occurred.

Fault block   A crustal block bounded on at least two sides by faults. Usually elongate; if it is down-dropped it produces a graben; if uplifted, it is a horst.

Fauna   The animal life of any particular area or of any particular time.

Fetch   1. Pertaining to the area of the open ocean over which the wind blows with constant speed and direction, thereby creating a wave system. 2. The distance across the fetch (wave-generating area) measured in a direction parallel to the direction of the wind.

Fishery assessment   The conduct of research on the ecological and economic factors related to a fishery and the application of this knowledge gained to its regulation.

Fjord   A long, narrow, deep, U-shaped inlet that usually represents the seaward end of a glacial valley that has become partially submerged after the melting of the glacier.

Flagellum   A whiplike living process used by some cells for locomotion.

Floe   A piece of floating ice other than fast ice or icebergs. May range in dimension from about 20 cm across to more than a kilometer.

Flood current A tidal current associated with increasing height of the tide, generally moving toward the shore.

Flora The plant life of any particular area or of any particular time.

Folded mountain range Mountain ranges formed as a result of the convergence of lithospheric plates. They are characterized by masses of folded sedimentary rocks that formed from sediments deposited in the ocean basin that was destroyed by the convergence.

Food chain The passage of energy materials from producers through a sequence of a herbivore and a number of carnivores.

Food web A group of interrelated food chains.

Foraminifera An order of planktonic and benthic protozoans that possess protective coverings, usually composed of calcium carbonate.

Forced wave A wave that is generated and maintained by a continuous force such as the gravitational attraction of the moon.

Foreshore The portion of the shore lying between the normal high and low water marks—the intertidal zone.

Fortnight Half a synodic month (29.5 days), or about 14.75 days. Normally used in reference to a period of time equal to two weeks. The time that elapses between new moon and full moon.

Fracture zone An extensive linear zone of unusually irregular topography of the ocean floor, characterized by large seamounts, steep-sided or asymmetrical ridges, troughs, or long, steep slopes. Usually represents ancient, inactive transform fault zones.

Free wave A wave created by a sudden rather than a continuous impulse that continues to exist after the generating force is gone.

Freezing point The temperature at which a liquid becomes a solid under any given set of conditions. The freezing point of water is 0°C under atmospheric pressure.

Fringing reef A reef that is directly attached to the shore of an island or continent. It may extend more than 1 km from shore. The outer margin is submerged and often consists of algal limestone, coral rock, and living coral.

Frustule The siliceous covering of a diatom, consisting of two halves (epitheca and hypotheca).

Fucoxanthin The reddish brown pigment that gives brown algae its characteristic color.

Gastropoda A class of molluscs, most of which possess an asymmetrical spiral one-piece shell and a well-developed flattened foot. A well-developed head will usually have two eyes and one or two pairs of tentacles. Includes snails, limpets, abalone, cowries, sea hares, and sea slugs.

Geostrophic current A current that grows out of the earth's rotation and is the result of a near balance between gravitational force and the Coriolis effect.

Gill A thin-walled projection from some part of the external body or the digestive tract used for respiration in a water environment.

Glacial epoch The Pleistocene epoch, the earlier of two divisions of the Quaternary period of geologic time.

During this time high latitude continental areas now free of ice were covered by continental glaciers.

Glacier A large mass of ice formed on land by the recrystalization of old compacted snow. It flows from an area of accumulation to an area of wasting where ice is removed from the glacier by melting.

Global plate tectonics The process by which lithospheric plates are moved across the earth's surface to collide, slide by one another, or diverge to produce the topographic configuration of the earth.

Gondwanaland A hypothetical protocontinent of the Southern Hemisphere named for the Gondwana region of India. It included the present continental masses Africa, Antarctica, Australia, India, and South America.

Graded bedding Stratification in which each layer displays a decrease in grain size from bottom to top.

Gradient The rate of increase or decrease of one quantity or characteristic relative to a unit change in another. For example, the slope of the ocean floor is a change in elevation (a vertical linear measurement) per unit of horizontal distance covered. Commonly measured in m/km.

Gran method A system involving the observation of the change in dissolved oxygen within paired transparent and opaque bottles containing phytoplankton and suspended in the ocean surface water to determine the base of the euphotic zone.

Granite A light colored igneous rock characteristic of the continental crust. Rich in nonferromagnesian minerals such as feldspar and quartz.

Gravity wave A wave for which the dominant restoring force is gravity. Such waves have a wavelength of more than 1.74 cm, and their speed of propagation is controlled mainly by gravity.

Groin A low artificial structure projecting into the ocean from the shore to interfere with longshore transportation of sediment. It usually has the purpose of trapping sand to cause the buildup of a beach.

Gross ecological efficiency The amount of energy passed on from a trophic level to the one above it divided by the amount it received from the one below it.

Gulf Stream The high-intensity western boundary current of the North Atlantic Ocean subtropical gyre that flows north off the east coast of the United States.

Guyot A tablemount, a conical volcanic feature on the ocean floor that has had the top truncated to a relatively flat surface.

Gyre A circular spiral form. Used mainly in reference to the circular motion of water in each of the major ocean basins centered in subtropical high pressure regions.

Habitat A place where a particular plant or animal lives. Generally refers to a smaller area than environment.

Hadal Pertaining to the deepest ocean environment, specifically that of ocean trenches deeper than 6 km.

Hadal zone Pertaining to the deepest ocean benthic environment specifically that of ocean trenches deeper than 6 km (3.7 mi).

Half-life The time required for half the atoms of a sample of a radioactive isotope to decay to an atom of another element.

Halocline A layer of water in which a high rate of change in salinity in the vertical dimension is present.

Headland A steep-faced irregularity of the coast that extends out into the ocean.

Heat Energy moving from a high temperature system to a lower temperature system. The heat gained by the one system may be used to raise its temperature or to do work.

Heat budget The equilibrium that exists on the average between the amount of heat absorbed by the earth and its atmosphere in one year and the amount of heat radiated back into space in one year.

Heat capacity Usually defined as the amount of heat required to raise the temperature of 1 g of a substance 1°C.

Heat energy Energy of molecular motion. The conversion of higher forms of energy such as radiant or mechanical energy to heat energy within a system increases the heat energy within the system and the temperature of the system.

Herbivore An animal that relies chiefly or solely on plants for its food.

Hermaphroditic Pertaining to the possession of functional male and female reproductive organs by an animal. It is rare for both systems to function at the same time.

Hermatypic coral Reef-building corals that have symbiotic algae in their ectodermal tissue. They cannot produce a reef structure below the euphotic zone.

Heterotroph Animals and bacteria that depend on the organic compounds produced by other animals and plants as food. Organisms not capable of producing their own food by photosynthesis.

High water (HW) The highest level reached by the rising tide before it begins to recede.

Higher high water (HHW) The higher of two high waters occurring during a tidal day where tides are mixed.

Higher low water (HLW) The higher of two low waters occurring during a tidal day where tides are mixed.

Holoplankton Organisms that spend their entire life as members of the plankton.

Homologous A basic similarity of structures in different organisms resulting from a similar embryonic origin and development, e.g., the foreflippers of a seal and the arms and hands of a person are homologous.

Hook A spit or narrow cape of sand or gravel with an end that bends landward to form a "hook."

Horse latitudes The latitude belts between 30° and 35° north and south where winds are light and variable, since the principal movement of air masses at these latitudes is one of vertical descent. The climate is hot and dry.

Hurricane A tropical cyclone in which winds reach speeds in excess of 120 km/h (73 mi/h). Generally applied to such storms in the North Atlantic Ocean, eastern North Pacific Ocean, Caribbean Sea, and Gulf of Mexico. Such storms in the western Pacific Ocean are called *typhoons*.

Hydrogenous sediment Sediment that forms from precipitation from ocean water or ion exchange between existing sediment and ocean water. Examples are manganese nodules, phosphorite, glauconite, phillipsite, and montmorillonite.

Hydrologic cycle The cycle of water exchange among the atmosphere, land, and ocean through the processes of evaporation, precipitation, runoff, and subsurface percolation.

Hydrothermal spring Vents of hot water found primarily along the spreading axes of oceanic ridges and rises.

Hydrozoa A class of coelenterates that characteristically exhibits alternation of generations with a sessile polypoid colony giving rise to a pelagic medusoid form by asexual budding.

Hypertonic Pertaining to the property of an aqueous solution having a higher osmotic pressure (salinity) than another aqueous solution from which it is separated by a semipermeable membrane that will allow osmosis to occur. The hypertonic fluid will gain water molecules through the membrane from the other fluid.

Hypotonic Pertaining to the property of an aqueous solution having a lower osmotic pressure (salinity) than another aqueous solution from which it is separated by a semipermeable membrane that will allow osmosis to occur. The hypotonic fluid will lose water molecules through the membrane to the other fluid.

Iceberg A massive piece of glacier ice that has broken from the front of the glacier (calved) into a body of water. It floats with its tip at least 5 m above the water's surface and at least ⅘ of its mass submerged.

Ice floe *See* floe.

Ice shelf A thick layer of ice with a relatively flat surface which is attached to and nourished by a continental glacier from one side. The shelf, which is for the most part afloat, may extend above water level by more than 50 m along its seaward cliff formed by the break-off of large tabular chunks of ice that become icebergs.

Igneous rock One of the three main classes into which all rocks are divided, i.e., igneous, metamorphic, and sedimentary. Rock that forms the solidification of molten or partly molten material (magma).

Inertia Newton's first law of motion. It states that a body at rest will stay at rest and a body in motion will remain in a uniform motion in a straight line unless acted on by some external force.

Infauna Animals that live buried in the soft substrate (sand or mud).

Infrared radiation Electromagnetic radiation lying between the wavelengths of 0.8 μm and about 1000 μm. It is bounded on the shorter wavelength side by the visible spectrum and on the long side by microwave radiation.

In situ In place, i.e., in situ density of a sample of water is its density at its original depth.

Insolation The rate at which solar radiation is received per unit of surface area at any point at or above the earth's surface.

Interface   A surface separating two substances of different properties, i.e., density, salinity, or temperature. In oceanography, it usually refers to a separation of two layers of water and with different densities caused by significant differences in temperature and/or salinity.

Internal wave   A wave that develops below the surface of a fluid, the density of which changes with increased depth. This change may be gradual or occur abruptly at an interface.

Intertidal zone   Littoral zone, the foreshore. The ocean floor covered by the highest normal tides and exposed by the lowest normal tides and the water environment of the tide pools within this region.

Invertebrate   Animal without a backbone.

Ion   An atom that becomes electrically charged by gaining or losing one or more electrons. The loss of electrons produces a positively charged cation, and the gain of electrons produces a negatively charged anion.

Island arc system   A linear arrangement of islands, many of which are volcanic, usually curved so the concave side faces a sea separating the islands from a continent. The convex side faces the open ocean and is bounded by a deep ocean trench.

Island mass effect   As surface current flows past an island, surface water is carried away from the island on the down-current side. This water is replaced in part by upwelling of water on the down-current side of the island.

Isohaline   Of the same salinity.

Isopoda   An order of dorso-ventrally flattened crustaceans that are mostly scavengers or parasites on other crustaceans or fish.

Isotherm   Line connecting points of equal temperature.

Isothermal   Of the same temperature.

Isotonic   Pertaining to the property of having equal osmotic pressure. If two such fluids were separated by a semipermeable membrane that will allow osmosis to occur, there would be no net transfer of water molecules across the membrane.

Isotope   One of several atoms of an element that has a different number of neutrons, and therefore a different atomic mass, than the other atoms, or isotopes, of the element.

Jellyfish   1. Free-swimming, umbrella-shaped medusoid members of the coelenterate class, Scyphozoa. 2. Also frequently applied to the medusoid forms of other coelenterates.

Jetty   A structure built form the shore into a body of water to protect a harbor or a navigable passage from being shoaled by deposition of longshore (littoral) drift material.

Kelp   Large varieties of Phaeophyta (brown algae).

Kelvin temperature scale (K)   0 K = − 273.16°C. One degree on the Kelvin scale equals the same temperature range as one degree on the Celsius scale. 0 K is the lowest temperature possible.

Key   A low, flat island composed of sand or coral debris that accumulates on a reef flat.

Kinetic energy   Energy of motion. It increases as the mass or speed of the object in motion increases.

Knot (kt)   Unit of speed equal to 1 nautical mile per hour, approximately 51 cm/s.

Krill   A common name frequently applied to members of crustacean order, Euphausiacea (euphausiids).

Lagoon   A shallow stretch of seawater partly or completely separated from the open ocean by an elongate narrow strip of land such as a reef or barrier island.

Laminar flow   Flow in which water, or any fluid, flows in parallel layers or sheets. The direction of flow at any point does not change with time. Nonturbulent flow.

Langmuir circulation   A cellular circulation set up by winds that blow consistently in one direction with speeds in excess of 12 km/h. Helical spirals running parallel to the wind direction are alternately clockwise and counterclockwise.

Larva   An embryo that is on its own before it assumes the characteristics of the adults of the species.

Latent heat   The quantity of heat gained or lost per unit of mass as a substance undergoes a change of state (liquid to solid, etc.) at a given temperature and pressure.

Latitude   Location on the earth's surface based on angular distance north or south of the equator. Equator, 0°; North Pole, 90° N; South Pole, 90° S.

Laurasia   A hypothetical protocontinent of the Northern hemisphere. The name is derived from Laurentia, pertaining to the Canadian Shield of North America, and Eurasia of which it was composed.

Lava   Fluid magma coming from an opening in the earth's surface, or the same material after it solidifies.

Leeward   Direction toward which the wind is blowing or waves are moving.

Levee   1. Natural levees are low ridges on either side of river channels that result from deposition during flooding. 2. Artificial levees are built by human beings.

Limestone   A class of sedimentary rocks composed of at least 80 percent carbonates of calcium or magnesium. Limestones may be either biogenous or hydrogenous.

Limpet   A mollusc of the class Gastropoda that possess a low conical shell that exhibits no spiraling in the adult form.

Lithogenous sediment   Sediment composed of mineral grains derived from the rocks of continents and islands and transported to the ocean by wind and running water.

Lithosphere   The outer layer of the earth's structure, including the crust and the upper mantle to a depth of about 200 km. It is this layer that breaks into the plates that are the major elements of the theory of plate tectonics.

Lithothamnion ridge   A feature common to the windward edge of a reef structure, characterized by the presence of the red algae, Lithothamnion.

Littoral zone   The benthic zone between the highest and lowest spring tide shorelines; the intertidal zone.

Lobster   Large marine crustacean used as food. *Homarus americanus* (American lobster) possesses two large chelae (pincers) and is found off the New England coast.

Panulirus sp. (spiny lobsters or rock lobsters) have no chelae but possess long spiny antennae effective in warding off predators. *P. argus* is found off the coast of Florida and in the West Indies, while *P. interruptus* is common along the coast of southern California.

Longitude Location on the earth's surface based on angular distance east or west of Greenwich Meridian (0° long.). 180° longitude is the International Date Line.

Longitudinal wave A wave phenomenon where particle vibration is parallel to the direction of energy propagation.

Longshore current A current located in the surf zone and running parallel to the shore as a result of waves breaking at an angle on the shore.

Longshore drift The load of sediment transported along the beach from the breaker zone to the top of the swash line in association with the longshore current.

Lophophore Horseshoe-shaped feeding structure bearing ciliated tentacles characteristic of the phyla Bryozoa, Brachiopoda, and Phoronidea.

Low water (LW) The lowest level reached by the water surface at low tide before the rise toward high tide begins.

Lower high water (LHW) The lower of two high waters occurring during a tidal day where tides are mixed.

Lower low water (LLW) The lower of two low waters occurring during a tidal day where tides are mixed.

Lunar day The time interval between two successive transits of the moon over a meridian, approximately 24 h and 50 min of solar time.

Lunar hour One twenty-fourth of a lunar day. About 62.1 min.

Lunar tide The part of the tide caused solely by the tide-producing force of the moon.

Magma Fluid rock material from which igneous rock is derived through solidification.

Magnetic anomaly Distortion of the regular pattern of the earth's magnetic field, resulting from the various magnetic properties of local concentrations of ferromagnetic minerals in the earth's crust.

Manganese nodules Concretionary lumps containing oxides of iron, manganese, copper, and nickel found scattered over the ocean floor.

Mantle The relatively plastic zone rich in ferromagnesium minerals between the core and crust of the earth.

Marginal sea A semienclosed body of water adjacent to a continent and floored by submerged continental crust.

Mariculture The application of the principles of agriculture to the production of marine organisms.

Marsh An area of soft, wet land. Flat land periodically flooded by salt water, common in portions of lagoons.

Meander A sinuous curve, bend, or turn in the course of a current.

Mean high water (MHW) The average height of all the high waters occurring over a 19-year period.

Mean low water (MLW) The average height of the low waters occurring over a 19-year period.

Mean sea level (MSL) The mean surface water level determined by averaging all stages of the tide over a 19-year period, usually determined from hourly height observations along an open coast.

Mean tidal range The difference between high water and mean low water.

Mechanical energy Energy manifested as work being done; the movement of a mass some distance.

Mediterranean circulation Circulation characteristic of bodies of water with restricted circulation with the ocean that results from an excess of evaporation as compared to precipitation and runoff. Surface flow is into the restricted body of water with a subsurface counterflow as exists between the Mediterranean Sea and the Atlantic Ocean.

Medusa Free-swimming bell-shaped coelenterate body form with a mouth at the end of a central projection and tentacles around the periphery. Reproduces sexually.

Meridian of longitude Great circles running through the north and south poles.

Meroplankton Planktonic larval forms of organisms that are members of the benthos or nekton as adults.

Mesoderm Primitive cell layer of embryo that develops between endoderm and ectoderm. In vertebrates, it gives rise to the skeleton, muscles, circulatory, and excretory systems, and most of the reproductive system.

Mesopelagic That portion of the oceanic province 200–1000 m deep. Corresponds approximately with the disphotic (twilight) zone.

Metamorphic rock Rock that has undergone recrystallization while in the solid state in response to changes of temperature, pressure, and chemical environment.

Metaphyta Kingdom of many-celled plants.

Metazoa Kingdom of many-celled animals.

Microplankton Net plankton. Plankton not easily seen by the unaided eye, but easily recovered from the ocean with the aid of a silk-mesh plankton net.

Mineral An inorganic substance occurring naturally in the earth and having distinctive physical properties and a chemical composition that can be expressed by a chemical formula. The term is also sometimes applied to organic substances such as coal and petroleum.

Mixed layer The surface layer of the ocean water mixed by wave and tide motions to produce relatively isothermal and isohaline conditions.

Mixed tide A tide having two high waters and two low waters per tidal day with a marked diurnal inequality. Such a tide may also show alternating periods of diurnal and semidiurnal components.

Mohorovicic discontinuity A sharp compositional discontinuity between the crust and mantle of the earth. It may be as shallow as 5 km below the ocean floor or as deep as 60 km beneath some continental mountain ranges.

Molecule The smallest particle of an element or compound that, in the free state, retains the characteristics of the substance.

Mollusca Phylum of soft unsegmented animals usually protected by a calcareous shell and having a muscular foot for locomotion. Includes snails, clams, chitons, and octopuses.

Molt Periodic shedding of exoskeleton by arthropods to permit growth.

Monera Kingdom of organisms that do not have nuclear material confined within a sheath but spread throughout the cell. Bacteria and blue-green algae.

Mononodal Pertaining to a standing wave with only one nodal point or nodal line.

Monsoons A name for seasonal winds derived from the Arabic word for season, *mausim*. The term was originally applied to winds over the Arabian Sea that blow from the southwest during summer and the northeast during winter.

Moraine A deposit of unsorted material deposited at the margins of glaciers. Many such deposits have become important economically as fishing banks after being submerged by the rising level of the ocean.

Mud Sediment consisting of silt and clay-sized particles smaller than 0.06 mm. Actually, small amounts of larger particles will also be present.

Mutualism A symbiotic relationship in which both participants benefit.

Mycota The kingdom of fungi. In the marine environment they are found living symbiotically with algae as lichen in the intertidal zone and as decomposers of dead organic matter in the open sea.

Nadir The point on the celestial sphere directly opposite the zenith and directly beneath the observer.

Nanoplankton Plankton less than 50 μm in length that cannot be captured in a plankton net and must be removed from the water by centrifuge or special microfilters.

Nansen bottle A device used by oceanographers to obtain samples of ocean water from beneath the surface.

Nauplius A microscopic free-swimming larval stage of crustaceans such as copepods, ostracods, and decapods. Typically has three pairs of appendages.

Neap tide Tides of minimal range occurring when the moon is in quadrature, first and third quarters.

Nearshore That zone from the shoreline seaward to the line of breakers.

Nektobenthos Those members of the benthos that can actively swim and spend much time off the bottom.

Nekton Pelagic animals such as adult squids, fish, and mammals that are active swimmers to the extent they can determine their position in the ocean by swimming.

Neritic province That portion of the pelagic environment from the shoreline to where the depth reaches 200 m.

Neritic sediment That sediment composed primarily of lithogenous particles and deposited relatively rapidly on the continental shelf, continental slope, and continental rise.

Net primary production The primary production of plants after they have removed what is needed for their metabolism.

Niche The ecological role of an organism and its position in the ecosystem.

Nitrogen fixation Conversion by bacteria of atmospheric nitrogen (N_2) to oxides of nitrogen (NO_2,NO_3) usable by plants in primary production.

Node The point on a standing wave where vertical motion is lacking or minimal. If this condition extends across the surface of an oscillating body of water, the line of no vertical motion is a nodal line.

Nonconservative property A property of ocean water attained at the surface and changed by processes other than mixing and diffusion after the water sinks below the surface. For example, dissolved oxygen content will be altered by biological activity.

North Atlantic Deep Water A deep-water mass that forms primarily at the surface of the Norwegian Sea and moves south along the floor of the North Atlantic Ocean.

Nudibranch Sea slug. A member of the mollusc class Gastropoda that has no protective covering as an adult. Respiration is carried on by gills or other projections on the dorsal surface.

Nutrients Any number of organic or inorganic compounds used by plants in primary production. Nitrogen and phosphorus compounds are important examples.

Oceanic crust A mass of rock with basaltic composition that is about 5 km thick and may or may not extend beneath the continents.

Oceanic province That division of the pelagic environment where the water depth is greater than 200 m.

Oceanic ridge A linear, seismic mountain range that extends through all the major oceans, rising 1–3 km above the deep-ocean basins. Averaging 1500 km in width, rift valleys are common along the central axis. Source of new oceanic crustal material.

Oceanic sediment The inorganic abyssal clays and the organic oozes that accumulate on the deep-ocean floor slowly, particle by particle.

Oceanic spreading center The axes of oceanic ridges and rises that are the locations at which new lithosphere is added to lithospheric plates. The plates move away from these axes in the process of sea-floor spreading.

Offshore The comparatively flat submerged zone of variable width extending from the breaker line to the edge of the continental shelf.

Omnivore An animal that feeds on both plants and animals.

Ooze A pelagic sediment containing at least 30 percent skeletal remains of pelagic organisms, the balance being clay minerals. Oozes are further defined by the chemical composition of the organic remains (siliceous or calcareous) and by their characteristic organisms (diatom ooze, foraminifera ooze, radiolarian ooze, pteropod ooze.)

Orthogonal lines Lines drawn perpendicular to wave fronts and spaced uniformly so equal amounts of energy are contained by the segments of the wave front lying between any two orthogonal lines in a series. The areas where energy is concentrated as the waves break on the shore can be identified by the convergence of the orthogonal lines.

Orthophosphate Phosphoric oxide (P_2O_5) can combine with water to produce orthophosphates ($3H_2O \cdot P_2O_5$ or H_3PO_4) that may be used by plants as nutrients.

Osmosis Passage of water molecules through a semipermeable membrane separating two aqueous solutions of different solute concentration. The water molecules pass from the solution of lower solute concentration into the other.

Osmotic pressure A measure of the tendency for osmosis to occur. It is the pressure that must be applied to the more concentrated solution to prevent the passage of water molecules into it from the less concentrated solution.

Osmotic regulation Physical and biological processes used by organisms to counteract the osmotic effects of differences in osmotic pressures of their body fluids and the water in which they live.

Ostracoda An order of crustaceans that are minute and compressed within a bivalve shell.

Oviparous Referring to embryological development in an egg that is deposited outside the mother's body where nutrition for development is provided by the yolk of the egg.

Ovoviviparous Referring to embryological development in the female reproductive tract where nutrition for development is provided by the yolk of the egg.

Oxygen compensation depth The depth in the ocean at which marine plants receive just enough solar radiation to meet their basic metabolic needs. It marks the base of the euphotic zone.

Oxygen utilization rate (OUR) The rate at which oxygen is being used by the demands of respiration and bacterial decomposition of dead organic material.

Pacific-type margin Leading edge of a continent that undergoes tectonic uplift as a result of lithospheric plate convergence.

Pack ice Any area of sea ice other than fast ice. Less than 3 m thick, it covers the ocean sufficiently that navigation is possible only by icebreakers.

Pancake ice Circular pieces of newly formed sea ice from 30 cm to 3 m in diameter that form in early fall in polar regions.

Pangaea A hypothetical supercontinent of the geologic past that contained all the continental crust of the earth.

Panthalassa A hypothetical protoocean surrounding Pangaea.

Parapodia Flat protuberances on each side of most segments of polychaete worms. Most possess cirri and setae (bristlelike projections); may be modified for special functions such as feeding, locomotion, and respiration.

Parasitism A symbiotic relationship between two organisms in which one benefits at the expense of the other.

Pelagic environment The open ocean environment which is divided into the neritic province (water depth 0–200 m) and the oceanic province (water depth greater than 200 m).

Pelecypoda A class of molluscs characterized by two more or less symmetrical lateral valves with a dorsal hinge. These filter feeders pump water through the filter system and over gills through posterior siphons. Many possess a hatchet-shaped foot used for locomotion and burrowing. Includes clams, oysters, mussels, and scallops.

Perigee The point on the orbit of an earth satellite (moon) that is nearest the earth.

Perihelion That point on the orbit of a planet or comet around the sun that is closest to the sun.

Permeability Capacity of a porous rock or sediment for transmitting fluid.

Phaeophyta Brown algae characterized by the carotinoid pigment fucoxanthin. Contains the largest members of the marine plant community.

Photic zone The upper ocean in which the presence of solar radiation is detectable. It includes the euphotic and disphotic zones.

Photophore One of several types of light-producing organs found primarily on fishes and squids inhabiting the mesopelagic and upper bathypelagic zones.

Photosynthesis The process by which plants produce carbohydrate food from carbon dioxide and water in the presence of chlorophyll, using light energy and releasing oxygen.

Phycoerythrin A red pigment characteristic of the Rhodophyta (red algae).

Phytoplankton Plant plankton. The most important community of primary producers in the ocean.

Plankton Passively drifting or weakly swimming organisms that are not independent of currents. Includes mostly microscopic algae, protozoa, and larval forms of higher animals.

Plankton bloom A very high concentration of phytoplankton, resulting from a rapid rate of reproduction as conditions become optimum during the spring in high latitude areas. Less obvious causes produce blooms that may be destructive in other areas.

Plankton net Plankton extracting device that is cone shaped and typically of a silk material. It is towed through the water or lifted vertically to extract plankton down to a size of 50 μm.

Polar emergence The emergence of low- and mid-latitude temperature-sensitive deep-ocean benthos onto the shallow shelves of the polar regions where temperatures similar to that of their deep-ocean habitat exist.

Pollution (marine) The introduction of substances or energy into the marine environment that result in harm to the living resources of the ocean or humans that use these resources.

Polychaeta Class of annelid worms that includes most of the marine segmented worms.

Polynya A nonlinear opening in sea ice.

Polyp A single individual of a colony or a solitary attached coelenterate.

Population A group of individuals of one species living in an area.

Porifera Phylum of sponges. Supporting structure composed of $CaCO_3$ or SiO_2 spicules or fibrous spongin.

Water currents created by flagella-waving choanocytes enter tiny pores, pass through canals, and exit through a larger osculum.

Potential (world) fishery The mass of fish that are supported annually by nutrients that originate outside of an ecosystem (world fishery ecosystems) and flow into it through upwelling or other processes. This amount of fish can be removed on an annual basis without degrading the ecosystem.

Precipitation In a meteorological sense, the discharge of water in the form of rain, snow, hail, or sleet from the atmosphere onto the earth's surface.

Primary productivity The amount of organic matter synthesized by organisms from inorganic substances within a given volume of water or habitat in a unit of time.

Prime meridian The meridian of longitude 0° used as a reference for measuring longitude. The Greenwich Meridian.

Progressive wave A wave in which the waveform progressively moves.

Propagation The transmission of energy through a medium.

Protein A very complex organic compound made up of large numbers of amino acids. Proteins make up a large percentage of the dry weight of all living organisms.

Protista A kingdom of organisms that includes all one-celled forms with nuclear material confined to a nuclear sheath. Includes the animal phylum Protozoa and the phyla of algal plants.

Protoplasm The complicated self-perpetuating living material making up all organisms. The elements carbon, hydrogen, and oxygen constitute more than 95 percent; water and dissolved salts make up 50–97 percent of most plants and animals with carbohydrates, lipids (fats), and proteins constituting the remainder.

Protozoa Phylum of one-celled animals with nuclear material confined within a nuclear sheath.

Pseudopodia An extension of protoplasm in a broad, flat, or long needlelike projection used for locomotion or feeding. Typical of amoeboid forms such as foraminifera and radiolaria.

Pteropoda An order of pelagic gastropods in which the foot is modified for swimming and the shell may be present or absent.

Purse seine A curtainlike net that can be used to encircle a school of fish. The bottom is then pulled tight much the way a purse string is used to close a baglike purse.

Pycnocline A layer of water in which a high rate of change in density in the vertical dimension is present.

Pycnogonid A spiderlike arthropod found on the ocean bottom at all depths. The more commonly observed nearshore varieties are usually less than 1 cm across, while deeper water varieties may reach spreads of over 1 m.

Pyrrophyta A phylum of microscopic algae that possesses flagella for locomotion—the dinoflagellates.

Radiata A grouping of phyla with primary radial symmetry—phyla Coelenterata and Ctenophora.

Radioactivity The spontaneous breakdown of the nucleus of an atom resulting in the emission of radiant energy in the form of particles or waves.

Radiolaria An order of planktonic and benthic protozoans that possess protective coverings usually made of silica.

Ray A cartilaginous fish in which the body is dorso-ventrally flattened, eyes and spiracles are on the upper surface, and gill slits are on the bottom. The tail is reduced to a whiplike appendage. Includes electric rays, manta rays, and stingrays.

Recruitment (fishery) The year-class (number of fish or mass of fish) of young adults added to a fishery following each spawning season.

Red tide A reddish brown discoloration of surface water, usually in coastal areas, caused by high concentrations of microscopic organisms, usually dinoflagellates. It probably results from increased availability of certain nutrients for various reasons. Toxins produced by the dinoflagellates may kill fish directly; or large populations of animal forms that spring up to feed on the plants, along with decaying plant and animal remains, may use up the oxygen in the surface water to cause asphyxiation of many animals.

Reef A consolidated rock (a hazard to navigation) with a depth of 20 m or less.

Reef flat A platform of coral fragments and sand on the lagoonal side of a reef that is relatively exposed at low tide.

Reef front The upper seaward face of a reef from the reef edge (seaward margin of reef flat) to the depth at which living coral and coralline algae become rare, 16–30 m.

Reflection The process in which a wave has part of its energy returned seaward by a reflecting surface.

Refraction The process by which the part of a wave in shallow water is slowed down to cause the wave to bend and tend to align itself with the underwater contours.

Relict beach A beach deposit laid down and submerged by rise in sea level. It is still identifiable on the continental shelf, indicating no deposition is presently taking place at that location on the shelf.

Residence time The average length of time a particle of any substance spends in the ocean. It is calculated by dividing the total amount of the substance in the ocean by the rate of its introduction into the ocean or the rate at which it leaves the ocean.

Respiration The process by which organisms utilize organic materials (food) as a source of energy. As the energy is released, oxygen is used and carbon dioxide and water are produced.

Reversing current The tide current as it occurs at the margins of landmasses. The water flows in and out for approximately equal periods of time separated by slack water where the water is still at high and low tidal extremes.

Rhodophyta Phylum of algae composed primarily of small encrusting, branching, or filamentous plants that receive their characteristic red color from the presence of the pigment phycoerythrin. With a worldwide distribution, they are found at greater depths than other algae.

Rip current A strong narrow surface or near surface current of short duration (up to 2 h) and high speed (up to 4 km/h) flowing seaward through the breaker zone at nearly right angles to the shore. It represents the return to the ocean of water that has been piled up on the shore by incoming waves.

Rise A long, broad elevation that rises gently and rather smoothly from the deep-ocean floor.

Rotary current Tidal current as observed in the open ocean. The tidal crest makes one complete rotation during a tidal period.

Sabellid A member of the annelid family Sabellidae that lives in a tube composed of shell fragments, sand, and a glutinous material. Featherlike gills and feeding structures filter food from the water above the tube opening.

Salinity A measure of the quantity of dissolved solids in ocean water. Formally, it is the total amount of dissolved solids in ocean water in parts per thousand by weight after all carbonate has been converted to oxide, the bromide and iodide to chloride, and all the organic matter oxidized. It is normally computed from conductivity, refractive index, or chlorinity.

Salpa Genus or pelagic tunicates that are cylindrical, transparent, and found in all oceans.

Salt Any substance that yields ions other than hydrogen or hydroxyl. Salts are produced from acids by replacing the hydrogen with a metal.

Sand Particle size of 1/16–2 mm. It pertains to particles that lie between silt and granules on the Wentworth scale of grain size.

Sargasso Sea A region of convergence in the North Atlantic lying south and east of Bermuda where the water is very clear deep blue in color and contains large quantities of floating Sargassum.

Sargassum A brown alga characterized by a bushy form, substantial holdfast when attached, and a yellow brown, green yellow, or orange color. Two species, *S. fluitans* and *S. natans,* make up most of the macroscopic vegetation in the Sargasso Sea.

Scaphopoda A class of molluscs commonly called *tusk shells.* The shell is an elongate cone open at both ends. The conical foot surrounded by threadlike tentacles extends from the larger end to aid the animal in burrowing in fine sand or mud.

Scarp A linear steep slope on the ocean floor separating gently sloping or flat surfaces.

Scavenger An animal that feeds on dead organisms.

Scyphozoa A class of coelenterates that includes the true jellyfish in which the medusoid body form predominates and the polyp is reduced or absent.

Sea 1. A subdivision of an ocean. Two types of seas are identifiable and defined. They are the *mediterranean seas,* where a number of seas are grouped together collectively as one sea, and *adjacent seas* that are connected individually to the ocean. 2. A portion of the ocean where waves are being generated by wind.

Sea anemone A member of the class Anthozoa whose bright color, tentacles, and general appearance make it resemble flowers.

Sea arch An opening through a headland caused by wave erosion. Usually develops as sea caves are extended from one or both sides of the headland.

Sea cave A cavity at the base of a sea cliff formed by wave erosion.

Sea cow An aquatic, herbivorous mammal of the order Sirenia that includes the dugong and manatee.

Sea cucumber A common name given to members of the echinoderm class Holotheuroidea.

Sea-floor spreading A process producing the lithosphere when convective upwelling of magma along the oceanic ridges moves the ocean floor away from the ridge axes at rates of from 1 to 10 cm (0.4 to 4 in.) per year.

Sea ice Any form of ice originating from the freezing of ocean water.

Seamount An individual peak extending over 1000 m above the ocean floor.

Sea snake A reptile belonging to the family Hydrophiidae with venom similar to that of cobras. They are found primarily in the coastal waters of the Indian Ocean and the western Pacific Ocean.

Seasonal thermocline A thermocline that develops due to surface heating of the oceans in mid- to high latitudes. The base of the seasonal thermocline is usually above 200 m (656 ft).

Sea state A description of the ocean surface that includes the average height of the highest one-third of the waves observed in a wave train, referred to a numerical code.

Sea turtle Any of the reptilian order Testudinata found widely in warm water.

Sea urchin An echinoderm belonging to the class Echinoidea possessing a fused test (external covering) and well-developed spines.

Sediment Particles of organic or inorganic origin that accumulate in loose form.

Sedimentary rock A rock resulting from the consolidation of loose sediment, or a rock resulting from chemical precipitation, i.e., sandstone and limestone.

Seiche A standing wave of an enclosed or semienclosed body of water that may have a period ranging from a few minutes to a few hours, depending on the dimensions of the basin. The wave motion continues after the initiating force has ceased.

Seismic Pertaining to an earthquake or earth vibration, including those that are artificially induced.

Seismic sea wave *See* tsunami.

Semidiurnal tide Tide having two high and two low waters per tidal day with small inequalities between successive highs and successive lows. Tidal period is about 12 h and 25 min solar time. Semidaily tide.

Serpulid A polychaete worm belonging to the family Serpulidae that builds a calcareous or leathery tube on a submerged surface.

Sessile   Permanently attached to the substrate and not free to move about.

Seta   Hairlike or needlelike projections on the exoskeleton of arthropods. Similar structures exist on some annelids.

Shallow water wave   A wave on the surface of the water whose wavelength is at least 20 times water depth. The bottom affects the orbit of water particles, and speed is determined by water depth.
$$S(m/s) = 3.1 \sqrt{\text{water depth (m)}}.$$

Shelf break   The depth at which the gentle slope of the continental shelf steepens appreciably. It marks the boundary between the continental shelf and continental rise.

Shoal   Shallow.

Shore   Seaward of the coast, extends from highest level of wave action during storms to the low water line.

Shoreline   The line marking the intersection of water surface with the shore. Migrates up and down as the tide rises and falls.

Sigma-T (σ_t)   A term used in place of density but derived from density of a water sample after the pressure has been reduced to one atmosphere. It is computed as $\sigma_t =$ (specific gravity − 1) 1000.

Silica   Silicon dioxide (SiO_2).

Sill   A submarine ridge partially separating bodies of water such as fjords and seas from one another or from the open ocean.

Silt   A particle size of $1/128$–$1/16$ mm. It is intermediate between sand and clay.

Siphonophora   An order of hydrozoan coelenterates that forms pelagic colonies containing both polyps and medusae. Examples are *Physalia* and *Velella*.

Slack water   Occurs when a reversing tidal current changes direction at high or low water. Current speed is zero.

Slick   A smooth patch on an otherwise rippled surface caused by a monomolecular film of organic material that reduces surface tension.

Solar tide   The partial tide caused by the tide-producing forces of the sun.

Solstice   The time during which the sun is directly over one of the tropics. In the Northern Hemisphere the summer solstice occurs on June 21 or 22 as the sun is over the Tropic of Cancer, and the winter solstice occurs on December 21 or 22 when the sun is over the Tropic of Capricorn.

Solute   A substance dissolved in a solution. Salts are the solute in salt water.

Solution   A state in which a solute is homogeneously mixed with a liquid solvent. Water is the solvent for the solution that is ocean water.

Solvent   A liquid that has one or more solutes dissolved in it.

Sonar   An acronym for *sound navigation and ranging*. A method by which objects may be located in the ocean.

Sounding   Measuring the depth of water beneath a ship.

Species diversity   The number or variety of species found in a subdivision of the marine environment.

Specific gravity   The ratio of density of a given substance to that of pure water at 4°C and at atmospheric pressure.

Specific heat   The quantity of heat required to raise the temperature of 1 g of a given substance 1°C. For water it is 1 cal.

Spermatophyta   Seed-bearing plants.

Spicule   A minute needlelike calcareous or siliceous form found in sponges, radiolarians, chitons, and echinoderms that acts to support the tissue or provide a protective covering.

Spit   A small point, low tongue, or narrow embankment of land commonly consisting of sand deposited by longshore currents and having one end attached to the mainland and the other terminating in open water.

Sponge   *See* Porifera.

Spring tide   Tide of maximum range occurring every fortnight when the moon is new and full.

Stack   An isolated mass of rock projecting from the ocean off the end of a headland from which it has been detached by wave erosion.

Standard laboratory bioassay   Tests conducted in a laboratory setting to determine the percentage of dying caused within a population by a given concentration of foreign material.

Standing stock (crop)   The biomass of a population present at any given time.

Standing wave   A wave, the form of which oscillates vertically without progressive movement. The region of maximum vertical motion is an *antinode*. On either side are *nodes* where there is no vertical motion but maximum horizontal motion.

Stenohaline   Pertaining to organisms that can withstand only a small range of salinity change.

Stenothermal   Pertaining to organisms that can withstand only a small range of temperature change.

Storm surge   A rise above normal water level resulting from wind stress and reduced atmospheric pressure during storms. Consequences can be more severe if it occurs in association with high tide.

Streamlining   The shaping of an object so it produces the minimum of turbulence while moving through a fluid medium. The teardrop shape displays a high degree of streamlining.

Subduction   The process by which one lithospheric plate descends beneath another as they converge.

Sublimation   The transformation of the solid state of a substance to a vapor without going through the liquid phase, and vice versa.

Sublittoral zone   That portion of the benthic environment extending from low tide to a depth of 200 m (656 ft); considered by some to be the surface of the continental shelf.

Submarine canyon   A steep V-shaped canyon cut into the continental shelf or slope.

Submergent shoreline   Shoreline formed by the relative submergence of a landmass in which the shoreline is on landforms developed under subaerial processes. It is characterized by bays and promontories and is more irregular than a shoreline of emergence.

Suboceanic province   Benthic environments seaward of the continental shelf.

Substrate   The base on which an organism lives and grows.

Subsurface chlorophyll maximum (SCM)   The concentration of chlorophyll in many places in the ocean is highest immediately below the oxygen compensation depth.

Subsurface current   A current usually flowing below the pycnocline, generally at slower speed and in a different direction from the surface current.

Subsurface oxygen maximum (SOM)   Generally observed within the lower euphotic zone, the SOM may represent supersaturations of oxygen of up to 120 percent as a result of photosynthetically produced oxygen.

Supralittoral zone   The splash or spray zone above the spring high-tide shoreline.

Tide   Periodic rise and fall of the surface of the ocean and connected bodies of water resulting from the gravitational attraction of the moon and sun acting unequally on different parts of the earth.

Tide wave   The long-period gravity wave generated by tide-generating forces described above and manifested in the rise and fall of the tide.

Tintinnid   A ciliate protozoan of the family Tintinnidae with a tubular to vase-shaped outer shell.

Tissue   An aggregate of cells and their products developed by organisms for the performance of a particular function.

Tombolo   A sand or gravel bar that connects an island with another island or the mainland.

Topography   The configuration of a surface. In oceanography it refers to the ocean bottom or the surface of a mass of water with given characteristics.

Total allowable catch   The permissiable annual catch of a given species of fish that will not degrade the fishery given the knowledge of the ecosystem from which the fishery is being removed.

Trade winds   The air masses moving from subtropical high pressure belts toward the equator. They are northeasterly in the Northern Hemisphere and southeasterly in the Southern Hemisphere.

Transform fault   A fault characteristic of oceanic ridges along which they are offset.

Transitional crust   Thinned section of continental crust at the trailing edge of a continent created by the breaking apart of an ancient continent over a newly formed spreading center.

Transitional wave   A wave moving from deep water to shallow water that has a wavelength more than twice the water depth but less than 20 times the water depth. Particle orbits are beginning to be influenced by the bottom.

Transverse wave   A wave in which particle motion is at right angles to energy propagation.

Trawl   A study bag or net that can be dragged along the bottom to catch fish or a variety that can be towed various depths above the bottom for the same purpose.

Trench   A long, narrow, and deep depression on the ocean floor, with relatively steep sides.

Trophic level   A nourishment level in a food chain. Plant producers constitute the lowest level, followed by herbivores and a series of carnivores at the higher levels.

Tropical tide   A tide occurring twice monthly when the moon is at its maximum declination north and south of the equator. It is in the tropical regions where tides display their greatest diurnal inequalities.

Tsunami   Seismic sea wave. A long-period gravity wave generated by a submarine earthquake or volcanic event. Not noticeable on the open ocean but build up to great heights in shallow water.

Tube worms.   *See* Sabellidae and Serpulidae.

Tunicates   Members of the chordate subphylum Urochordata, which includes sacklike animals. Some are sessile (sea squirts) while others are pelagic (salps).

Turbidite   A sediment or rock formed from sediment deposited by turbidity currents characterized by both horizontally and vertically graded bedding.

Turbidity   A state of reduced clarity in a fluid caused by the presence of suspended matter.

Turbidity current   A gravity current resulting from a density increase brought about by increased water turbidity. Possibly initiated by some sudden force such as an earthquake, the turbid mass continues under the force of gravity down a submarine slope.

Turbulent flow   Flow in which the flow lines are confused, heterogeneously due to random velocity fluctuations.

Typhoon   A severe tropical storm in the western Pacific.

Ultraplankton   Plankton for which the greatest dimension is less than 5 μm. Very difficult to separate from the water.

Ultrasonic   Sound frequencies above those that can be heard by humans (above 20,000 cycles per second).

Ultraviolet radiation   Electromagnetic radiation shorter than visible radiation and longer than X rays. The approximate range is 1–400 nanometers (nm).

Upwelling.   The process by which deep, cold, nutrient-laden water is brought to the surface, usually by diverging equatorial currents or coastal currents that pull water away from the coast.

Valence   The combining capacity of an element measured by the number of hydrogen atoms with which it will combine.

van der Waals force   Weak attractive force between molecules resulting from the interaction between the nuclear particles of one molecule and the electrons of another.

Vector   A physical quantity that has magnitude and direction. Examples are force, acceleration, and velocity.

Ventral   Pertaining to the lower or under surface.

Vertebrata   Subphylum of chordates that includes those animals with a well-developed brain and a skeleton of bone or cartilage; includes fish, amphibians, reptiles, birds, and animals.

Viscosity   A property of a substance to offer resistance to flow. Internal friction.

Water mass   A body of water identifiable from its temperature, salinity, or chemical content.

Wave   A disturbance that moves over the surface or through a medium with a speed determined by the properties of the medium.

Wave-cut bench   A gently sloping surface produced by wave erosion and extending from the base of the wave-cut cliff out under the offshore region.

Wave-cut cliff   A cliff produced by landward cutting by wave erosion.

Wave height   Vertical distance between a crest and the preceding trough.

Wavelength   Horizontal distance between two corresponding points on successive waves, such as from crest to crest.

Wave period   The elapsed time between the passage of two successive wave crests past a fixed point.

Wave steepness   Ratio of wave height to wavelength.

Wave train   A series of waves from the same direction.

Weathering   A process by which rocks are broken down by chemical and mechanical means.

Westerly winds   The air masses moving away from the subtropical high pressure belts toward higher latitudes. They are southwesterly in the Northern Hemisphere and northwesterly in the Southern Hemisphere.

Windrows   Rows of floating debris aligned parallel to the direction of the wind that result from Langmuir circulation.

Windward   The direction from which the wind is blowing.

Zenith   That point on the celestial sphere directly over the observer.

Zooplankton   Animal plankton.

INDEX

Wirt, Leslie S., 177
WLVEC (Water Low-Velocity Energy
    Converter), 155
Woods Hole Oceanographic Institution,
    15, 16
Woolly sculpin, 307

World Health Organization, 349
Worms, 312–13

Yellowfin tuna, 288, 339
Yellowtail, 288

Zenith, 185
Zeolites, 86
Zobell, Claude, 322
Zooplankton, 249, 250, 271